DETERMINATION OF STRUCTURAL SUCCESSIONS IN MIGMATITES AND GNEISSES

DETERMINATION OF STRUCTURAL SUCCESSIONS IN MIGMATITES AND GNEISSES

A. M. Hopgood

Reader
Department of Geology
University of St Andrews
Scotland

A manual for geologists concerned with the interpretation of complex structural relationships in highly deformed deep crustal rocks

SPRINGER SCIENCE+BUSINESS MEDIA, B.V.

Library of Congress Cataloging-in-Publication Data

ISBN 978-94-010-5902-2   ISBN 978-94-011-4427-8 (eBook)
DOI 10.1007/978-94-011-4427-8

Printed on acid-free paper

DEDICATION

To J. J. Sederholm whose insight and pioneering work on migmatites led to studies which provide so much of our present understanding of deep crustal processes.

CONTENTS

(h) 'Cryptic' structures
Look for evidence (such as 'anomalous' orientation of the structure) for structures whose presence is not immediately obvious (e.g. very open folds).
(i) Variation of structural expression
Recognize changes in expression in structures of the same set.
(j) Introduced structures
Note the presence of minor intrusions, agmatites, cleavages etc. potentially useful for distinguishing subsequent deformation (Chapter 4).
(k) Orientated specimens
Record localities where there is suitable material for collecting orientated specimens.
(l) Key structures
Record distinctive structures likely to be suitable for key structures.
(m) Relationships to key structures
Note relationships between structures and those likely to be suitable for key structures.
(n) Asymmetry of folds relative to large structures: vergence
Note relationships between large-scale and apparently related small-scale structures.
(o) Record of exposures showing clear structural relationships
Note localities where structural relationships are particularly clear.
(p) Ending reconnaissance
Circumstance governing the termination of the reconnaissance.

(a) Introduction
Reminders of some basic field techniques.
(b) Geometrical analysis using stereographic projection
Geometrical analysis, limitations of method and need for direct observation of structural relationships in the field.
(c) Orientated specimen collection
Procedure for collection and preparation.
(d) Sketches of structural relationships
Value of field sketches of structural relationships.
(e) Orientated photographs
Importance of recording observational data, including time and direction, when taking photographs.
(f) Field notes
Some important aspects of keeping field records.
(g) Structural relationships check
Need to clarify and confirm uncertain structural relationships continually.
(h) Consistency of successions check
Need for continual check and confirmation of structural relationships.
(i) Good field practice
(j) Summary

PREFACE

This book has been written in response to requests from a number of colleagues in the earth sciences in different parts of the world. Prominent among these are geologists whose interests lie in the fields of isotopic and economic geology and who have a particularly keen appreciation of the importance to their work of a thorough understanding of the structural relationships in rocks, especially where such rocks, like migmatites, have a long and often complex developmental history.

What these geologists asked for was a guide to the methods employed in resolving the structural complexity of repeatedly deformed rocks (i.e. those affected by 'polyphase' or multiple deformation), especially in Precambrian basement terranes. These rocks normally comprise gneisses and migmatites in which, because of the reasons given later, the combined effect of lithological complexity and an involved deformational history can present a daunting structural picture. The intricacy of their structure means that methods traditionally used in the investigation of less complexly-deformed rocks are seldom suitable for structural analysis in migmatite terranes. These analytical difficulties can be overcome by adopting a different approach to understanding and coping with the structural complexity of migmatites and related rocks, and the methods presented here provide a rigorous means of structural interpretation in the field. They emphasize the importance of accurate observation, particularly of structural detail, and aim to alter the perception of observers who might be discouraged by highly complex structural relationships so that they can deal systematically and confidently with analysis of the structure.

While considerable effort has been made to ensure that the coverage is comprehensive, it is not necessarily exhaustive.

The importance attached to the structural analysis of migmatites (and other complex structural associations) stems from the fact that resolution of their structural complexity has such a significant bearing on many aspects of geology, both academic and economic. Examples of some of these applications are listed in Chapter 14. Two particularly important aspects are that the structural characteristics so identified can be used (1) to correlate between separated rock units with comparable deformational histories, including segments of fragmented supercontinents, and (2) to reliably identify the relative positions in the structural succession of rock units suitable for isotopic dating.

This analytical approach follows the development over many years of methods used for the interpretation of the extreme structural complexity developed at deep crustal levels by repeated deformation throughout a long tectonic history. In the past, and even now, this aspect of structural geology has tended to be largely neglected, understandably perhaps, considering the complexity of the structural patterns that can be seen: some observers simply dismiss the patterns as being 'wild' and lacking any underlying regularity. Nevertheless, interpretation of structural relationships in metamorphic complexes is of fundamental importance, both for the understanding of crustal development and for the recognition of changes that have taken place through geological time. Although challenging, such structural complexity can be resolved, as has been shown by the results of several studies like that of the early Proterozoic Svecofennian migmatites of southern Finland (Hopgood *et al.*, 1983).

It is well known that because of their nature and composition, gneisses and migmatites affected by multiple deformation present special problems when it comes to determining their structural succession. Gneisses typically lack distinctive lithological units suitable for use as key, or marker, layers which are normally very useful, if not essential, for delineating major structures which affect, or may be affected by, other structures. Problems in migmatites are created by the complexity caused by partial melting and local intrusion. Another factor is the rarity of carbonates and other rock types particularly responsive to deformation and sensitive to metamorphism which can provide mineral assemblages related to individual tectonothermal events and characteristic of particular fold sets. Their absence also serves to increase the difficulty of distinguishing between the different sets of structures. This, coupled with the complexity of deformational history, as well as effects arising from migmatization, palingenesis and anatexis, further accentuates the difficulty of analysis aimed at determining a structural succession and subsequently establishing a deformational sequence. It is for these reasons that it is usually necessary to adopt a special approach to structural analysis in migmatite terranes.

Techniques have not yet been devised that enable complex structures to be dealt with in a mechanized or computerized manner comparable to that used in making a chemical analysis of a rock, e.g. by feeding powdered samples into automated analytical equipment (Figure P.1). This is partly because no two structures are strictly identical, and the geologist is still required to make the initial observations of structures and their relationships and to assess these in each case. However, it **is** possible to establish and use certain rules which can be adapted to meet each set of circumstances. Almost invariably this means working on the outcrop – if necessary backing this up by laboratory work on orientated specimens collected in the field – so that it is important, and almost always essential, to resolve the structural complexity prior to leaving the field.

The approach is a direct result of involvement over a number of years with, firstly, the study of the structural complexity of basement rocks, often in the company of colleagues (from whom there has been considerable input and invaluable criticism), and secondly, the teaching of the structural analytical techniques to resolve this complexity. It is based on field work carried out on basement rocks in various parts of the world, including southwestern

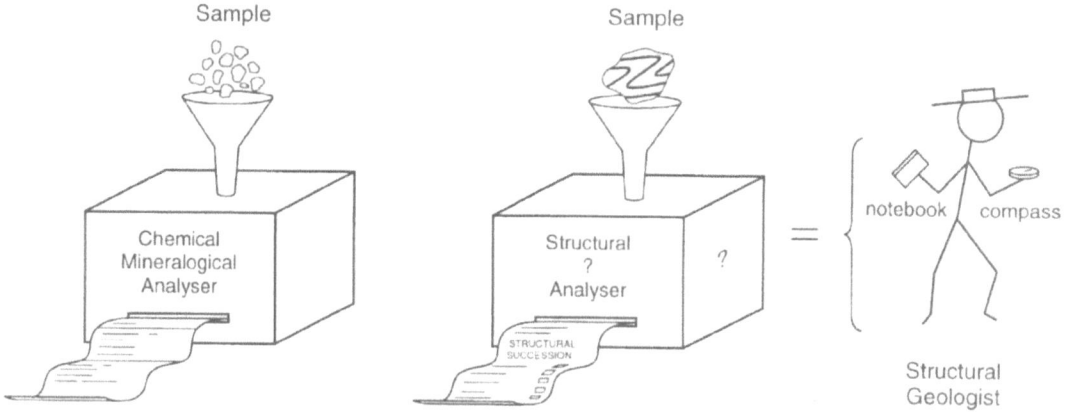

Figure P.1 Automated geochemical analysis contrasted with structural analysis.

Australia, southern China, the Czech Republic, Finland, southwestern Greenland, India, Scotland, eastern Siberia, Uganda and Zimbabwe, but principally on the superbly exposed, ice-polished Proterozoic Svecofennides of the Gulf of Finland, where in all probability the exceptionally high quality of the exposure is unsurpassed anywhere else in the world.

The methods presented are those that have been used to demonstrate ways of resolving structural complexity, both in the field (the practice) and in the laboratory (including the theory), to undergraduates and graduates as well as to experienced geologists. While at the outset the attitude of some in the last group was often one which could at best be described as 'sceptical' (this scepticism apparently stemming from the sheer number of 'events' involved), their initial reservations (or prejudices) were rapidly overcome once they had seen the methods demonstrated, or had themselves applied them successfully in the field. After being applied to the resolution of complex structural relationships over several years, and in many different terranes, the techniques discussed here have been shown to yield consistently reproducible results, not only by the same observer but also by different observers tackling the same problems.

The completion of this book has taken considerable time since its inception as notes put together during a summer in the early 1970s when bad weather interrupted field work on the edge of the Frederikshåbs Isblink in West Greenland (Figure P.2). What began merely as a collection of ideas has evolved through several sets of lectures presented to various groups in different parts of the world, from Finland to Australasia, and has undoubtedly benefited considerably from suggestions from those who received the lectures as well as from colleagues in branches of earth science other than structural geology, and indeed in subjects other than the earth sciences.

There have been times when, with the increasing appreciation of the value of the methods by colleagues involved with the reso-

Figure P.2 Terrain and outcrop typical of exposures of deep crustal rocks in high latitudes. Camp at Kangiussap near the Inland Ice, West Greenland, 1971.

lution of complex structures throughout the world, it seemed that this book was unnecessary. However, to my continuing surprise, personal experience and discussion with other colleagues in the intervening years has shown that the need to instil an understanding of the basic principles and their application remains as strong as ever. The importance of this is no less in 1997 than it was in 1977, so that although preparation of the book was often set aside, it was taken up again as a result of my having been persuaded, not only from personal experience but also by others who had seen and successfully applied the methods in the field, that the need is still very much in evidence. While much has been written on the description of structures in high-grade terranes since the classic work of Sederholm (1967) in the 1900s, and more recently on the interpretation of these structures (e.g. Passchier, Myers and Kröner, 1990), no textbook aimed specifically at the analysis of the complex structural relationships observed with the intention of determining structural successions has been published previously.

It is intended that the discussion can be followed by anyone with a basic knowledge of structural geology, and this includes advanced undergraduates and research students as well as experienced professional geologists. Besides attempting to present the concepts discussed in as simple a manner as possible (not always easy away from the outcrop), the use of specialist terms that have not been well-established is kept to a minimum, or avoided as far as possible. There is a widely-held impression that the subject of structural geology has a reputation for the proliferation of new terms. There is some repetition of subject matter and concepts and this is intentional. Not only is it intended to remind the reader of concepts discussed earlier, but it is also employed to emphasize particular aspects, to present some topics in different contexts and, by the adoption of a different approach in some instances, it is intended to remove any lingering doubts and uncertainties. Experience has shown that varying the approach to a difficult concept is an effective way of clarifying misconceptions. Although some of the methods discussed are exacting, and although the initial impression might be that some are even tedious, practice and experience will eventually lead to their being used with little more conscious effort than that required by the experienced commuter in following a complicated yet familiar city route to work. In time, the rules for the resolution of structural complexity should grow to be instinctive and their successful application will become close to 'second nature'.

Chapters 12 and 13 summarize the analytical approach and can be referred to directly by those with experience in working with structures derived from multiple deformation who do not wish to read about the theory and philosophy behind the principles at this stage. There are several sections of text in close-spaced print that provide additional explanation not strictly essential for following the techniques described, and these can be bypassed if so desired without losing the train of the discussion.

While the principles described apply equally to all structures, the presentation here concentrates largely on the structure relating to folds and folding. This is because, of all structures, folds generally contribute most to the complexity of rocks such as migmatites. Correspondingly fold analysis presents one of the principal challenges to the understanding of structural complexity, and the resolution of superimposed fold structures often forms the basis for resolving the overall structural complexity.

Although in most of the illustrations, fold profiles (fold cross-sections) rather than three-dimensional representations have been used in the interests of simplicity, this does not represent a conflict with the analytical approach used here because complex three-dimensional structural relationships are commonly resolved by determining relationships on several essentially two-dimensional surfaces.

Finally, in response to comments and suggestions from geologists other than structural geologists who have seen the application of this work in the field, it may be worth pointing out that because the methods are closely linked to observation as well as analysis, some aspects are also broadly applicable to any aspect of geology that depends on observation.

THE AIM OF THE BOOK
['What the book is about']

This book provides the reader with facts, techniques and examples collected over a period of approximately 40 years' experience of working in complexly deformed rocks, mainly gneisses and migmatites.

1. It seeks to provide a systematic approach to the resolution of the complex structural relationships found particularly in migmatites and other deep crustal rocks.

2. It sets out to:

   - provide some basic background to the theory behind the development of structural imprints on rocks in response to stress, i.e. deformation or tectonism;
   - outline the basic principles underlying the resolution of these structural imprints or patterns on the basis of interpreting **overprinting** relationships between structures;
   - discuss some of the particular aspects and problems associated with the field study of structures in migmatite terranes;
   - consider the factors contributing to the structural complexity of profoundly deformed deep crustal rocks such as migmatites, and the importance of these factors in resolving the structure;
   - demonstrate, using progressively complex examples, the basic methods and principles employed in unravelling the structure of rocks affected by **multiple** (**polyphase**) deformation;

   - explain the basis of the method using the concept of **fold sets** and their **overprinting** relationships; the involvement of **recognition** of folds in fold sets and the importance of the **identity** (identification) of folds in **correlating** between folds within sets; fold set **nomenclature** and the use of **key** (datum) structures in correlation;
   - consider the problems and apparent paradoxes relating to the establishment of fold successions;
   - illustrate the principles used in determining structural **successions** (succession of fold sets) in crustal segments involving correlation between **local subsuccessions** and their integration;
   - provide photographs, explained by diagrams, of real examples to illustrate the determination of structural successions;
   - present and discuss the practical aspects involved in determining structural successions, field work, collection of data and material for laboratory analysis; problems and 'pointers';
   - summarize the whole approach covered in the preceding sections. For geologists with some experience of studying multiple deformation in migmatite terranes, this could obviate the need to read all of some of the previous sections in detail.

THE NEED FOR THE BOOK

To describe the methods used in the following:

- finding the age of specific events in the deformational history of the rocks, using isotopic methods to date rocks emplaced at specific times in the structural succession;
- determining the nature of the development of the structural history of deep crustal rocks;
- determining the physical conditions (temperature and pressure etc.) associated with the development of deep crustal rocks such as migmatites;
- determining the *rates* of orogenic processes on the basis of dating specific events in the structural succession;
- establishing the time span of orogenic events on the basis of having determined the dates of specific events throughout the structural succession;
- identifying particular events (deformational or intrusive) in the succession that are related to rocks of economic importance.

ACKNOWLEDGEMENTS

A great deal is owed to my geology teachers at Auckland University, especially to A. R. Lillie for fostering a self-critical approach and to R. N. Brothers for introducing me to the study of complex structural relationships in melanges. My thanks are due also for their encouragement and support which ensured the continuation of my study in terranes beyond New Zealand.

I am also indebted to the following:

D. R. Bowes for discussion and companionship in the field in many terranes and over many years and for tolerating my persistent refusal to ignore the apparently intractable. T. J. Koistinen for discussion and constructive criticism in the field in Finland and for critical assessment of the manuscript. P. Joubert, S. Moorbath and O. van Breeman whose appreciation of the effectiveness of the methodology helped ensure its wider acceptance in the geoscience community, and particularly O. v B. who, after close collaboration in the field suggested that the book be written. They and many other geologists have continued to encourage me to produce the book.

Geologists and others who have provided hospitality in the field and at meetings in many countries, in particular during frequent visits to Finland and Czechoslovakia since the early 1970s. The Geology Department at Auckland University for hospitality and access to their Library during the final stages of preparation. Colleagues, friends and acquaintances who offered suggestions or, through chance remarks, provided inspiration unwittingly. All have helped to contribute to this book as well as making my work rewarding, at times entertaining, and always interesting.

R. H. Hopgood, a scientist with no geological background, for giving a 'lay' scientist's opinion of the manuscript. J. Allan, J. R. Ball, D. S. Campbell, D. Maclean and W. Nightingale for advice and technical assistance, and I. Rolf for the original source of the illustration used in Figures 5.7 and 12.5.

ARCO Great Britain and the Russell Trust whose contributions towards the cost of drafting the illustrations enabled the publication of the book to proceed. For grants and awards which enabled me to undertake much of the research on which this book is based, in particular from the following: The Royal Society of London and Carnegie Trust for the Universities of Scotland, The British Council, The Fulbright Foundation.

Finally, my wife Jenny for her forbearance and unstinting support during the final stages, for providing a lay person's test of the explanations of difficult concepts, and for companionship in some remarkable (often difficult) field locations throughout the world.

GENERAL CHARACTERISTICS OF MIGMATITES AND MIGMATITE TERRANES

1.1 INTRODUCTION

Structures and structural relationships resulting from superimposed deformation have long been recognized by geologists, and references to this perception can be found in the literature at least as far back as the late nineteenth century (Clough, *The Geology of Cowal* (Gunn *et al.*, 1897, pp. 23, 24), and here, under 'Fold (Structure) sets', section 5.3), and even earlier than this, in the eighteenth century, James Hutton in his *Theory of the Earth*, Chapter IV, p. 19 (1795), described overprinting relationships between granite, 'schistus' and a later transecting granite vein observed on a boulder in Glen Tilt, Scotland (see also section 8.1.2). The systematic study and geometrical analysis of superimposed folds, i.e. the effects of overprinting (*Überprägung*), were formalized in the early 1900s by the work of the Austrian geologists Sander (1930, 1970) and Schmidt (1932). The publication of Sander's work, particularly that in English in the 1930s (Knopf, 1933) greatly stimulated the study of structural relationships in folded rocks, a study which increasingly employed stereographic projection as an analytical tool. This in turn led to an increase in studies tending to concentrate on two major aspects of structural geology; one concerned with the geometrical relationships and behaviour of structures during refolding by different mechanisms, and the other involving statistical analysis of refold relationships with Schmidt nets, using Lambert equal area projection, of data concerning refold relationships collected in the field (see Turner and Weiss 1963; Phillips 1971; and Figures 2.10, 2.11, 2.12).

However, the detailed study of the structure of gneisses and migmatites[1], which comprise much of the Precambrian shield (cratonic) areas of the world, has consistently received considerably less attention than that of other highly deformed rocks. This apparent lack of interest stems only partly from the complexity of the structural geometry of migmatites and is most likely due to other factors which contribute to this overall complexity (see also references to migmatites under 'Need for a systematic approach' in section 2.1 and again under 'Influence of rock type on the structure

---

[1] The definition of 'Migmatite' in the 'Glossary' of the *Cambridge Encyclopaedia of Earth Sciences* (1981) as, 'A very high-grade metamorphic rock in which extremes of temperature and pressure have induced partial melting so that the rock has taken on some of the characteristics of igneous texture', is more closely akin to the definition of an 'anatexite' and is inconsistent with the account by J. B. Wright under, 'Migmatites and melting' on p. 92 of the encyclopaedia which more closely accords to Sederholm's 1967 definition of migmatites and describes them in the following terms, '...migmatites, composite rocks in which high grade gneisses are injected by granitic magma derived from partial melting of nearby rocks. Plastic deformation is common in migmatites, which thus become difficult to interpret. The greater the degree of magma generation and deformation, the greater the loss of original texture, until anatexis finally destroys all trace of pre-existing structure, and metamorphism grades into igneous processes.' See also Ashworth, pp. 1–3 (and Table 1.1), 1985 for a discussion of the difficulties in framing a modern definition of migmatite. (See also comments under, 'Influence of rock type on the structure of migmatites', section 1.1.3.)

of migmatites' below. Such factors include the effects of polyphase metamorphism and ultrametamorphism ('granitization'), quartzo-feldspathic veining, agmatization and the frequency of successive emplacement of igneous bodies. The combined effect of these means that, apart from complexly folded primary banding, it is possible to observe on only a single exposure, features such as (1) agmatites containing isolated 'floating' hinges of disrupted folds or tectonic 'fish', (2) anastomosing basic, intermediate and leucocratic veins, either concordant or discordant to the main foliation, (3) blurring or complete obliteration of foliation by leucocratic partial melt material (neosome), (4) several sets of cleavages, and (5) complicated shear and fracture patterns. These and the general absence of distinctive rock layers which could serve as reference 'horizons', together with the differential response to deformation of units of different competence, can result in extremely complicated and apparently irregular structural patterns (Figure 1.1). The impression, at times bewildering, which the structural pattern of such rocks presents at first sight has given rise to such terms as 'wild folds' and 'wild migmatites' (see Berthelsen *et al.*, 1962) and has frequently led to the early abandoning of any systematic attempt at resolution of the structure. This is either because the structural pattern was regarded as too complex, or, since it *appeared* to lack a consistent pattern, because it was not considered to be amenable to structural analysis. Such an assessment (i.e. that the structure of gneisses and migmatites is generally so complex as to defy attempts at resolution by the generally accepted methods of structural analysis) has often meant that the understanding of the structure, and hence the deformational history of these terranes has, at best, been inadequate. Because of this imperfect understanding of their structure, attempts to present a history for these rocks has often called for considerable ingenuity in order to provide a plausible explanation for the phenomena observed.

Figure 1.1 Folded migmatite. Patterns such as these, which appear to lack regularity, have given rise to the expression 'wild folds'. Svecofennian migmatites. Beach, Pohja, Finland.

1.1.1 STRESS TRANSMISSION DURING MIGMATITE DEVELOPMENT AND THE RECORD OF DEFORMATION – A PARODOX?

The close similarity between the appearance of some tectonic structural interference patterns formed in deep crustal rocks and that of superficial structures such as those seen on surface films like oil slicks on water (Figure 1.2) seems to have prompted the view among some observers less familiar with these rocks that their structure differs fundamentally from that of other tectonites[2]. It seems to be thought that the complex interference patterns prevalent in these rocks, which deformed at deep crustal levels where conditions of high ductility obtain, stem from random or 'wild' deformation, perhaps because of a preponderance of a fluid rather than a solid continuum. Presumably it is considered that in such an environment where melt is a

---

[2] There is an inherent danger in drawing analogies between distinctly different processes on the basis of the superficial similarity of their products as in the case of the fold-like forms of disturbed thin films such as oil slicks (essentially from two-dimensional deformation) and true folds caused by deformation in three dimensions. While there is a superficial resemblance between folds and the organic growth patterns seen on the surfaces of sawn timber no one would suggest the latter are the product of tectonism!

common or predominant constituent of the rock mass, the stress transmission necessary to produce consistently regular structures is not possible, having been inhibited or precluded entirely by the melt fluid. In fact it has been found that similar, apparently irregular structures looking like 'stirred porridge' can be computer-generated in a geometrically regular fashion (see the discussion in Ragan (1985), '13.4 Wild folds', pp. 252, 253).

Such misinterpretation of the significance of structural relationships in migmatites stems from, or is reinforced at least in part by, the evidence for the presence of quantities of melt at some stage(s) during migmatite development. Although partial melting (anatexis) is an integral part of much migmatite formation, the presence of isolated discrete patches of fluid material, or of extremely ductile material, does not preclude the pervasive development of penetrative structures. Observation shows that structural sets form in migmatites just as they do in other, lower-grade metamorphic tectonites and that refolding relationships between sets can be determined in these rocks as in other tectonites, as can crosscutting relationships of superimposed planar fabric elements. The fact that apparently irresolvable structural 'wildness' does not seem to be generally characteristic of migmatite complexes has been demonstrated in Finland (Hopgood 1984; Koistinen 1981), Scotland (Hopgood and Bowes 1972), India (Halden *et al.* 1982; Naha *et al.* 1990) and elsewhere. The explanation for the apparent paradox of pervasive fold structural sets in rocks containing melt material is a simple one. During anatexis the melt is developed irregularly,

(a)

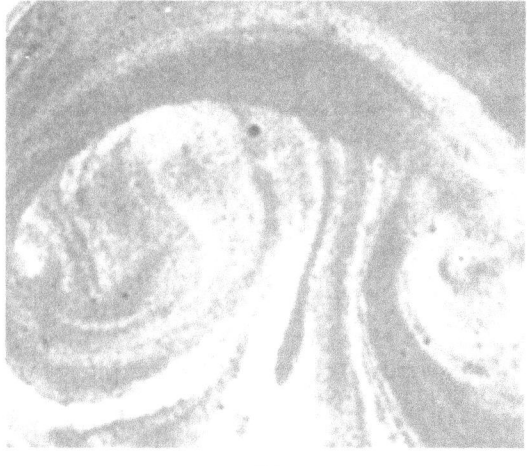

(b)

Figure 1.2 'Mushroom' interference pattern. (**a**) The shape caused by intersection with the exposure plane of the three-dimensional structure resulting from refolding. Migmatites, Pohja, Finland. Compare this shape with that of (**b**), a 'mushroom' structure' caused by mechanical stirring of a layer with composition different from that of the supporting fluid, in this case formed in the surface layer of a cup of soup (**c**).

(c)

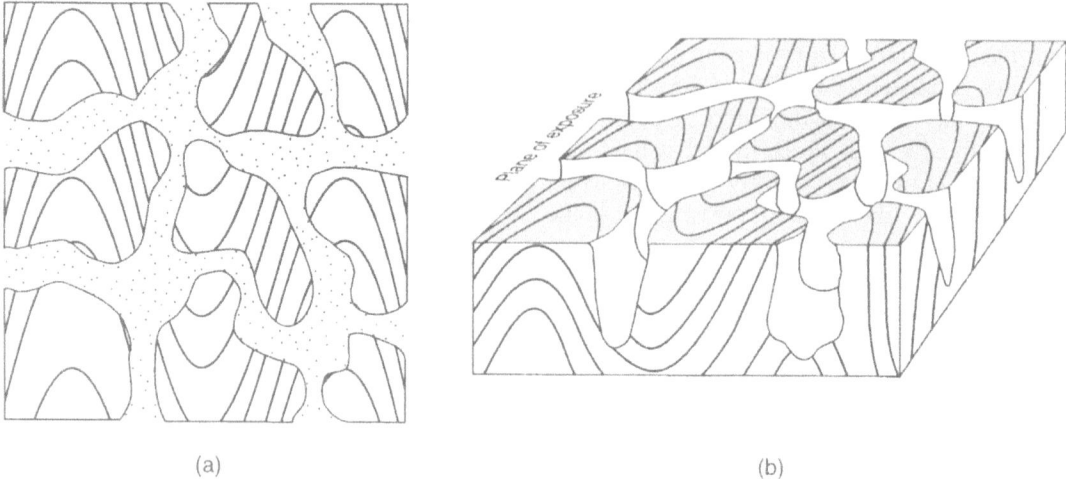

(a)   (b)

Figure 1.3   (**a**) Non-random, apparently discrete palaeosome 'blocks' viewed in two-dimensions showing *structural* continuity between 'blocks' even though they appear to be physically unconnected. (**b**) When viewed three-dimensionally the apparently isolated palaeosome blocks are seen to be linked together as parts of a continuous rigid framework so that the structural integrity is preserved, as in the example of Figure 11.12.

and usually **discontinuously** initially, rather than as a continuum. Even when the pockets of melt are interconnected, the relationship between the melt and the host tectonite is one of fluid between rigid interconnected septa rather than that of disconnected rafts in a fluid host, so that the stress can be transmitted throughout the entire rock mass by way of the rigid interconnected framework to produce a penetrative structural pattern, at least in the palaeosome (Figure 1.3). Only in those cases where there are rafts of rock isolated by melt is it likely that the stress will fail to be transmitted throughout the rock mass to produce a pervasive structural pattern. Fortunately this situation is not likely to lead to misinterpretation because such agmatites derived from visibly deformed rocks are, like metasedimentary breccias (or immature conglomerates (Figure 1.4a) which commonly show other 'tell-tale' sedimentary features), readily recognizable by the random attitude of the palaeosome blocks in the neosome (Figure 1.4b). Figure 1.4c shows an agmatite in which there is a transition between palaeo-

some blocks exhibiting structural and physical continuity (as in Figure 1.3) to one in which the blocks show random orientation, as in Figure 1.4c. Thin section examination might be necessary to confirm this condition in those cases where the scale of the structure is microscopic.

In essence this means that in general:

- deformed rocks **do** preserve a 'systematic' record of their deformational history and
- it **is** possible to resolve the structural complexity so recorded, even in the case of the highly complex structure in deep crustal rocks such as migmatites.

1.1.2   MIGMATITES – WHAT THEY IMPLY TO THE STRUCTURAL GEOLOGIST

The developmental complexity of migmatites must inevitably govern their structural aspect on the outcrop and this aspect is commonly one of complex fold patterns further complicated by ultrametamorphic and intrusive

(a)

(b)

(c)

Figure 1.4 (a) Immature conglomerate for comparison with an agmatite. Svecofennian complex, Kisko, Finland. (b) Random blocks of palaeosome in agmatite. Little remains of the 'pre-agmatite' structural continuity suggested by the disposition of the elongate foliated blocks at upper left, right and lower centre. Sharyzhalgay complex, Lake Baikal, eastern Siberia. (c) Agmatite showing progression from structural (and physical) continuity (foreground) to the condition where, with breakdown of this continuity, palaeosome blocks with random orientation 'float' in a neosome continuum (background). In the background agmatite, stress transmission was inhibited (or precluded) by melt-formation which caused the blocks of palaeosome to be isolated from one another by the fluid neosome, so preventing the transmission of stress from one block to another. The absence of a continuous rigid rock framework prevented the stress field from producing a pervasive (penetrative) structural pattern throughout the whole rock mass until such time as the neosome was no longer fluid. Nötö, southern Finland. Compare these examples with Figures 1.3, 1.11, 1.12, 1.13a, 4.23, 10.17, where the structural integrity is preserved during anatexis.

phenomena. It is this superficially irregular, or 'wild' appearance that has often deterred the geologist from beginning a structural investigation likely to involve a large amount of time. This is in the belief that the outcome will almost certainly be unrewarding, or at best, produce a very poor return in terms of new information.

It is essential for the geologist involved in a study of a migmatite terrane to realize at the outset one important aspect of these rocks. It is this: **Every feature expressed represents an event which contributed to the character of the observed rock**. In this respect it is useful to recall the comment attributed to Ernst Cloos in 1946 and quoted at the front of Ramsay and Huber, Volume 2 (1987): '. . . every structure in a rock is significant, none is unimportant, even if, at first sight, it may seem irrelevant'.

Hence, paradoxical as it might seem, although the structure of migmatites is complex because of the great number of features contributed during all the events comprising their developmental history, it is the very existence of such complicating factors that allows the structure to be resolved. (See the discussion on structures introduced into the succession in Chapter 4).

Therefore, if the geologist can

(i)   recognize each of these features, and
(ii) relate them one to the other.

an ordered succession of structural features can be built up. Once that has been achieved, an attempt can be made to establish the tectonic sequence responsible for the development of the rocks observed.

That, at least, is the theory of the field approach and we will see how this compares with what is entailed in practice.

1.1.3   INFLUENCE OF ROCK TYPE ON THE STRUCTURE OF MIGMATITES

To the observer, even one well versed in the principles of geometrical structural analysis, faced for the first time with attempting to resolve the structural complexity of deformed gneisses and migmatites, the practicality seems to bear little resemblance to the theory. There are two reasons for this. One is that the structure is usually more complicated than it is in the examples used to demonstrate the principles of geometrical analysis. The other reason is that the rock types are such that deformation has affected a succession which, to begin with, is neither straightforward nor discretely layered, but, as stated in the preface, one that often lacks distinctive 'marker', or reference layers (Figures 1.5 and 11.16) so that even when the structural succession is relatively simple, the imposed structures may not themselves be regular or clearly defined. It was not without good reason that in the early 1900s Sederholm using a word derived from the Greek word for 'mixture', proposed the term 'migmatites': '. . . I would suggest for them the name migmatites, from μιγμα,

(a)

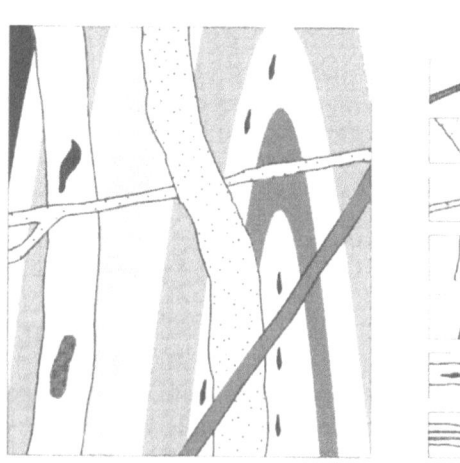

(b)

Figure 1.5 (a) Pre-Ketilidian migmatites, West Greenland. (b) Sketch showing the typically complex interrelationships between components of a migmatite comparable to that shown in (a) with the succession of components comprising the overall structure shown in column form (youngest at top).

a mixture', (Sederholm, 1967, p. 86). In this respect it is important to note that some authors, when speaking of migmatite, sometimes appear to be referring largely or solely to the partial melt (anatectic) component of the rock rather than migmatite *sensu stricto* (e.g. Mehnert, 1968). A clear definition of the word 'migmatite' in modern petrological terms appears to present some difficulty, as pointed out by Ashworth, 1985, pp. 1–3 and Table 1.1 (see also footnote in section 1.1), whereas Sederholm's original definition neatly encapsulates the characteristics of migmatites and is appropriate in the context of the study of structural relationships.

The absence of marker layers in gneiss and migmatite (Figure 1.6) also poses a problem, particularly in mapping larger-scale structures (see Figure 1.16) and in migmatites this (by definition) is especially the case, because of the diversity of relationships between the different rock units comprising them, so they are one of the most difficult groups of rocks to deal with structurally. Early-formed structures have been modified or obliterated because they have been affected by concordant and discordant intrusions including emplacement of very coarse-grained rocks sometimes with megacrysts several centimetres across (Figure 1.7). They have also been subjected to partial melting and anatexis, to agmatization and to the development of nebulites. Differential movement between the neosome and palaeosome in agmatites, or between the blocks in conglomerates and agglomerates, can obscure original structural relationships (Figure 1.8a) as will shearing, whether it is localized, on discrete surfaces (Figure 1.8b) or pervasive. Continued deformation leading to transposition structures (Figure 1.9) further complicates the structure and, together with repeated deformation and emplacement of new rock bodies, the early-formed structures can be modified so drastically as to leave very few clues to their existence and almost none concerning their relationships.

Figure 1.6 Migmatites showing absence of clear, continuous reference layers. Pre-Ketilidian Complex, West Greenland.

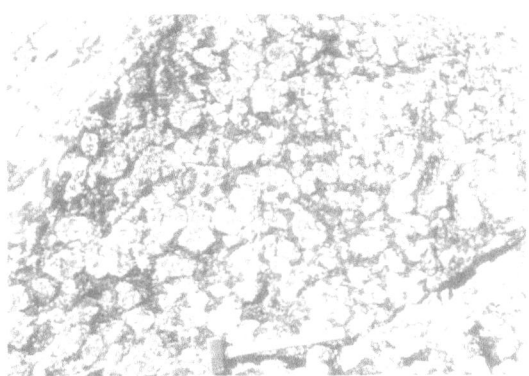

Figure 1.7 Feldspar megacrysts in anorthosite. Pre-Ketilidian migmatite, West Greenland.

In extreme cases little may remain of the earliest fold sets other than a few intrafolial hinges superficially indistinguishable from one another (Figure 1.10). In such cases even the suspicion that some folds belong to different sets should form the basis for separating them **provisionally** from others. If at the end of the study no firm evidence has been found in support of this distinction then the folds can be grouped together, but with the recognition that they may represent more than one set (see section 5.2.3, 'Importance of early provisional classification of structures', *et seq.*).

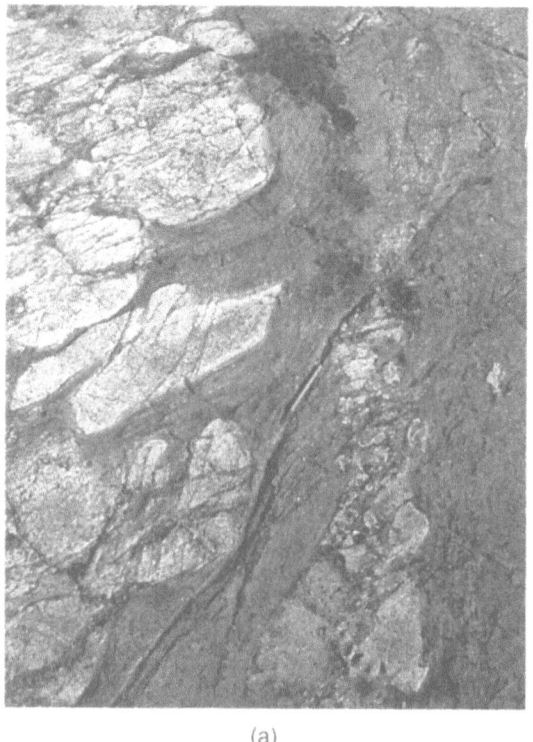

(a)

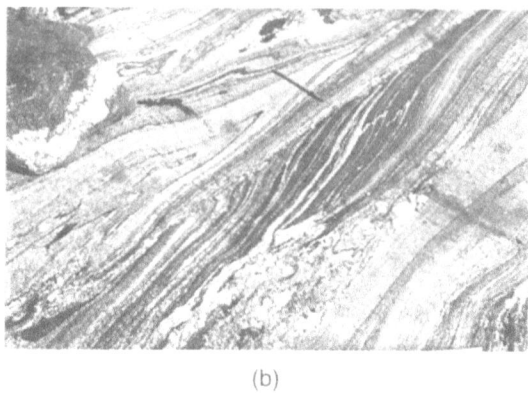

(b)

Figure 1.8   (**a**) Agglomerate with acidic tuff matrix showing displacement and separation of blocks by differential movement between them and tuffaceous matrix. Svecofennian migmatites, Orijärvi, Finland. (**b**) Effect of differential movement by shear in migmatite. Pre-Ketilidian migmatite, West Greenland.

(a)

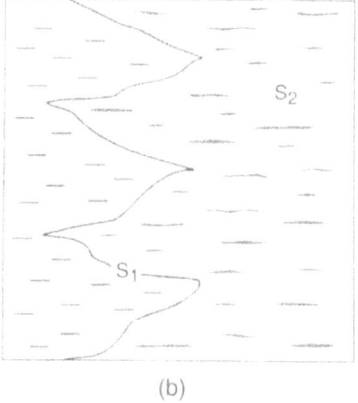

(b)

Figure 1.9   (**a**) Foliation transposition from the earlier lithological contact (S_1) to the later foliation (S_2). (**b**) Explanatory sketch of (**a**). Pre-Ketilidian migmatite, West Greenland.

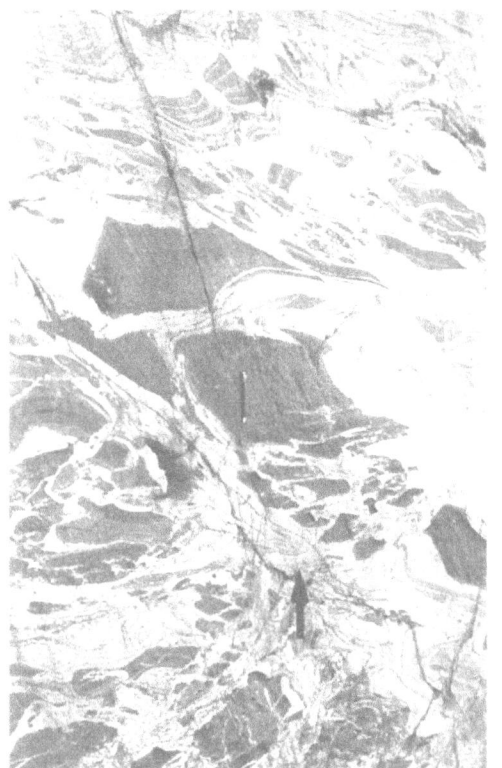

Because lithostructural mapping is seldom possible, individual structures can often be defined only on the basis of deformed 'form surfaces' (Figure 1.11) especially on a large scale, and synclines and anticlines can rarely be recognized because of the lack of stratigraphical information. It may be possible only to infer the presence of large-scale structures on the basis of observed small-scale structures (Figure 1.12). (The words 'large' and 'small' scale are purely relative so that there is some degree of imprecision in the use of the word 'scale' which is not absolute.)

In some cases anatectic recrystallization might have been so extreme as to have destroyed the foliation entirely, producing a

Figure 1.10 A small tight fold hinge (arrow) 'floating' in leucocratic neosome enclosing a palaeosome comprising the fragmented remnants of a larger, more open fold defined by the amphibolite blocks (under the pencil). The smaller hinge is not obviously related to any other fold set. Svecofennian agmatite, Jussarö area, southern Finnish archipelago.

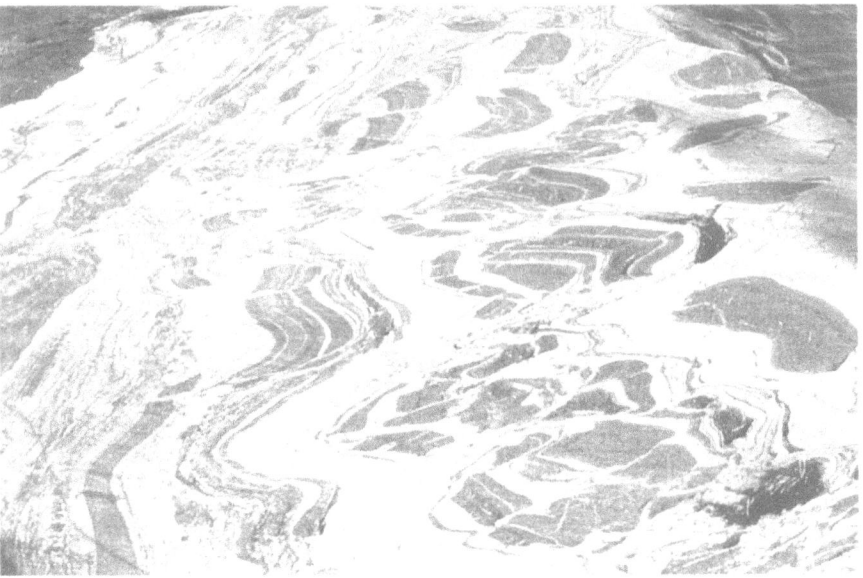

Figure 1.11 Form lines. The broad trend of the gross foliation can be traced (as form lines) between and parallel to the fragmented darker lithological layers. Svecofennian agmatite, Orrholmarna, Jussarö area, southern Finnish archipelago.

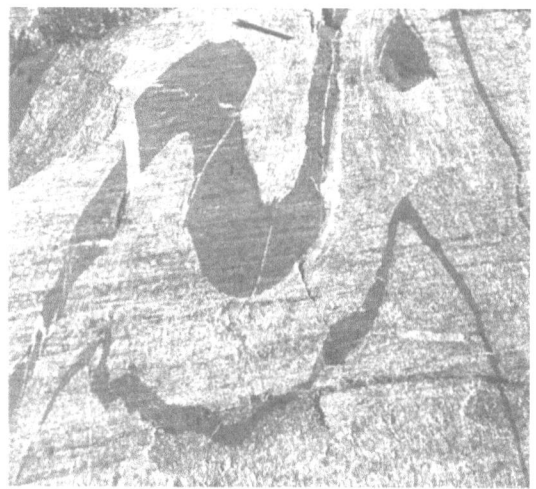

Figure 1.12 Small-scale folds such as this one could indicate the existence of related large-scale structures. Svecofennian migmatite, Jussarö area, southern Finland.

nebulite lacking in any but the faintest relict structure, or even an anatectic granitoid with almost uniform texture (Figure 1.13).

Despite this, it can still be possible, on a larger scale, to define zones parallel to the original foliation in which refractory 'resisters' ('restite' or 'palaeosome') of say hornblendite (Figures 1.14a, b) are predominant, which can be mapped on a lithostratigraphical basis as lithological zones.

The gradual disappearance of the structure by such 'granitization' is a phenomenon commonly observed in cratonic areas where deep crustal rocks are well exposed (Figure 1.15).

Nevertheless, even in spite of such extreme effects as in these cases, it might still be possible to learn a considerable amount about the structure. The effects of ultrametamorphism (anatexis or 'granitization' producing

(a)

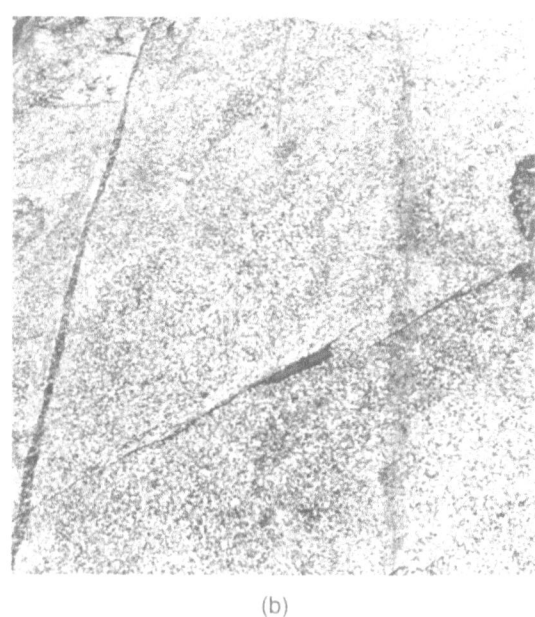

(b)

Figure 1.13 (a) Nebulite development, showing transition from folded faintly foliated gneiss (left) to almost macro-structureless granitoid (right). (b) Detail of granitoid texture showing the absence of the gross foliation present at the left of (a). Pre-Ketilidian migmatite, West Greenland.

(a)   (b)

Figure 1.14   Folds preserved in agmatite as relict palaeosome blocks (resisters) in leucocratic neosome.
(a) Replacement of fold core by neosome. Kinarsani, Andhra Pradesh, India. (b) Massive replacement of
foliation by neosome. Svecofennian migmatite. Åland Islands, Finland. Compare Figure 1.15d.

nebulites, migmatites and agmatites) which
tend to obscure all structures such as folds,
foliation etc. is not necessarily a serious
problem, especially on a large scale, because
in many cases the granitization has been selective, having been controlled lithologically,
so that the distribution of the granite reflects
that of the original rock type – the protolith.
Hence, while the foliation might have been
obliterated and the lithological boundaries
might be blurred and overlapped, on a large
scale it is often still possible to use lithology in
a general way to show the structure on a map
covering an area of several square kilometres.
Such is the case in southern West Greenland
(Figure 1.13; also Figure 1.14) where tonalites
and dioritic gneisses are so poorly banded that
it would not have been possible to produce a
structural map were it not for the fact that the
'granitization' was lithologically controlled
(Figure 1.16).

It is often the case that close examination of
nebulites shows relicts of structures which

deformed the 'original' foliation and which
are earlier than the new melt (neosome) of
the nebulites themselves. With careful observation the relationships between even these
structures can be resolved (Figure 1.17).

Furthermore, mechanical masking or obliteration of structure, e.g. by attenuation and
shearing-out (referred to earlier) of fold limbs
to leave rootless, intrafolial hinges need not
necessarily mean that the relationship of these
folds cannot be established.

In a region where foliation transposition
has taken place because of the later, strong
development of a foliation with a new trend,
perhaps parallel to the axial plane of tight
folds and oblique to the lithological layering (especially in poorly, or intermittently
exposed ground), accurate mapping of the
structure on a large scale may be difficult,
if not 'hazardous' because the trend of the
new foliation might be mistaken for that
of the old. Failure to recognize the effects
of this transposition can lead to incorrect

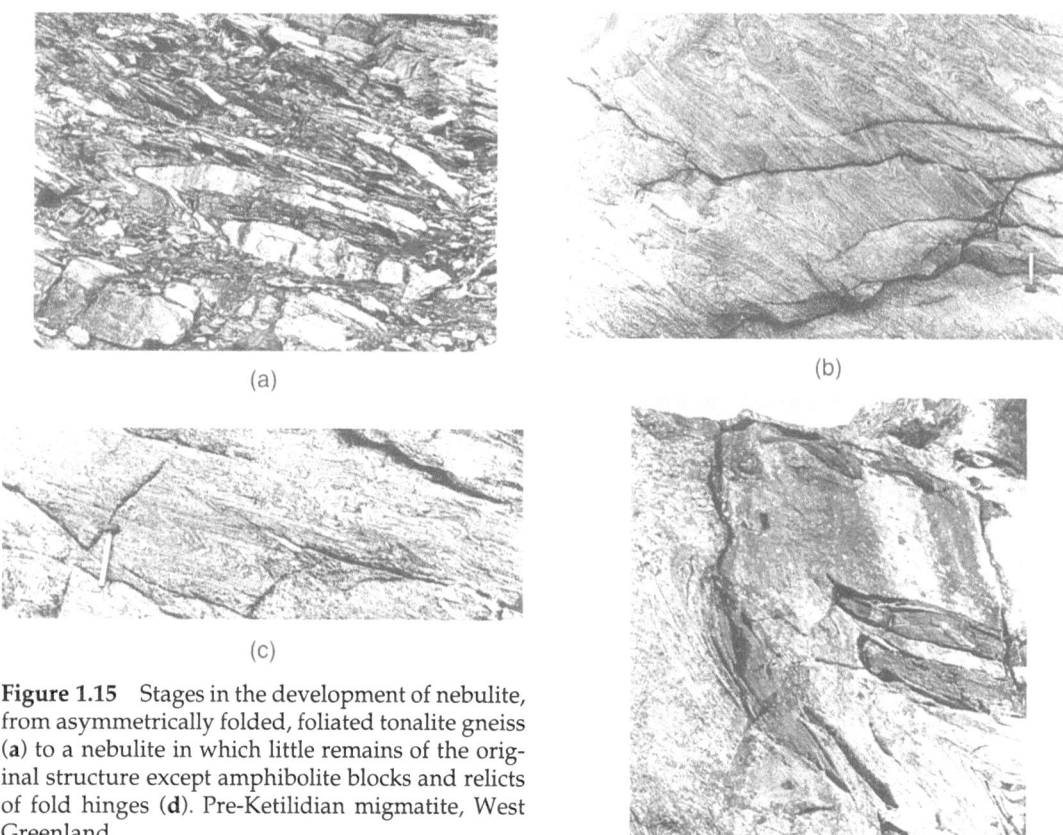

(a)

(b)

(c)

Figure 1.15   Stages in the development of nebulite, from asymmetrically folded, foliated tonalite gneiss (**a**) to a nebulite in which little remains of the original structure except amphibolite blocks and relicts of fold hinges (**d**). Pre-Ketilidian migmatite, West Greenland.

(d)

interpretation of the structure. On the other hand, recognition of this effect is valuable in that it provides evidence of another deformation stage and if the new foliation was associated with a distinctive mineral growth its value would be further enhanced (Figures 1.18a, b, c).

Even in the absence of marker horizons it is still possible to use the presence of small-scale structures to infer the presence of large structures, and the sense of vergence of small folds will serve to establish the positions of limbs and hinges of these large structures (see the discussion of 'Vergence and symmetry aspects of large folds', section 10.1.3.

1.1.4   INFLUENCE OF EXPOSURE TYPE ON THE STRUCTURAL STUDY OF MIGMATITES

The places where structure studies of this kind are likely to be undertaken have a wide geographical distribution, from polar to equatorial, but they will tend to be confined to cratonic, i.e. 'shield' areas, and usually to continental areas, viz. Africa, Asia, India, Australia, Greenland, Northwest Britain, North and South America, the Fennoscandian Shield (Finland, Sweden, Norway). However, this is not necessarily so. For example, the complex fold structure of Mesozoic rocks in New Caledonia bordering the convergent margin between the Pacific and Indo-Australian lithospheric plates also stems from multiple deformation (Figure 1.19).

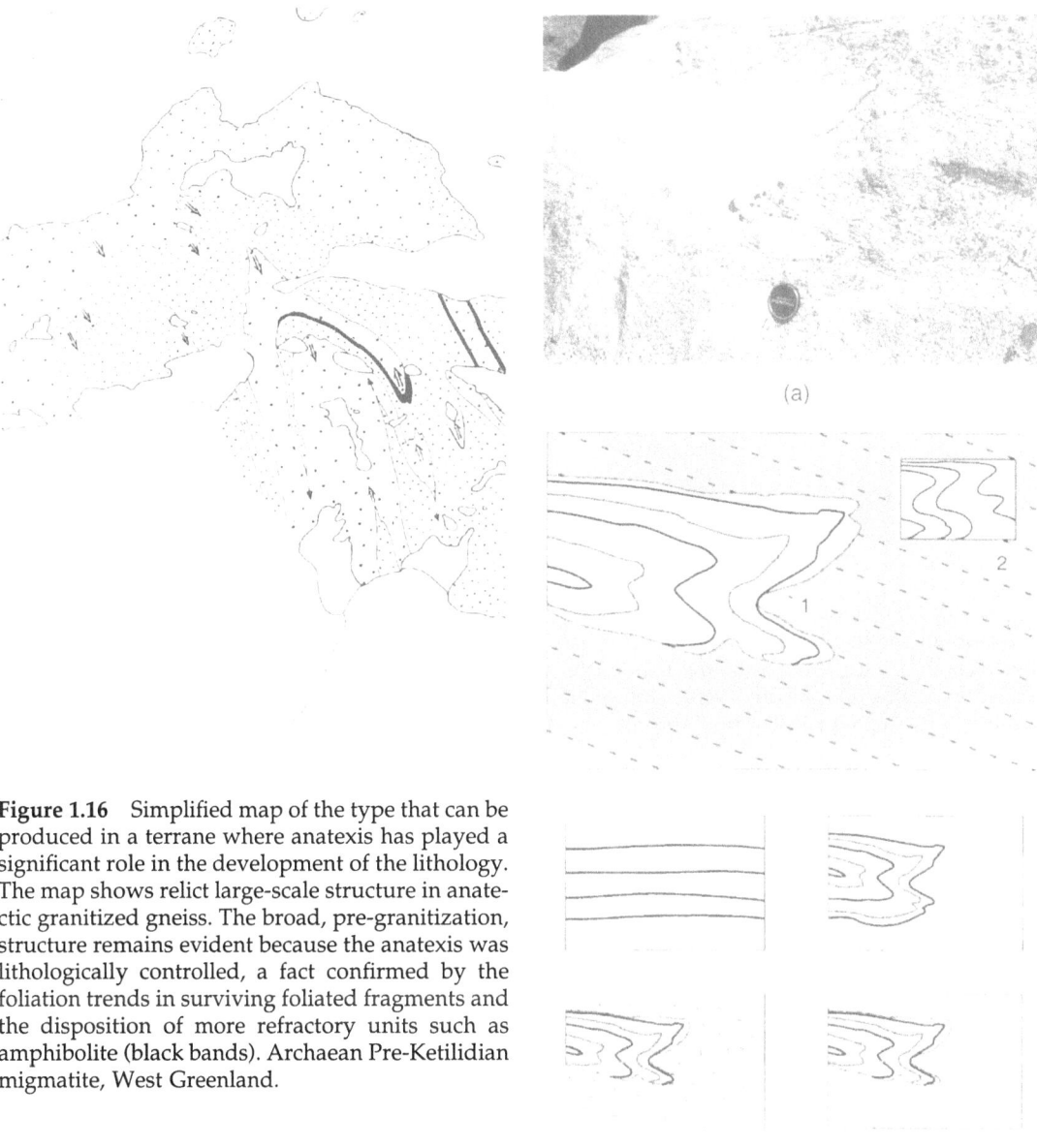

(a)

(b)

Figure 1.16 Simplified map of the type that can be produced in a terrane where anatexis has played a significant role in the development of the lithology. The map shows relict large-scale structure in anatectic granitized gneiss. The broad, pre-granitization, structure remains evident because the anatexis was lithologically controlled, a fact confirmed by the foliation trends in surviving foliated fragments and the disposition of more refractory units such as amphibolite (black bands). Archaean Pre-Ketilidian migmatite, West Greenland.

Figure 1.17 (a) Photograph of a palaeosome block in neosome containing traces of relic palaeosome structure, a nebulite. (b) Sketch of (a) showing the trace of the later, post-agmatization foliation in the neosome (dashed lines) discordant to the early foliation and four smaller sketches showing the stages (1–4) in the development of the present structure. The neosome encloses, at (1), a clearly-defined enclave (palaeosome) containing relict folds formed in early foliation prior to the neosome and at (2), faint 'ghost' remnants of early folds not completely replaced by the neosome. Hangö, Finland.

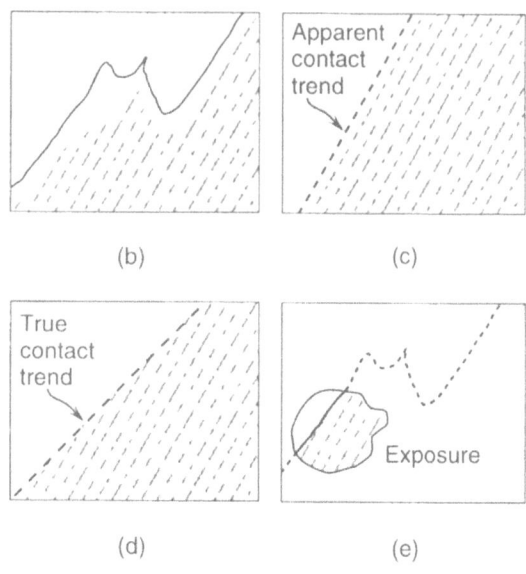

(a)

Figure 1.18 (**a**) Photograph of the contact between foliated leucocratic gneiss on the left and strongly foliated amphibolite showing (**b**) broad angular discordance between the foliation and contact and (**c**) a likely (but incorrect) interpretation of the relationships between the foliation and the contact in the case of limited exposure (**e**), and the correct interpretation (**d**). In (**a**) the true relationships are obvious only because of the high quality of the exposure. Pre-Ketilidian complex, West Greenland.

Figure 1.19 Two sets of intrafolial isoclinal folds refolded around a round-hinged tight fold. The earlier isoclinal hinge closes at the broken rock (top, below the hammer shaft at the letters 'ER' of 'UNIVERSITY') and the later hinge is below the hammer head. East side of Pam Peninsula, New Caledonia.

Figure 1.20 General view of the terrain on the northeast side of Col d' Amos, New Caledonia, showing the quality of the outcrop.

Figure 1.21 (**a**) Exposure resulting from the effects of glaciation. Lewisian complex, Isle of Rona, Inner Hebrides, Scotland. (**b**) Shoreline exposure. Folded gneiss and amphibolite, Lewisian complex. Isle of Barra, Outer Hebrides, Scotland.

Figure 1.22 Exposures typical of rock outcrops affected by tropical and sub-tropical weathering. (**a**) Mazoe River, northeastern Zimbabwe. (**b**) Hilltop exposure. Wheatbelt, Western Australia.

Consequently the degree and type of exposure will vary considerably, from nearly one hundred percent in glaciated areas such as Greenland to tropical areas where the exposure may be very limited indeed, or where there are intermittent outcrops in river beds or along shorelines. Examples of some of these are shown in Figures 1.21–1.25.

Tropical weathering resulting in differential weathering and erosion often produces well-developed, three-dimensional outcrops of larger structures. The gross structural features tend to weather out well, although the finer detail is commonly lost by abrasion of the softened rock by wind-blown sand, or obscured by weathering products (Figure 1.25). However, because differential weathering depends on the lithological contrast between the deformed layers, sometimes the converse is true and the detail of the structure is **enhanced** (see Figure 1.27).

In all of these areas (even in the most extensive exposure such as in Greenland), when attempting to build up a structural succession,

Figure 1.23 Antiform crest. Pre-Ketilidian complex. Dalagers Nunatak, Southwest Greenland. Recently glaciated outcrops afford good widespread exposure of large-scale structural features but are of less value for showing fine detail because the rock surfaces are commonly obscured by glacial detritus, rock flour and lichen.

Figure 1.24 Island shoreline exposure in the Fennoscandian Shield. Exposure originally caused by glaciation and recently modified by wave action, as in the Finnish Archipelago, is excellent for revealing even the finest structural detail. Svecofennian migmatite, Skåldö, southern Finland.

Figure 1.25 Gross three-dimensional structure accentuated by differential weathering in coastal exposure of gneiss. In this case some of the finer structural detail stands out because of the contrast in colour between the lithological units. Sugar Loaf, Western Australia.

Figure 1.26 Isoclinally folded basic layer in leucocratic host rock. Compare the clarity of this structure where there is a strong colour contrast between the fold and the host rocks, with the barely visible structure within the host rock at the left of the picture where there is little contrast between the folded layers. Belemorides, Soviet Karelia.

Figure 1.27 Fine structural detail enhanced by differential erosion of quartzo-feldspathic gneiss. Open, angular 'recumbent' folds affecting intrafolial isoclinal folds. Mazoe River, Zimbabwe.

Figure 1.28 Gneiss etched by wind-blown sand. Differential abrasion by sand-blasting has accentuated the rectilinear trace (parallel to the pencil) of a late fracture cleavage crossing the gneissic foliation. Lewisian complex, Tangusdale, Isle of Barra, Outer Hebrides, Scotland.

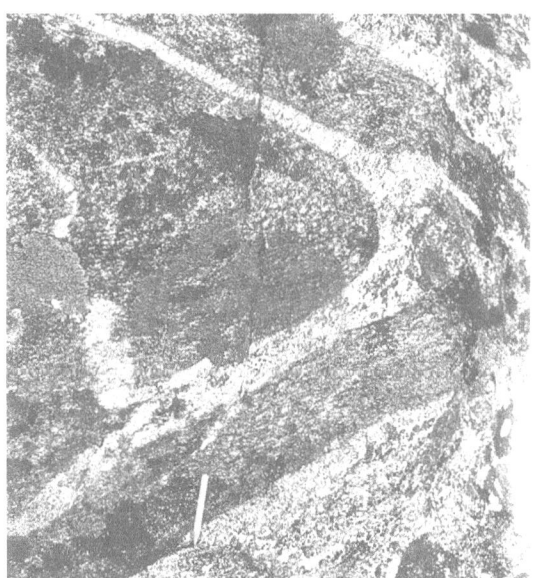

Figure 1.29 Fold in migmatite exposed on a glaciated surface partly covered by lichen. Although the broad shape of the fold profile can still be seen, the associated fine structural detail is not clear because it is obscured by the lichen. Pre-Ketilidian migmatites, West Greenland.

Figure 1.30 The first stage in studying the structure in migmatites accentuated by the leaching effect of plant acid. Rolling back the vegetation cover. Forest near Orijärvi, southern Finland.

an important objective is the comparison and correlation of isolated exposures whether on lake shores and islands (as in Finland) or nunataks (in Greenland) or in dry river beds such as in tropical zones. There will be a difference in the type of information one can obtain from different exposures. In Greenland, for example, although exposure is very often nearly one hundred percent, the rocks are commonly covered by glacial dust (rock flour) or by lichen (Figure 1.23). This means that in such terrain only large-scale structures are obvious and even then it is only the boldest and clearest of those formed in strongly contrasting lithological layering that stand out (Figure 1.26).

However, in most cases it is possible to extract a considerable amount of structural information from the rocks, even when they are strongly weathered (as is the case in southwestern Australia), because here the weathering tends to accentuate the three-dimensional aspect of the exposure (as happens in micaceous layers) by removal of the most easily altered minerals. Such is the case on the exposure at the Sugar Loaf in southwestern Australia, shown in Figure 1.25.

On the other hand, as noted earlier, tropical differential weathering sometimes enhances even the very finest structural detail as can be seen in Figure 1.27.

Beach exposure can be very informative because wind-blown sand causes differential abrasion which accentuates the fine structural detail (Figure 1.28). This also applies to river bed exposures if sand is present (Figure 1.27) because there also delicate structures can be highlighted by differential abrasion by wind, or water-borne sand.

While lichen can present a problem in structural interpretation because it obscures fine structural detail in some recently glaciated terrains like those of Greenland (Figure 1.29), in others, such as Scandinavia, the thin vegetation cover of moss and shrubs can be peeled or rolled off to expose rock surfaces on which the structural detail has been etched differentially by plant acids (Figure 1.30). Initially only the coarsest structural features are visible, but after the exposure has been left for a period to allow rain to wash off the detritus leaving the surface clean, this reveals the fine structure which can be studied in detail.

PRINCIPLES OF STRUCTURAL ANALYSIS OF MIGMATITES 2

2.1 COMPLEX STRUCTURES CONSIDERED AS COMPONENTS OF A SUCCESSION

2.1.1 INTRODUCTION – THE NEED FOR A SYSTEMATIC APPROACH

It should be a source of encouragement to the inexperienced observer that, in most cases, the lack of success when attempting to resolve the structural complexity of migmatites and other rocks affected by multiple deformation stems largely from a failure to adopt a systematic, sequential approach to the study. Presumably this is because, having initially been overawed by the complexity of the structure, the observer has forgotten, or has not appreciated (or accepted), the fact that the structure is the result of overprinting and is therefore amenable to systematic analysis. This first impression is understandable and has undoubtedly been experienced by everyone faced with such complexity for the first time. Nevertheless, the reasons for the structural pattern observed and the consequent fact that it can be resolved should never be forgotten.

Paradoxically, it is this complexity, which at the beginning of this century led Sederholm (1967) to propose for these rocks the name 'migmatites' (section 1.1.3), that can often be utilized to unravel their complicated history. This has been shown time and again by many studies which have resulted in the successful resolution of the structural complexity in Precambrian migmatite terranes throughout the world (e.g. Hopgood, 1976 – Greenland,

1984 – Finland; Hopgood and Bowes, 1990 – Siberia, 1995 – southwestern Australia; Joubert, 1971 – southern Africa; Laing, et al., 1978 – Australia; Naha et al., 1990 – India; Suo et al., 1982 – China; Tobisch et al. 1970 – northwest Britain). Apart from the restrictions imposed by time when analysing this structure, the limiting factor in most cases is the amount and type of exposure. The most successful approach to solving the structural complexity of migmatites is one that, perhaps surprisingly, requires no special equipment or techniques, relying as it does solely on the application of principles fundamental in geology. It does however, require **objectivity** on the part of the geologist and careful **observation**, and each new study should be approached without preconception or prejudice regarding the likely structure and tectonic history or its causes.

While a structural study of deformed rocks may have several objectives depending on the circumstances of the investigation, where the structure is the result of multiple deformation one of the principal aims will be (or is likely to be) the establishment in the first instance of the structural succession for which a reasonable deformational sequence or history can be proposed. The recognition of the structural succession and an understanding of the deformational history that produced it may be required for any number of reasons, economic or academic, such as studies involving the determination of the structural control of ore bodies, geochronology, geochemistry, igneous and/or metamorphic petrology, deformational mechanisms, major tectonic processes, deep crustal development and so on.

In some cases the deformational history may have been relatively simple so that the structural succession to be determined is short and uncomplicated whereas in others, particularly in basement complexes, the structure of the terrane under consideration may be very complex indeed. It is with the latter case, involving elucidation of structural successions stemming from multiple deformation, that the discussion in this book is primarily concerned.

The structural succession

A primary objective in the study of a complexly-deformed terrane is to establish a **succession** (of structures). These structures include structural elements and their relationships as well as relationships with igneous intrusions and the products of metamorphic episodes such as mineral growths (which may

or may not be syntectonic). Commonly a subsequent objective will be the determination of the **sequence** (of events) that produced the successions of structures.

As a simple example of what determining a structural succession means in principle, consider the case of foliation that has been folded asymmetrically with the fold limbs offset by shear along the fold axial plane. Parallel to the axial plane, and more or less replacing one limb of the fold, there is an unsheared leucocratic vein containing hornblende crystals exhibiting preferred alignment oblique to the fold axial plane. A structure comparable to this is shown in Figure 2.1a and explained in Figure 2.1b.

The succession of structural features in this case (Figure 2.1b) is as follows:

5.   Aligned hornblende crystals;
4.   Axial planar leucocratic vein;
3.   Axial planar shear surfaces;
2.   Asymmetrical fold;
1.   Foliation.

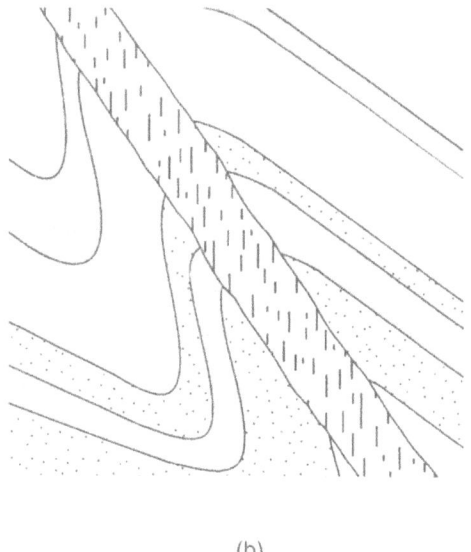

(a)   (b)

Figure 2.1   (**a**) Asymmetrical fold with a lineated, leucocratic axial planar vein. Pre-Ketilidian migmatite, West Greenland. (**b**) Sketch of structure comparable to that shown in (**a**) showing offset of the fold limbs by sinistral shear parallel to the fold axial plane and an axial planar vein containing crystals (of hornblende) in preferred orientation. The succession of structures is shown in the text.

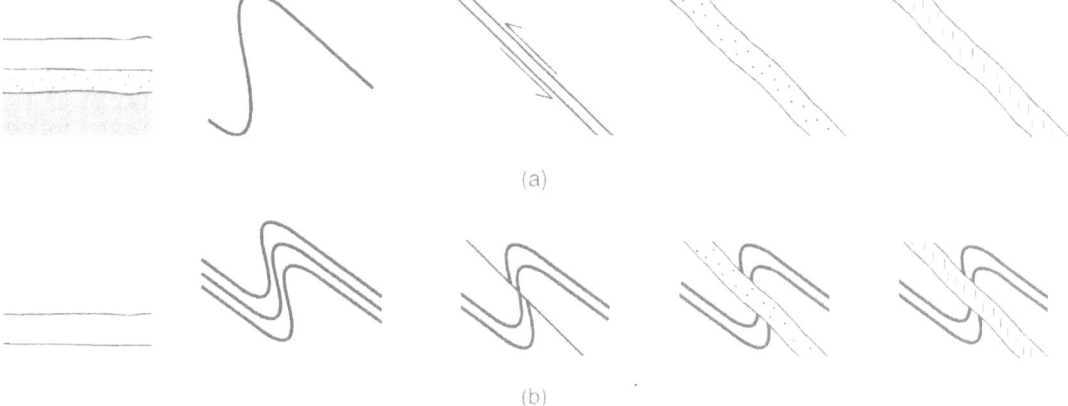

(a)

(b)

Figure 2.2 (**a**) Sequence of events (1–5), leading to (**b**) the succession of structures (1–5) which together comprise the structure shown in Figure 2.1.

These structures were produced by the sequence of events shown in Figure 2.2.

Thus the sequence of events leading to the development of these relationships would be something like the following:

1. Sedimentation?
2. Metamorphism producing foliation
3. Asymmetrical folding
4. Offset of fold limbs by shear parallel to the fold axial plane
5. Leucocratic vein emplacement within the shear zone
6. Syntectonic metamorphism to produce aligned hornblende crystals.

Table 2.1 compares the succession of structures with the sequence of events which produced them.

Table 2.1

Succession of structures	Sequence of events
5. Hornblende growth	6. Metamorphism + deformation
4. Vein	5. Vein emplacement
3. Shear surface	4. Offset
2. Fold	3. Folding
1. Foliation	2. Metamorphism (? + deformation)
	1. Sedimentation?

2.1.2 IMPORTANCE OF THE STRUCTURAL SUCCESSION IN MIGMATITE TERRANE INTERPRETATION

In determining the relative chronology of successively-formed structures one is also establishing the basis for determining the relative chronology of metamorphic and igneous episodes. This in turn forms a basis for isotopic studies enabling time spans to be established for tectonothermal episodes as well as allowing one to determine the timing of the intervals between the events responsible for structure sets within these episodes. The structural chronology so established also provides a basis for demonstrating the changing pressure and temperature conditions within that part of the orogen being studied, as well as changes in these parameters from place to place within the orogen. In addition, the nature of structures developed can provide evidence relating to the movement of lithospheric plates. For example, the evolution of the Svecofennian migmatite complex of southern Finland appears to have taken place in response to alternating compression and slip, suggesting a combination of lithospheric plate collision and strike-slip fault movement (Hopgood, 1984). Integration of neosome emplacement episodes with this structural evolution further provides a basis for tracing the changing pattern of the nature and depth of magma sources and so of assisting in the understanding on a broad scale of the overall development of orogenic regimes.

A number of important points follow from this.

1. The structural succession, particularly where it is complex, could be used as a basis for correlation between formerly contiguous crustal segments separated by major tectonism such as thrusting and transcurrent faulting, provided the separation took place **after** the imprint of the structural patterns being studied (e.g. in the Lewisian complex of northwestern Scotland, Hopgood and Bowes 1972, and in the central European Hercynides of Bohemia, Hopgood *et al.*, 1995). On a larger scale, structural successions could be used in the matching of crustal terrane segments such as those of the east coast of South America and the west coast of Africa or between the southwest coast of Australia, and the east coast of India and Antarctica (Hopgood and Bowes, 1995). See Chapter 14, 'Broad Applications', and section 6.1.3, 'Structural correlation of tectonostratigraphic terranes'. Such structural matching is comparable to matching pieces of a three-dimensional jigsaw puzzle. However, because it requires the coincidence of a much more complex 'pattern' (involving not only the matching of the physical shapes of the structures but also matching of the products of metamorphic and igneous events), it is likely to be more rigorous. See Figure 8.8 in section 8.1.3, and Figure 12.8 in section 12.5.

2. The structural succession can be used for establishing the position of ore bodies in the structural succession making it possible to determine the distribution and timing of the ore formation. This serves as a basis for understanding the nature of the structural control on the emplacement of the ore bodies which may be highly modified subsequently (Koistinen 1981). This in turn can assist in the exploration for other similarly controlled ore

bodies in the same tract but outside the area studied initially, especially when this knowledge is used in conjunction with geophysical (magnetic and gravitational) data.

3. The study of structural successions in Precambrian terranes and the sequences of deformation causing them can not only explain the development of interference patterns produced by multiple deformation but can also provide clues to the nature of early crustal deformational processes (Hopgood, 1984).

4. Resolution of the structure of deeply eroded terranes provides an insight into the nature of early deep crustal history and the processes responsible for crustal development. From the structural succession it may be possible to determine the relative time and pressure–temperature relationships influencing migmatization; metamorphism, 'granitization' and partial melting (anatexis), intrusion and deformational processes.

5. Isotopic dating of igneous and metamorphic events in the sequence based on the established succession enables the ages of the deformational events to be determined. It also provides the time spans involved for individual deformational, igneous and metamorphic events and the lengths of the intervals between these (e.g. Hopgood *et al.*, 1983).

2.2   RESOLVING STRUCTURAL COMPLEXITY IN MIGMATITES

2.2.1   INTRODUCTION TO THE APPROACH

It is the intention to show how, using simple basic principles coupled with careful, objective observation, complicated structural patterns arising from extensive sequences of multiple deformation can be interpreted in the field and the component structures so identified placed in their correct order in the succession. Because an extensive knowledge of fold mechanisms is not, for the most part, essential for this approach to the resolution of structural successions, an exhaustive coverage of folding and fold mechanisms is not included here. These aspects will be discussed only where they have a bearing either on the relationship between the behaviour of different structures in response to deformation, or on the control they have on the nature of the resultant structure seen in the field. For more detailed and advanced treatments of folding and other types of deformation the reader is referred to textbooks such as those by Ghosh (1993), Hobbs, Means and Williams (1976), Park (1989), Ramsay (1967), Ramsay and Huber (1983, 1987) and Turner and Weiss (1963). However, so that the reader can feel competent to undertake a structural investigation without recourse to extensive further reading, a very brief discussion of those aspects considered to be fundamental to the analysis of structures formed by multiple deformation is included (section 2.2.5 'Fold Mechanisms') before the specialized treatment which is the main subject of this book.

The methods are illustrated by examples drawn from various migmatite and gneiss terranes throughout the world, starting with moderately simple cases where the deformational sequence has been comparatively short (section 10.1.9 'Structures associated with one or two folds'). These are followed by examples of greater complexity. In this way it is hoped that the reader will progressively develop sufficient confidence to tackle structures of increasing complexity, and ultimately will be undaunted by even the most complex structural patterns of migmatite terranes, such as those of the Svecofennian exposures in the archipelago of the Gulf of Finland (see Figures 10.1–10.4 in section 10.1.2, Figure 10.38 in section 10.1.10, and the examples discussed in Chapter 11).

While systematic analysis of the structure of deformed rocks is theoretically possible using stereographically-plotted orientation data that have been recorded non-selectively during the survey of terranes affected by limited multiple deformation (say two or three episodes or 'phases'), this is seldom, if ever, possible in practice if the deformation is more complex. The resolution of the complexity must be carried out on the basis of observations of structural relationships on the outcrop.

As has been stressed previously, and will be again, two important qualities are demanded of the structural geologist in the field: firstly the capability for critical, objective observation, and secondly, patience and a willingness to take the time to look closely and thoroughly at the field relationships. If this is not done, the early stages of the study will almost certainly lead to the conclusion that the structure lacks order and cannot therefore be resolved. Lack of critical observation may well be the reason for statements in the literature (generally unsupported by evidence) suggesting that structural successions of more than a few sets (section 5.3) cannot be established, and that recognition of more than about four 'generations' of structures is generally quite speculative because it is based on assumptions relating to style and correlation. But in the same context it must also be said that the acceptance of the existence of structural successions of fewer than five or six structure (fold) sets that have been established using the same criteria implies tacit acceptance by these critics of the basic principles used. It would seem therefore that while there has been acceptance of the basic principles of succession analysis there is a psychological barrier to accepting that these principles can be applied to extensive, complex successions. In other words, there is a reluctance to accept that the order of succession can be resolved in such cases solely because of the sheer number of components, perhaps because it is considered to be impracticable. This perception serves to reinforce the need for a systematic and thorough, rather than superficial, examination of the structure before conclusions are drawn regarding its development. One must be prepared to take the time and trouble to 'get down on one's knees', as one geologist expressed it, and study the outcrop closely.

While it is hoped that this book will be useful to the geologist who has little or no practical experience of working on the structure of repeatedly deformed gneisses and migmatites, it must be said that an illustrated written explanation of the methods can be a guide only. It cannot be expected satisfactorily to replace explanation and discussion on the outcrop where experience has shown that an understanding of the techniques can be successfully taught in just a few hours. Nevertheless, this presentation of an approach which has already been tested by researchers in highly deformed terranes, coupled with the examples described, should help even the inexperienced geologist to tackle the investigation of the most complex structural relationships and, with perseverance, enable them to resolve these successfully.

The presentation has been arranged so as to enable the reader with some experience of this kind of structural geology to proceed with a minimum of time spent on explanations of already familiar principles and examples. The blocks of explanation in close-spaced text can simply be omitted. On the other hand, for those who want it, this explanation is immediately available in context, so that there is no need to stop to refer to other sources and so lose the train of the discussion.

2.2.2 PRINCIPLES OF STRUCTURAL ANALYSIS

According to Sander (1911) a record of every aspect of the deformation (from the largest scale down to sub-microscopic) is preserved in the deformed rock as its **structure** (in this respect see also the reference to the statement attributed to E. Cloos in section 1.1.2). Where the deformation and the consequent structure of the deformed rock (the **tectonite**) are pervasive, the structure is said to be **penetrative**. Subsequently, both observational and experimental evidence have supported this thesis so

that it does indeed appear consistently to be the case that the tectonite contains, or (strictly **is**) a structural record of the deformation. This being so, the structural record so formed is potentially recognizable if the structures can be identified and their order of succession can be determined.

When endeavouring to resolve structural relationships in rocks affected by superimposed deformation it is first necessary to be able to recognize evidence of overprinting (Sander's '*Überprägung*'). In the case of superimposed **folds** this can often be found in the **relationships between their axial directions** which commonly exhibit angular discordance to those they overprint. Less common is the angular discordance between the **axial planes** of superimposed folds. Hence consistent non-parallelism of the hinges of folds seen on the outcrop is a clear indication that the rocks could have been affected by repeated, or multiple folding. However, it is theoretically possible to have very little or no angular discordance between either axial attitudes or axial planar attitudes of overprinted folds in which case there would be little or no visible evidence of superimposition. In such a case other evidence must be sought for the relationships between the fold sets and this will be discussed later (Chapter 5).

Nevertheless, it should be noted when considering non-parallelism of fold hinges as evidence for multiple deformation, angular discordance can arise in ways other than by tectonic overprinting (sections 3.2.1 and 3.2.5) and these must, of course, be distinguished from tectonically overprinted structures (Hobbs, *et al.* 1976, Section 3.6, pp. 156–9).

The geometry of folds and the nature of their re-folded relationships depends to some extent on the type of fold mechanism, e.g. whether flexure or slip (section 2.2.5, 'Fold mechanisms'), and this is also likely to affect the angular relationship both between fold hinges and between axial planes (Figure 2.3). However, it is unlikely that any particular mechanism would result in the consistent **absence** of

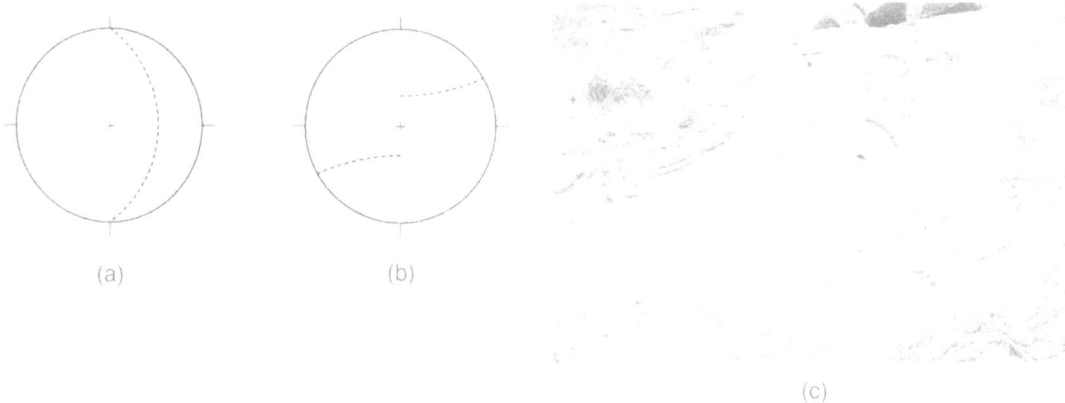

Figure 2.3 Stereo plots showing the distribution of reorientated linear data. (**a**) Reorientated by slip. Originally north-trending, horizontal lineation dispersed in a plane (east-dipping dashed great circle). (**b**) Reorientated by flexure. Originally horizontal, northeast-trending lineation dispersed on a conical surface (dashed small circle) as in the example shown in (**c**). (**c**) Linear structure parallel to the pencils curved by flexure across open folds with hinges parallel to the handle of the hammer in the background. Lewisian migmatites, Mingulay, Outer Hebrides, Scotland. (See note in section 2.2.5 on stereographic projection of structural data.

clearly recognizable overprinting relationships. Rather, it is more likely that the fold mechanism will have a systematic controlling influence on the variation in angular discordance between the elements (e.g. hinges) of one fold set and those of another between which there is an overprinting relationship. Hence, if the nature of the angular variation can be established it may be possible to use this to determine the mechanism which caused the refolding (section 2.2.5).

2.2.3   IDENTIFYING GROUPS OF RELATED FOLDS

When dealing with groups of (related) folds in a series, the aim of the method is first to identify those groups of folds that are related, i.e. the fold **sets** (section 5.3), and second to place these groups in order of **succession**. Although the distinction is probably artificial in many cases, it is convenient to regard each set as being the result of a separate deformational **event** or **phase**, and therefore belonging to a single **generation**. A second order aim is to establish a deformational **sequence** from which the structural succession is considered to have been derived.

2.2.4   EFFECTS OF MULTIPLE (POLYPHASE) DEFORMATION

The use of the word 'polyphase' in the expression 'polyphase deformation' – the repeated deformation of rocks – is analogous to its use in the term 'polyphase metamorphism', i.e. '. . . two or more successive metamorphic events that have left their imprint . . .' (American Geological Institute *Dictionary of geological terms*, 1976). This concept is discussed further in sections 5.1, 5.3 and 8.1.5.

Here the word 'phase' and the expression 'polyphase deformation' will be used in the sense of an **event**, on the one hand and **process(es)** on the other, having the effect of producing a succession of fold (structure) sets which can be recognized and correlated consistently throughout the region under consideration. Such processes result in successively-formed structures which are themselves deformed in turn by those produced later in the sequence. The concept of structure

sets (section 5.3) is independent of genetic criteria because the nature of the deformation causing the sets, and the conditions relating to the genesis of the structures are not inherent in the definition. Therefore questions such as those concerning the mechanism type and whether or not the structures formed in response to individual pulses, episodes, or phases or as part of a continuous process are not directly relevant to the sequential structural analysis and need not be considered at this stage. Genesis is a different order of problem and one that can be tackled once the succession of structures has been established.[1]

As stated earlier, when applied to the relationships between folds, recognition of overprinting depends essentially on the identification of angular discordance between fold axial, and axial planar attitudes arising from interference between successive folds, i.e. it depends on the ability of the observer to (1) recognize on the outcrop the intersection of fold hinge directions and/or their axial planar traces, and (2) determine the sequential relationships between the folds observed.

Intersecting fold directions can stem from a number of causes, both tectonic and non-tectonic (e.g. Ramsay 1967, Chapter 10; and the comments on the internal response to external stress fields in section 7.1.1 herein following the fourth of the 'assumptions' regarding the development of structures in a structurally complex terrane). Intersecting tectonic fold directions may be caused by one or more of the following conditions.

1. Simultaneous folding in more than one direction, either by extension perpendicular to the layer that is folding with concomitant contraction within the layer by constrictional strain (Figure 2.4), or as has been shown by Ghosh and Ramberg (1968), as a result of the anastomosing of axial directions within a single generation of **statistically** cylindrical folds (Figure 2.5). These are **not** examples of superimposed (sequential) deformation and the first of these is discussed further in 'Structures arising from constrictional deformation', section 3.2.1.

2. Refolding associated with progressive deformation during which frictional or viscosity differences within the layers caused curvature of the initially formed hinge, as

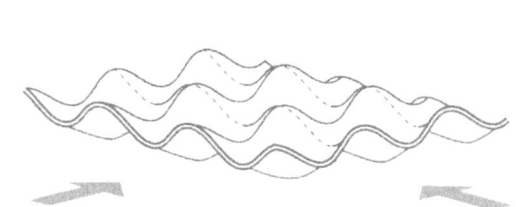

Figure 2.4 Layer-parallel contraction by folding in response to perpendicular compression within the layer.

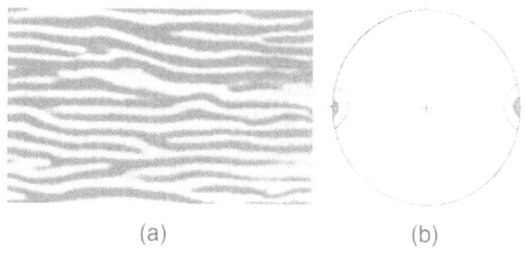

(a)   (b)

Figure 2.5 (**a**) Statistically parallel anastomosing folds produced experimentally by Ghosh and Ramberg. (**b**) Sterographic projection of the fold hinges produces a point maximum. From Figure 2, Hopgood, 1980. Reproduced with the permission of the Royal Society of Edinburgh.

---

[1] It is also important to bear in mind that deformational and metamorphic episodes are not necessarily synchronous: a deformational event ('phase') does not necessarily coincide with a metamorphic peak (or 'phase'). Indeed a deformational event is not necessarily accompanied by 'significant' metamorphism and diachronism between metamorphism and deformation, both within the same 'region' and between different 'regions', can have an important bearing on the relationships between structure and metamorphic grade. This is discussed in detail in metamorphic textbooks and its importance in relation to structural successions is referred to again later (see also Table 8.2).

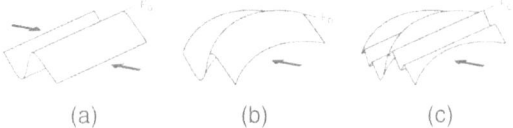

(a)   (b)   (c)

Figure 2.6 Development of cross-folds (F_c) on a pre-existing ('original') fold (F_o) during continuous progressive deformation under the same directions of horizontal compressive stress. From Figure 3, Hopgood, 1980. cf. Figure 2.7. See also Figure 8.9.

later, more or less rectilinear, sub-parallel hinges were overprinted penecontemporaneously in response to the continuation of the same stress system (Figure 2.6). This effect appears to be shown on a larger scale in Heim's 1919 map of the Jura Mountains, shown as Figure XIII – 8A, p. 422 in Hills (1963). The map (Figure 2.7) suggests superimposition of later folds with approximately NE-trending, near-rectilinear axial trends on the Jura Fold Arc. Conditions such as these are likely to be a common cause of superimposed folding.

3. Refolding stemming from crossed orogenic belts. An example is the superimposition of E–W Hercynian folds on northeast-trending Caledonian folds in Britain.

4. Refolding in response to successive deformation pulses in a stress field whose orientation changes with respect to the existing structures.

The treatment of structural resolution considered here is concerned mainly with the analysis of the complex relationships between superimposed folds resulting from cases (2) and (4) above, viz. from progressive deformation and from successive deformational pulses. It concerns folds that are mostly small-scale or 'mesoscopic' (Turner and Weiss 1963, p. 15). These are structures that can be seen more or less in their entirety on the outcrop. Nevertheless, other structures besides folds are important and in this respect reference should be made to the comments (and Figure 3.11) in section 3.1.3 relating to one-, two-, and three-dimensional structures.

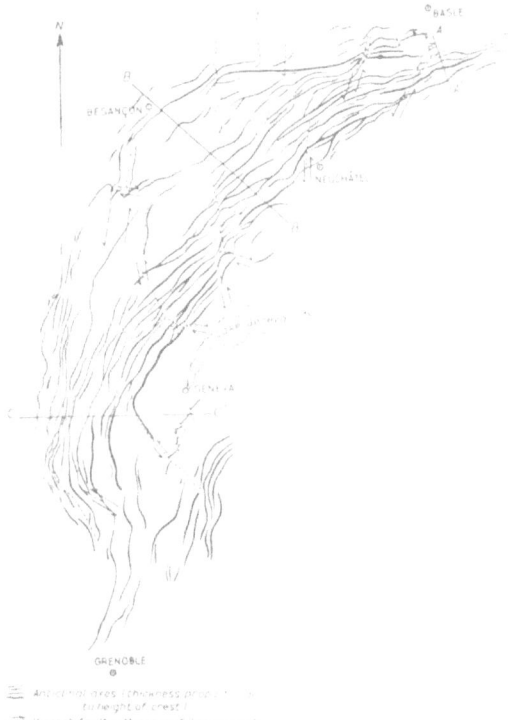

Figure 2.7 Arcuate fold belt of the Jura mountains. Although the fold belt is curved, there is a preponderance of more or less NE-trending, approximately rectilinear hinges suggesting that they are cross-folds superimposed on the arcuate structure (e.g. north of Neuchatel and south of Geneva). Adapted from Figure XIII-8A, Hills, 1963, after Heim, 1919. Compare Figure 2.6.

Even a single refolding can cause a wide range in angular discordance between hinges, from perpendicularity to near parallelism (Figures 2.8a, 2.8b), so that reorientation of fold structures in response to repeated deformation can result in widely divergent hinge attitudes whose behaviour during subsequent refolding cannot necessarily be predicted, and whose ultimate relationships may be difficult to analyse. This is because, in the simplest terms, two fundamentally different mechanisms of folding (but more likely a combination of these) can be responsible for the reorientation. These are, (a) passive folding (e.g. slip

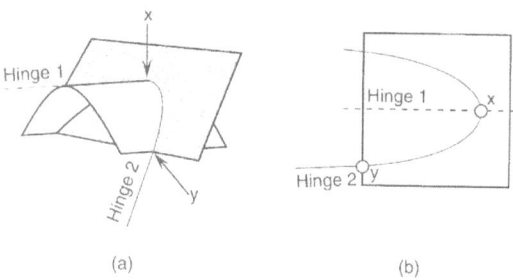

(a)   (b)

Figure 2.8 Variation in trend (and plunge) of the hinge of a fold with an inclined axial plane superimposed on an earlier (upright) fold. (**a**) Fold hinge 2 resulting from the superimposition of an inclined axial plane on an upright fold. (**b**) In plan, hinge 2 can be seen to trend perpendicular to hinge 1 at x whereas at y it trends nearly parallel to hinge 1.

Figure 2.9 Ice fall with a surface pattern which illustrates a natural example of slip folding. The traces of the folded surfaces are represented by transverse ridges in the ice fall. These are curved in response to differential rates of flow of the ice (the slip direction) which is reduced by friction at the margins of the ice fall and is therefore greater at the centre. West Greenland.

folding), producing **similar** folds in which the layers do not undergo bodily rotation with respect to fixed reference axes during deformation, and (b) folding during which the layers rotate bodily as they bend or buckle (e.g. flexure or flexural slip folding), producing **concentric** or **parallel** folds. Each process produces a different effect on existing planar structures or, as in the case of fold hinges, linear structures.

For a full up-to-date treatment of the relationships between successive folds the reader is referred to Chapter 5 of Ghosh (1993), Davis and Reynolds (1996) or any other modern structural geology textbook.

2.2.5 FOLD MECHANISMS

Slip folding involves parallel movement within a plane (cf. Figure 2.9). This plane is the fold axial plane, a great circle in stereographic projection, so that the new attitudes of a fold hinge which existed prior to the slip folding are all coplanar. Flexure, on the other hand, causes the new attitudes of a prefolding hinge to lie on a conical surface (on a small circle, or nearly so, in stereographic projection) rather than in a plane. For a full discussion of fold mechanisms see e.g. Davis and Reynolds (1996); Ghosh (1993); Hobbs *et al.*, (1976).

The analysis of the structure arising from multiple deformation (and hence the determination of a structural succession and deformational sequence) by interpretation of diagrams produced by the stereographic plotting of **non-selective** orientation data of fold hinges and axial planes is strictly limited and unlikely to be successful. This is because the confused distribution of data on the resultant 'scatter diagrams' fails to provide unique patterns characteristic of particular reorientation relationships so that it is not possible to identify folds of a particular set and hence determine their age relative to other sets (Figure 2.10).

On the other hand, stereographic plotting of **selective** orientation data (i.e. of measured attitudes of **identified** fold hinges) using different symbols for each set, is a useful way of presenting data to display their angular relationships (Figure 2.11b). As can be seen from Figure 2.11b, the distribution of the data for the latest fold set will be the most tightly clustered, tending towards being dispersed parallel to a diameter on the plot (e.g. the NE diameter in the figure). This diameter corresponds to the vertical plane parallel to the trend and the distribution along it represents the variation in plunge caused by the superimposition of the set on earlier folded surfaces. The data for earlier fold sets will become increasingly widely dispersed with age as a consequence of reorientation by later fold sets, with the distribution of axial attitudes of the earliest sets (and hence those subjected to the greatest amount of reorientation) likely to be spread throughout the plot and therefore unlikely to present anything resembling a regular pattern on the scatter diagram (see also the discussion in section 7.4, relating to the 'Variation with time of total structural complexity'.

Although it is unlikely that measured hinge attitudes affected by multiple folding can be resolved into sets simply by analysing stereo plots, it is sometimes possible to distinguish between only

two (or even three) sets of structures in terms of the degree of regularity of their distribution on a scatter diagram. In this case, if the pattern on the diagram can be seen to comprise two (or three) distinct groups this implies that two (or three) sets exist, with those having the closer, more regular grouping (i.e. the least scatter) being the latest (Figure 2.11b).

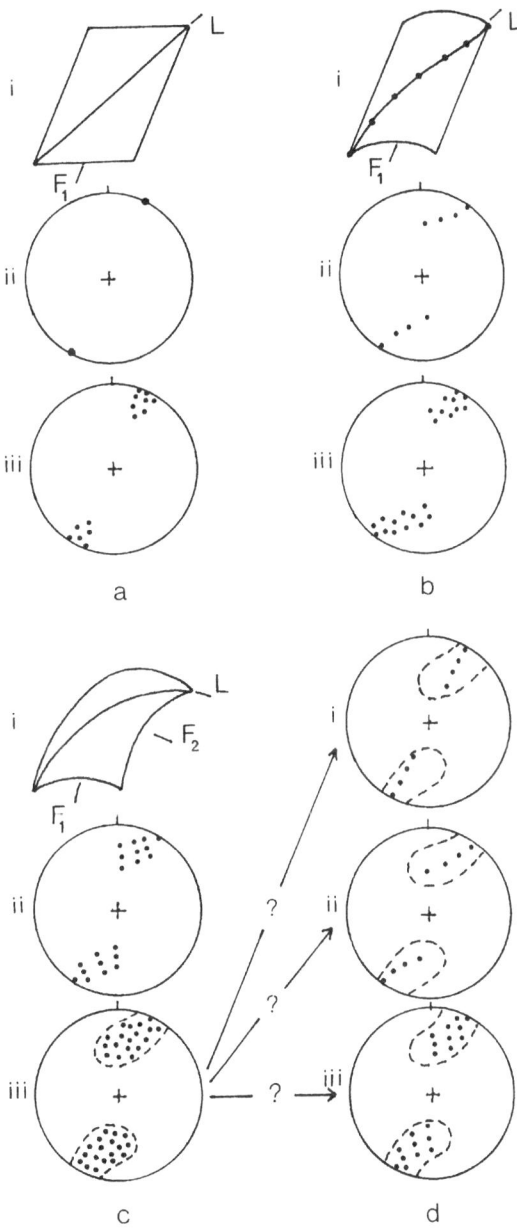

Figure 2.10 Sketches showing the behaviour of a linear structure (L) in a plane that is folded (F$_1$) and then refolded (F$_2$) together with plots (scatter diagrams) showing the distribution of the linear structure plotted in stereographic projection at different stages and the development of the scatter diagrams from an initial point maximum. These are compared with the type of scatter diagrams that would result from actual measurements of such data (**aiii**, **biii** and **ciii**) and show how the interpretation of such scatter diagrams is almost certain to be ambiguous.
(**ai**) Sketch of L lying in the plane prior to folding. (**aii**) Plot of L. (**aiii**) Plot of several measurements made in the field of a comparable structural relationship.
(**bi**) Sketch of L and the plane after the first folding showing selected attitudes of L which are plotted in (**bii**) where they lie on small circles. (**bii**) The data of (**aii**) after folding by F$_1$. (**biii**) Plot of several measurements comparable to (**aiii**).
(**ci**) Sketch of L and the plane after the second folding. (**cii**) The data of (**bii**) refolded by F$_2$. (**ciii**) Scatter diagram of several measurements of a lineation comparable to (**aiii**) and (**ciii**).
(**di–iii**) show that unambiguous interpretation of the recorded linear data of (**ciii**) is almost impossible. From Figure 5, Hopgood, 1980. Reproduced with the permission of the Royal Society of Edinburgh.

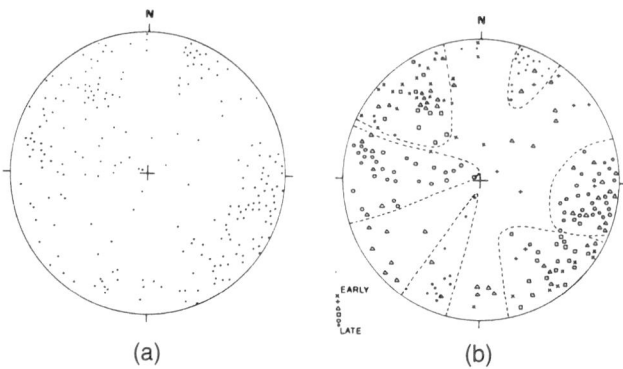

(a)   (b)

Figure 2.11   Two identical scatter diagrams of the axial attitudes of six sets of refolded folds plotted in equal area stereographic projection. (**a**) The data are undifferentiated, hence the axes of individual sets are indistinguishable from one another. (**b**) The data are identified and grouped according to fold sets. The latest three groups are enclosed by dashed lines. This demonstrates the relationship between distribution on the plot and degree of reorientation and shows that the distribution of the earlier sets is more widespread as a result of having been subjected to a greater degree of reorientation by later folding, whereas the latest are restricted to a narrow zone trending, in this case, NNE. Lewisian complex, Outer Hebrides, Scotland. From Figure 6, Hopgood, 1980. Reproduced with the permission of the Royal Society of Edinburgh. See also Figures 5.9 and 5.10.

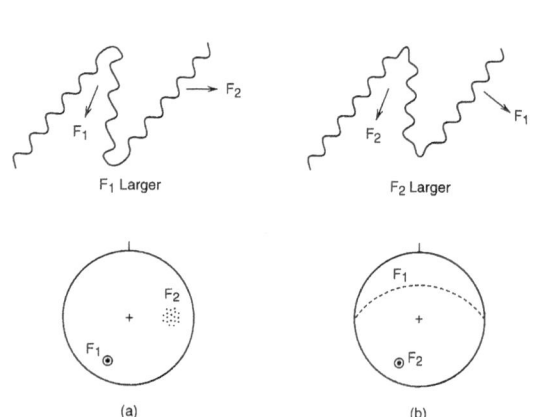

(a)   (b)

Figure 2.12   Effect of fold scale on grouping and dispersion of points on a scatter diagram (stereo plot) of refold data. (**a**) The first fold (F_1) is larger and F_2 is smaller. Both F_1 and F_2 are closely grouped on the plot (compare Figure 11.4a). (**b**) F_1 is smaller and F_2 is larger. Whereas F_2 data are closely grouped as a point maximum, F_1 are widely dispersed along a great circle maximum.

The number of fold sets (3 or 4) that can be unambiguously distinguished by analysis of stereo plots cannot be specified as a firm rule. It will vary with circumstance, depending particularly on the angular relationships between the axial directions of successive fold sets. If, for example, the axial attitudes of three sets (or even only two sets) of folds are within only a few degrees of one another, then plotting of these stereographically is almost certainly not going to help to distinguish them because the groupings of points on the scatter diagram representing each fold set will overlap (in part because of measurement error) and the clustering of the earlier structures becomes even less distinct as a result of later folding (cf. Figures 12.10 and 2.11). The relative scale of the earlier and later sets will also affect the grouping on the scatter diagram (Figure 2.12).

2.3   ASSUMPTIONS INHERENT IN THE APPROACH USED

In the undertaking of this type of investigation certain broad assumptions (1–5 below) may be made regarding the origin of the structures (see also the assumptions listed in section 7.1.1). These assumptions, some of which follow Sander's thinking (1911), are regarded as axiomatic.

It may be difficult or impossible to prove the assumptions rigorously. Nevertheless, in general they appear to hold consistently. Even when in some cases they do not hold, this does not necessarily invalidate the method. For example, non-tectonic structures such as those described by Hendry and Stauffer (1977) can generally be distinguished because of inconsistencies in geometry and attitude, and because of their restricted occurrence.

These assumptions are as follows.

1. Stresses responsible for the structures were regional and caused penetrative structures with consistent attitudes. Any changes in orientation are for the most part gradual and can be monitored.

Accordingly there is, in general, no sudden change in the attitude of folds of a particular set, except as a consequence of refolding, or because of local distinct inhomogeneity or deformation partitioning. Such variation as there is can be monitored from one locality to the next as has been shown in the Lewisian complex of the Outer Hebrides, Scotland (Hopgood and Bowes, 1972 and Figure 12.17). Any exceptions to these conditions are recognized as the investigation progresses.

2. Many folds were initiated with near-vertical axial planes and were then further deformed into inclined or recumbent attitudes so that the latest structures tend to remain more upright.

3. Earlier open folds have usually been tightened by continued or repeated deformation with reduction of their inter-limb angle, in some cases to zero.

In evolutionary terms this commonly means that folds first formed as open structures on upright axial planes, gradually became tighter, inclined, recumbent, overturned and ultimately isoclinal as deformation continued. In this way, later (more open) structures were imposed on a complex made relatively rigid in consequence of having already been highly folded during earlier deformational phases so that progressively more contorted structures resulted. Also, because the steep axial planes of later structures have been less affected by subsequent deformation than those of earlier-formed structures, they remain more upright so that their axial trends are more consistent throughout the orogenic belt (See Chapter 7, 'Characteristic features of structure successions'.

4. Examples of folds of every set are likely to have been preserved to some extent, having survived (if only in a modified form) the effects of subsequent deformation. This assumption is less certain than others and may be impossible to prove or disprove, especially in those cases where early sets of folds were initially only weakly developed, for example as very open structures. In such cases it is unlikely that the folds will be recognized if later deformation is intense, involving tight folding of possibly more than one episode. (See the discussion under 'Open folds', section 10.1.2.) Again in those cases where the immediately succeeding folding took place on parallel axial planes the two sets may be indistinguishable. The effects of successive folding on parallel axes but non-parallel axial planes will, however, be recognizable. Although extreme shearing may strongly modify or obliterate early structures, such an effect is likely to be local and confined to relatively narrow zones (section 8.1.9, 'Effects of strain partitioning') and, unless the area under investigation lies wholly within such a zone, early structures will be recognized in the intervening segments of low shear strain.

5. Because of the unique conditions (stress, pressure and temperature) responsible for their development, folds belonging to a particular generation are likely to have a sufficient number of characteristics to enable them to be identified unequivocally (section 5.2). However, the use of style (section 6.2), especially when only part of a fold is exposed, should be

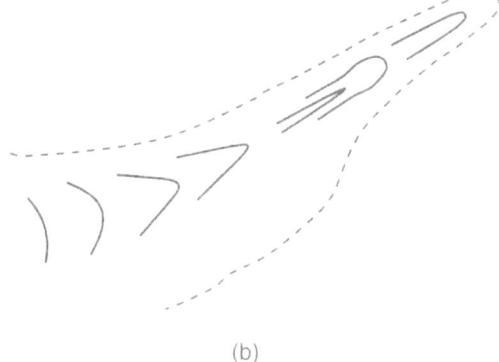

(a)

Figure 2.13 Photograph (**a**) and sketch (**b**) of the profile of a fold ranging from open to isoclinal. This shows how the form of a single fold can vary in profile, from open at left to isoclinal intrafolial at right, with the hinge shape ranging from round at the left, through sharply angular at the centre to round at the extreme left. Hence profile, particularly when only part of a fold is seen, cannot necessarily be regarded as diagnostic of a particular fold set. Pre-Ketilidian amphibolite facies migmatite, Southwest Greenland. See also Figure 6.26.

(b)

treated with caution (Williams, 1970) because the profile of even a single fold can vary considerably (Figure 2.13), as can the wavelength of a fold set (Watkinson and Cobbold, 1978), and modification by subsequent deformation may also differ between folds of the same set because of their differing attitudes

(Figure 2.14) at the time of the superimposed deformation.

Nevertheless, fold profile, taken with all the other evidence, should be sufficient at least to limit the degree of choice of categories or even to establish the identity of the structure in some cases. It is inferred also that the folds

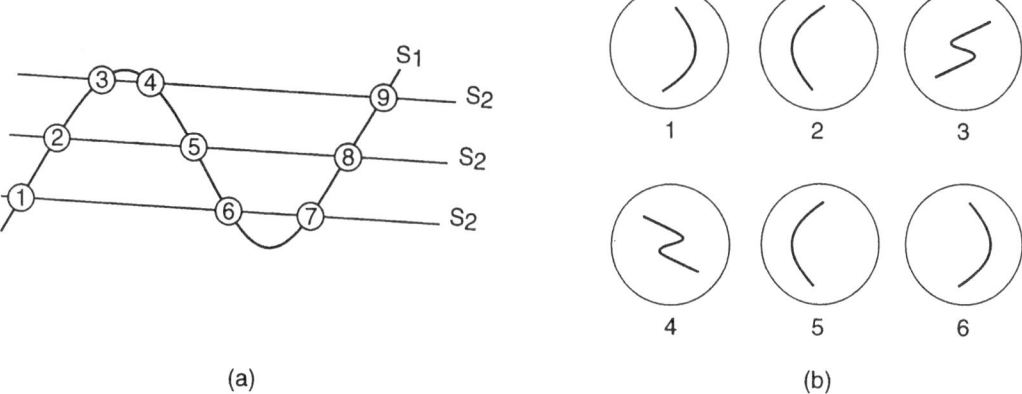

(a)   (b)

Figure 2.14 (a) Folds (1–8) of the same set (folds 7–9 are equivalent to folds 1–3) resulting from the overprinting of a second axial planar direction (S₂) on a pre-existing larger fold. All have different profiles as shown by the sketches of folds 1–6 in (b). The variation in profile is a consequence of difference in attitude of the pre-existing (folded) surface (S₁) with respect to the axial planes (S₂) of the later fold set. See also section 6.2.2.

observed can have larger-scale counterparts which could be identified if continuous reference horizons can be traced (Figure 2.15 and Figure 1.12, section 1.1.3).

It is necessary to bear in mind the possible effects caused by changes in physical conditions in the orogenic belt and hence in the response of the rocks to the imposed stress. Thus folds may vary between the products of deformation that on the one hand are 'brittle' and on the other highly 'ductile'. However, if such variation in conditions is time-dependent then its effect can be an advantage in that it could result in the formation of distinctive morphological features related to each successive set of folds, thereby adding to the sum of characteristics for each set.

On the other hand spatial (rather than temporal) variation in these conditions, for example with depth in the orogenic belt, would mean that cogenetic folds might vary considerably in appearance (Holland and Lambert, 1969). **Their refolding relationships with structures of other sets would nevertheless still be the same**. (See also section 5.2.12, section 7.7, Figure 7.37 which compares different successions and Figure 7.38 which contrasts similarities and differences of successions).

The assumptions listed, while not fundamental to the approach used here, are nevertheless very useful indicators of the likely status in the deformational sequence of particular fold structures observed during the investigation. However, it must be emphasized that these are no more than assumptions and that they form the basis of no more than an initial working hypothesis. At the outset they may be the only criteria available for deciding that the structures observed provide an indication of multiple deformation, at least until this can be shown unequivocally by refolding relationships. Also they form the basis for interim correlation of structures between different localities examined until such time as more certain evidence is adduced. These initial premises are of course subject to continual modification in the light of evidence stemming from overprinting relationships between folds.

An objection that has sometimes been raised against the significance of a fold sequence determined for a particular orogenic belt is that the 'same sort of sequence' can be deter-

(a)

(b)

Figure 2.15 The role of continuous reference horizons in allowing the recognition of large-scale counterparts of small-scale structures. (**a**) Small-scale folds observed in separate outcrops. (**b**) The larger fold inferred from the existence of the smaller structures, each incorporating the same reference horizon.

mined for any orogenic belt. This objection appears to be based on the notion that there is a limit to the number of ways in which any layered succession can deform, whether it comprises rock or any other material. For example, similar sorts of folds to those found in deformed rocks can be found in deformed layered ice (Figure 2.16). However, this argument has little relevance for gneisses and especially migmatites because of the fact that other aspects of an orogenic process contribute to the structure. These include metamorphic and igneous activity, both syntectonic and 'intertectonic', as well as fold episodes. They are

Figure 2.16 Fold in layered ice showing well-developed axial planar cleavage. The structure is very closely similar to folds formed in rocks. The figure at the centre shows the scale of the structure. Inland Ice, West Greenland.

important contributory factors to a comprehensive succession of structural features, a succession that is unique and characteristic of each orogenic belt (Hopgood 1973). In respect of this, see the discussion in the section on the characterization of fold successions (section 7.1) and that on the apparent similarity of fold successions (section 7.5).

3.1 OVERPRINTING

3.1.1 PRINCIPLE OF OVERPRINTING

Regardless of the mechanism involved in the deformation, the basic principle regarding refold relationships stated earlier always applies, namely that where overprinting (Sander's 'Uberprägung', 1911) has taken place, the structural features that are overprinted are earlier than those that overprint them. This rule, fundamental in geology, has as its counterpart in stratigraphy, the 'Law', or *'Principle of Superposition'* enunciated by William Smith (1789–1839) and based on observations by other geologists (Wyatt, 1986, in Challinor's *Dictionary of Geology*, 6th Edition, p. 179, and especially p. 309). See also the reference to Hutton (1795) in Chapter 1. The rule is the basis of the approach described here for resolving the structural complexity caused by repeated deformation, in particular multiple folding, an approach that depends on the direct observation of overprinted relationships on the outcrop. Two basic principles can be stated.

1. **Refolded folds are earlier than the folds that fold them**. Figure 3.1 shows a refolded fold on the left. Its structure is the result of two folding events involving folds like those shown on the right, an earlier tight fold and a later upright open fold. The tight fold subsequently becomes the refolded fold after being folded by the upright fold (the refolding fold).

2. **Intersected structures are earlier than the structures (e.g. veins, dykes, cleavages, fractures etc.) that cut them**. Figure 3.2 explains the relationships between (a) structures resulting from overprinting between a fold intersected by a discordant vein and (b) overprinted relationships between a layered succession intersected by a discordant vein.

This applies equally to all structures tectonically imposed on the rock including, besides veins and folds etc., other features common in migmatites such as agmatites (Hopgood and

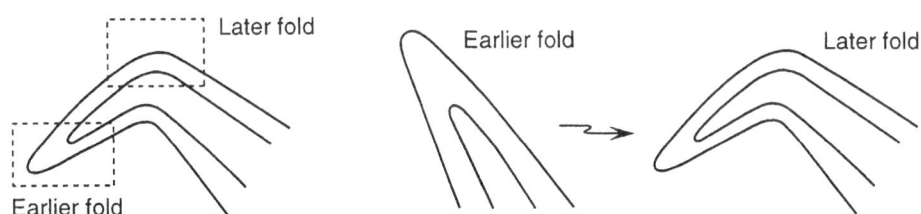

Figure 3.1 Sketch of refold relationships between two folds. The two structures, an earlier (refolded) tight fold and a later open, upright fold are shown on the right.

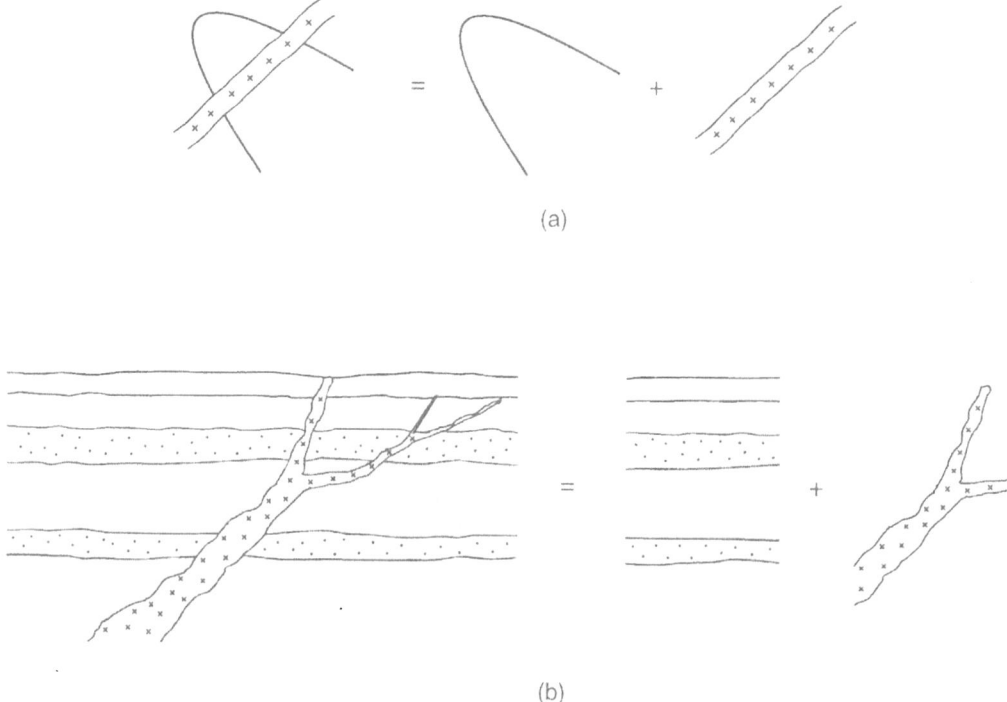

(a)

(b)

Figure 3.2   Intersected structures. (**a**) Fold intersected by a vein. The derivation of the observed relationship between the fold and the discordant vein is shown on the right, viz. the fold is overprinted by the (discordant) vein. (**b**) Succession of layers intersected by a vein. The relationship observed between the layered succession and a vein on the left are explained on the right as overprinting by a later (crosscutting) vein.

Bowes, 1978). See also section 2.2.4 and section 8.1.2.

3.1.2   RELATIVE TIMING OF EVENTS: OVERPRINTING

Once evidence of multiple deformation has been recognized, the next stage is to establish the relative order of the structures in the succession and hence the order of events. This entails identifying overprinted structural relationships of several kinds and various degrees of complexity, such as the following eight simple examples.

1. Folded primary foliation. The relationship between a planar structure and a three-dimen-sional structure. One of the simplest cases, that of folded foliation, such as bedding, is illustrated in Figure 3.3. Here, foliation in its unfolded state (Figure 3.3a) clearly represents an earlier state than that of Figure 3.3b where the foliation is folded.

2. Crenulation of a pre-existing foliation. This condition, a variant of (1) above is illustrated in Figure 3.4 which shows crenulation of a secondary foliation, a schistosity (in this case the pre-existing foliation), which commonly leads to the later foliation, a crenulation cleavage.

3. Refolded folds. The relationship between folds (three-dimensional structures) is less simple than the previous examples and is illustrated by the simple sketches of Figure 3.5

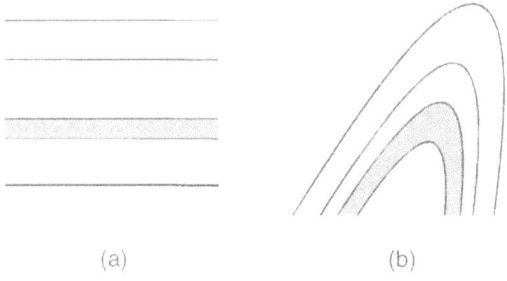

(a)   (b)

Figure 3.3 Simple relationship between (**a**) un-folded foliation (earlier) and (**b**) folded foliation (later).

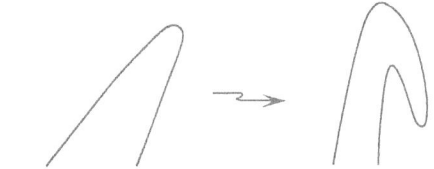

Figure 3.5 Sketch of a refolded fold. An inclined tight fold (left) folded by a second tight fold with an upright axial plane (right).

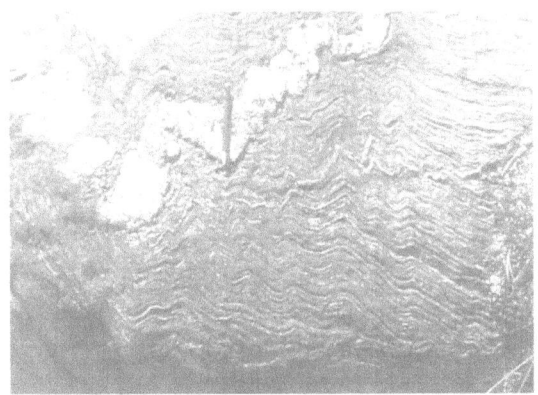

Figure 3.4 Crenulation of pre-existing foliation. Upright, small, chevron-like crenulations of foliation (schistosity). The fold axial planes are parallel to the pencil. Hammaslahti, Karelia, Finland. Compare Figures 6.40, 10.30 and 11.4.

which show a tight fold folded by a second tight fold with an axial plane that is now upright. Compare Figure 3.1.

4. Crenulations (or minor folds) affecting a fold such that the structure is not symmetrical with respect to the fold axial plane of the 'host' fold (Figure 3.6). Instead of 'S' folds on one limb being reflected on the other by 'Z' folds there will be approximately symmetrical minor folds on both limbs as can be seen immediately adjacent to the dashed '2nd' trace in Figure 3.6c. The symmetry of the minor structures on each limb of the affected fold shows these to be later structures, unrelated to the 'host' fold.

5. Fold or foliation affected by an undeformed discordant intrusive vein. This relationship, explained in Figure 3.2a is shown again here in Figure 3.7.

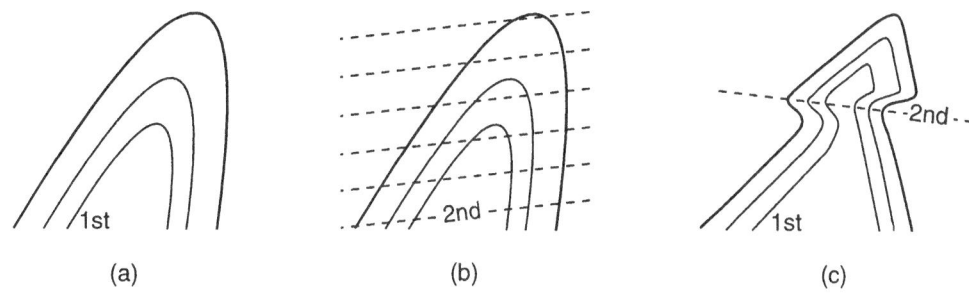

(a)   (b)   (c)

Figure 3.6 An earlier ('host') fold (**a**) cut by foliation or crenulations (**b**) or minor folds (**c**). In (**c**) the asymmetry sense of the later minor folds with respect to the earlier fold axial plane is different on each limb; they are more or less symmetrical (concave to the right) on both limbs. (cf. Figure 11.13).

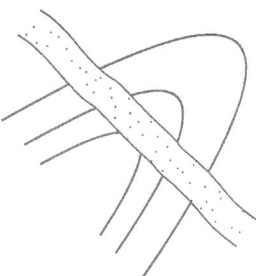

Figure 3.7   Discordant, undeformed intrusive vein obviously later than the fold. See also Figure 4.1.

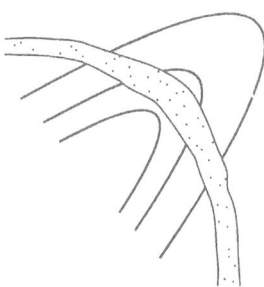

Figure 3.8   Deformed, originally planar tabular body discordant to a fold. On the basis of the reasonable assumption that the vein was tabular at the time of emplacement, the fold must therefore be a composite structure, formed in part prior to the intrusion of the vein and in part after the intrusion. See also Figures 4.1, 4.18 and 4.19.

6. Deformed originally planar tabular intrusive body (Figure 3.8). An extension of case (5) above, the intrusive vein clearly post-dates the fold and is itself deformed later to form an open fold. At the same time the earlier fold would have been refolded, apparently on the same axial plane as previously, and tightened. Time relationships between at least three events are recorded in this structure.

7. Foliation developed obliquely to pre-existing layering. Examples include foliation resulting from differential recrystallization (Figure 3.9), cleavage (caused by fracture, crenulation and recrystallization) and transposed foliation.

Figure 3.9   Foliation oblique to dominant layering. The foliation is defined by aligned crystals and elongate aggregates of andalusite and cordierite inclined at a high angle to the bedding, and apparently axial planar to the open fold and minor folds visible in the picture. Vetio, Kisko, Finland.

Figure 3.10   Rotated foliation in a small clast. The fabric of the clast is clearly discordant to that of the host rock showing that the clast has been rotated. Pre-Ketilidian migmatite, West Greenland. Compare Figure 4.24.

8. Evidence of rotation of the early foliation so that it is oblique, or truncated against, a later (or earlier-modified) foliation (Figure 3.10) and in thin section especially, rotation of porphyroblasts such as 'snowball' garnets.

There are of course other examples of over-printed relationships, such as those involving dislocation of pre-existing structural features. The amount of information that can be obtained from overprinted structures varies considerably but one of the factors influencing this is the type of structure involved and whether it is one-, two- or three-dimensional.

3.1.3 RELATIVE VALUE OF FOLDS COMPARED TO FOLIATIONS, AND LINEATIONS AS INDICATORS OF OVER-PRINTING RELATIONSHIPS

The structural complexity resulting from polyphase deformation is undoubtedly much greater where folds (which are three-dimensional structures) are involved in overprinting than it is where only planar (two-dimensional) or linear structures are involved. But the determination of a structural succession involving the resolution of overprinting relationships between fold sets can be accomplished more easily, and the results can be regarded with greater certainty, than they can where only planar or linear structures form. This is because there exists within the composite geometry of folds a greater potential for the development of features which, taken together, are characteristic of the particular set. Accordingly, the likelihood of distinguishing between different sets of structures, where these are folds, is considerably greater than where the structures are either foliations or lineations alone.

Because folds are three-dimensional structures the effects of overprinting on or by them are more likely to be recognized as a result of the certainty that structural discordance is no more than a possibility if only planar or linear structures are involved. This is because it is nearly certain (except in the case of isoclinal folds) that at least one of the elements of a fold will be favourably orientated to show discordant intercepts, whether this is with earlier structures or with later potential super-imposed structures and whether these are planar or linear or fold structures. See also section 5.3.

The chance that overprinting will take place is also controlled by both the spacing ('size') and (in the case of planar and linear structures) the direction (orientation) of the earlier and later structures (Figure 3.11). Again there is a greater likelihood of overprinting between three-dimensional structures such as folds than there is between folds and planar structures, and progressively less between folds and linear structures, planar structures, planar and linear, and linear structures.

3.1.4 EVIDENCE OF REFOLDING: INTERFERENCE PATTERNS

Structural analysis of the terrane will have been prompted at the outset by evidence of multiple deformation. On the outcrop (where the study involves mesoscopic scale structures) this evidence will be in the form of **interference patterns** resulting from the inter-action between two or more sets of folds. The intricacy of this pattern is dependent on (i) the number, (ii) the angular relationships, and (iii) the geometry, of the folds and also (iv) on the nature of the exposure (i.e. whether it is planar or irregular) and on its angular relationship (if it is approximately planar) to the fold elements (axis, axial plane and limbs). In thin section, evidence of polyphase deformation may be provided by the relationship between mineral growths and/or microstructures (Figures 3.9 and 3.10).

Different types of interference patterns have been formalized, described and discussed in textbooks such as those by Ghosh (1993), Ramsay (1967), Hobbs, Means and Williams (1976) and others.

Examples of simple patterns are shown in Figures 3.12 and 3.13 which show partial or complete closures ('eye' folds) indicative of the effects of cross-folding. The foliation of Figure 3.12 is highly irregular and shows simple examples of such eye folds, especially on the left of the view, as well as asymmetrical folding on the right of the already isoclinally-folded foliation (an isoclinal hinge shows

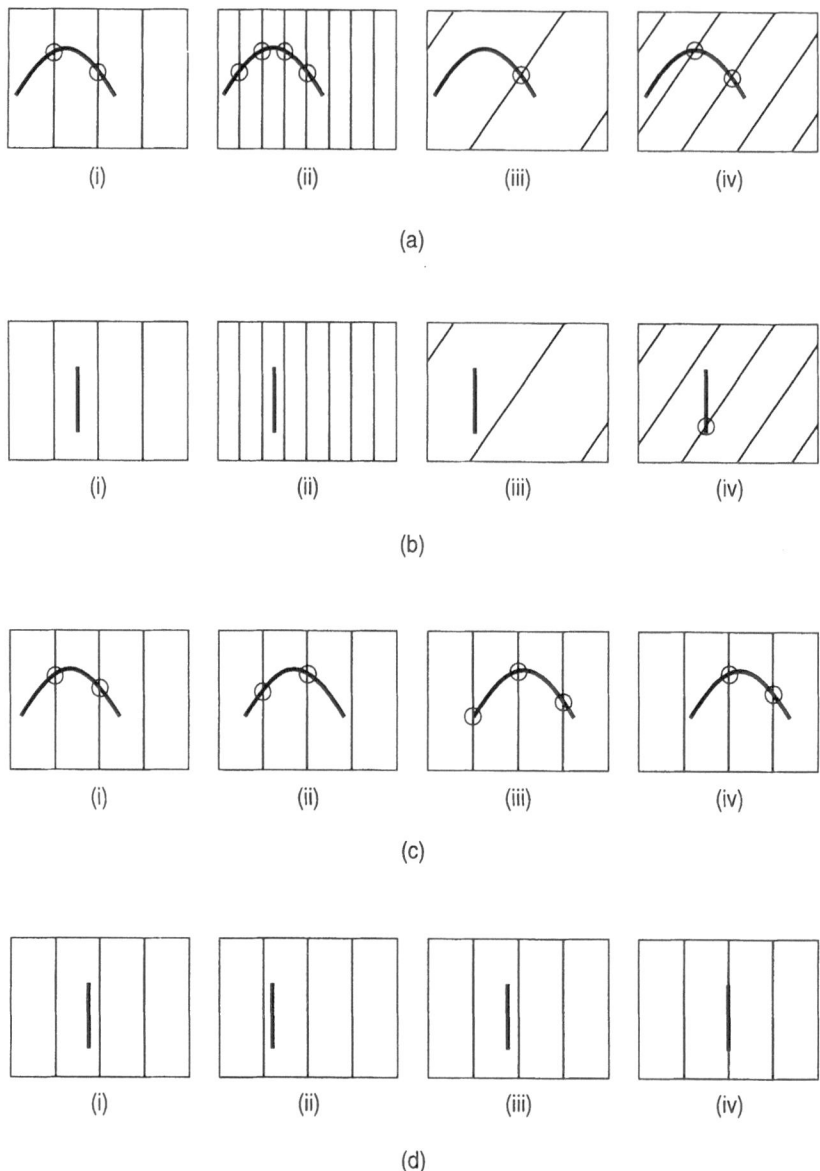

Figure 3.11   Cross-sections of planar structures and folds showing the variation in chance intersection between differently-spaced and differently-orientated superimposed two-dimensional (**a** and **c**) and one-dimensional (**b** and **d**) structures and corresponding sets of existing one-dimensional structures (vertical and inclined spaced lines in (**a**)–(**d**), (**i**)–(**iv**). Intersection (circled) with the superimposed two-dimensional structures is very much more likely than with the one-dimensional structures. There is corresponding variation in the likelihood of intersection between three-, two- and one-dimensional structures.

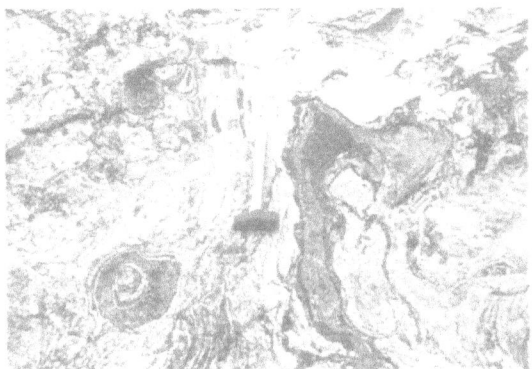

Figure 3.12 Interference patterns in anorthosite. Pre-Ketilidian complex, West Greenland.

Figure 3.13 Interference patterns in migmatites. Sand River, Republic of South Africa.

above and slightly to the left of the end of the hammer shaft). This implies that the foliation has been affected by at least three fold sets. While a more distant view of the foliation in Figure 3.13 gives the superficial impression that the structure is not particularly complicated, it is clear from closer inspection that, like that in Figure 3.12, it too has been strongly deformed. In this case there are asymmetrical folds with both a dextral and a sinistral rotation sense, and the double eye folds suggest cross-folding both perpendicular and parallel to the dominant foliation trend. Note the small white 'mushroom' interference structure (see Figure 3.18) along the strike from the double eye fold and to the left of the person standing on the right.

The presence of more than one fold set is clearly shown in the exposure of Figure 3.14a where at least three sets are visible and the time relationships between each is easily determinable.

As the previous figures show, the effects of refolding of one fold by another can normally be recognized fairly easily on the exposure because of the modified form of the fold shapes on the exposure caused by the interaction between the two folds (the interference pattern). However, while evidence of refolding may be easily recognizable, its true nature (i.e. the **relationships** between the fold sets

involved) may be less easily appreciated because of the factors noted above viz. (i) the geometry of the two folds, (ii) the orientation of one with respect to the other, and (iii) the attitude of the exposure surface with respect to the refold. With repeated (multiple, or polyphase) folding the interference pattern obviously becomes increasingly complex and when the outcrop is a non-planar surface, distortion of the interference pattern further complicates its appearance (Figure 3.15).

Simple interference patterns

In some cases it is possible to tell the order of superposition between folds solely from the interference pattern, but this is by no means always so, such as where the interference pattern results from overprinting between two similar fold sets with perpendicular axial planes (Figure 3.16). When the structural history is a complex one, three-dimensional exposure of the structure may be necessary in order to establish the refold relationships. On the other hand, where the axial traces of one fold can be seen to be folded by the other, the order of superposition is apparent from inspection as in the cases of the double zigzag and mushroom patterns of Figures 3.17 and 3.18.

The simple interference patterns of Ramsay's types 1–3 (Ramsay 1967) illustrated above are only occasionally seen clearly in the field because it is only rarely that the surface of the exposures is

(a)

Figure 3.14 (a) Refolded foliation in gneiss. Lewisian complex, South Harris, Outer Hebrides, Scotland. (b) The first fold (hinge 1), now isoclinal, has been folded by the tight fold (hinge 2) and both have been refolded asymmetrically (hinge 3).

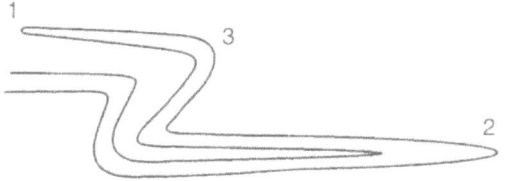

reasonably planar. What are most often seen are modifications of true interference patterns, including 'interference patterns' that are apparent rather than real (e.g. Figure 3.19).

In terranes where the structural succession comprises more than three fold sets the interference patterns are likely to be very complicated (Figure 3.20). In such cases there is less likelihood that the observer will be able to interpret the fold type on the basis of the interference pattern, especially if the scale of each of the interfering fold sets is approximately

the same (Figure 3.16). There is further discussion of this problem, in section 10.1.2.

Nevertheless, it is always worth making the attempt to learn something more about the structures than just their order of succession, factors such as the mechanism causing them and the stress field responsible.

As has been seen, complicated outcrop patterns on approximately planar exposures provide an indication of the existence of superimposed folds. However, whether these are interference patterns resulting from multiple deformation in response to stresses originating

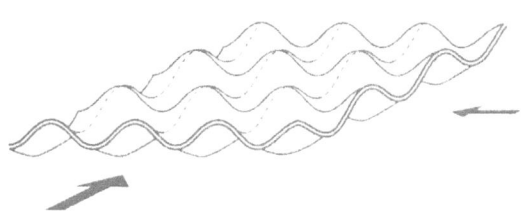

Figure 3.16 Cross-folding caused by compression parallel to a plane resulting in 'dome and basin' interference structures. Here the sequence of refolding is not apparent (compare the examples of constrictional deformation in Figures 3.22 and 3.23).

Figure 3.15 Distortion of the fold outcrop pattern where the exposure is non-planar (at left). The three-dimensional exposure does, however, allow measurement of the attitudes of the fold elements such as fold axial plunge. Pre-Ketilidian migmatite, West Greenland. Compare Figure 3.19.

(a)

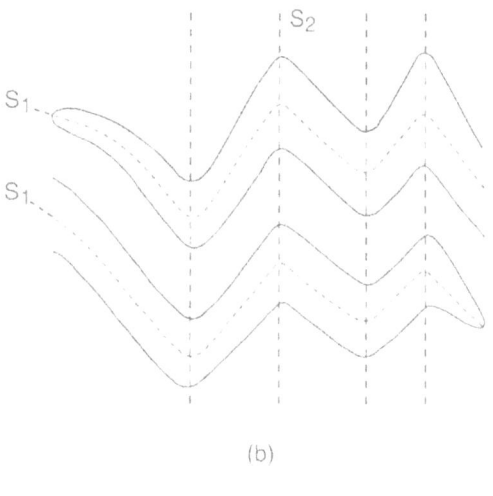

(b)

Figure 3.17 (a) Refold. 'Double zigzag' fold interference pattern between earlier isoclinal folds and later upright chevron folds. Hinges of the earlier, isoclinal fold set close to the right at the pencil top and to the left immediately above. Svecofennian migmatites, Trutlandet, Jussarö area, Finland. (b) Sketch of the double zigzag refold of (a) showing relationships between the earlier and later fold sets. Compare Figure 3.14. Harris, Outer Hebrides, Scotland.

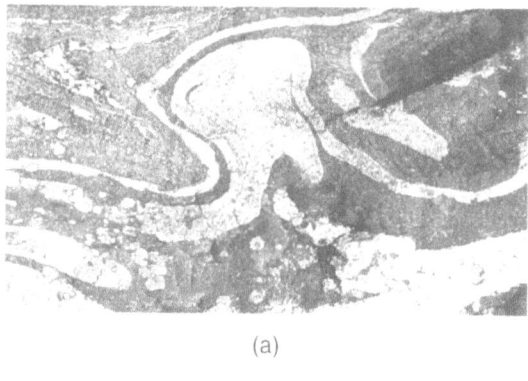

(a)

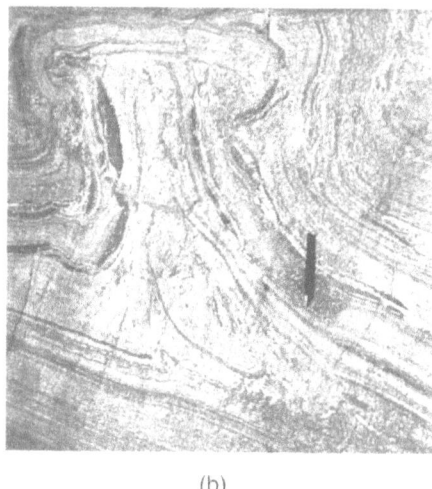

(b)

Figure 3.18 'Mushroom' interference patterns. (**a**) In Svecofennian migmatites, Pohja, Finland. (**b**) In Svecofennian migmatites, Kvarnskär, Åland Islands, Finland. The head of the 'mushroom' comprises the earlier, tighter fold and the stalk is parallel to the axial plane of the later fold set. Compare the photographs of Figures 3.12, West Greenland and 3.13, Sand River, South Africa.

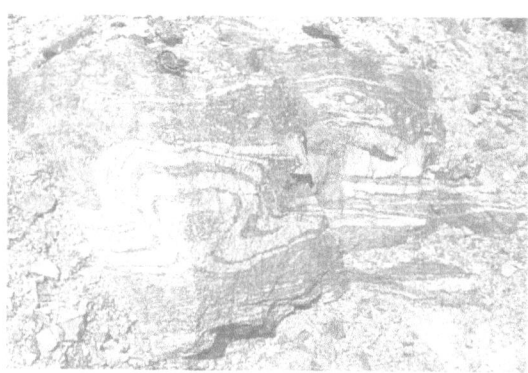

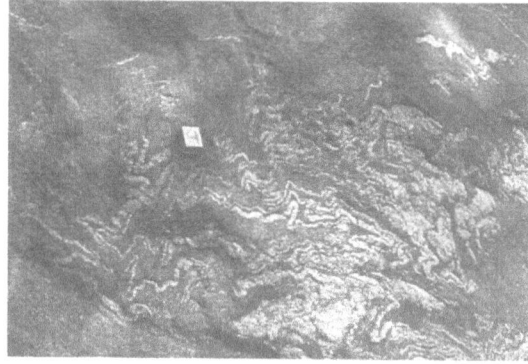

Figure 3.19 True and apparent interference patterns compared. Interference pattern resulting from overprinting of an isoclinal fold by an asymmetrical fold (centre and right) contrasted with an *apparent* interference pattern (a 'pseudo-conjugate' fold pattern) at the left and centre left resulting from the intersection of the asymmetrical fold with two non-parallel surfaces. Compare the structure on the right side of the photograph with the structure of Figure 3.14. Lewisian complex, South Harris, Outer Hebrides, Scotland.

Figure 3.20 Interference patterns. This complicated structure includes patterns similar to those of Figure 3.17. Mazoe River, Zimbabwe. Compare Figure 12.1.

tectonically or whether they are the meta-morphosed derivatives of non-tectonic structures (section 3.2.5), stemming, for example, from the effects of slump or water-expulsion from unconsolidated sediments (Figure 3.38) or movement of ice over fluvio-glacial sediments prior to lithification and metamorphism (Figures 3.36 and 3.39), must be confirmed by further observation. Systematic investigation will establish whether there exists the regular pattern associated with penetrative folding in response to a pervasive stress field or whether the structural pattern is non-penetrative and developed from some other cause. Mushroom-like, non-penetrative structures can be caused by diapirism, the result of differential movement because of competence differences arising from partial melting and anatexis during ultrametamorphism, or in response to stress differences acting on sediments prior to lithification where competence differences stemmed from differential consolidation (Figure 3.21).

3.2 DEPARTURES FROM CONSISTENT OVERPRINTED STRUCTURE PATTERNS

3.2.1 STRUCTURES ARISING FROM CONSTRICTIONAL DEFORMATION

Fold patterns arising from constriction are not uncommon in highly deformed rocks. Although often illustrated, they have seldom been discussed in terms of systematic geometrical analysis (Ramsay and Huber, 1983, p. 66). However, they have been the subject of extensive study and the classic experiments by S. K. Ghosh and colleagues (Figure 3.22).

Interference patterns arising from pure constrictional deformation differ markedly from patterns arising from overprinting and do not show the consistent fold patterns that arise from overprinting because of their less systematic interference geometry. Although such structures are not normally to be confused with overprinted structures because of the lack of consistent relationships shown between refolded pairs, or groups of fold sets,

Figure 3.22 Fold pattern formed by constriction. Note that there is no consistent relationship between the hinge trends of adjacent folds (cf. Figure 3.23). Copied from Figure 2a, Ghosh, S. K., Khan, D. and Sengupta, S., 1995. Interfering folds in constrictional deformation. *J. Struct. Geol.* **17**, 1361–1373, with kind permission from Elsevier Science Ltd, the Boulevard, Langford Lane, Kidlington OX5 1GB, UK.

Figure 3.21 Cross-section of structures formed by differential response to pressure of overburden (probably ice) in fluvio-glacial muds analogous to the domes formed by diapiric uprise of granitoid melt. Qeqetok, Midgaard, southern West Greenland.

care must be taken nevertheless to ensure that they are not misinterpreted in terms of overprinting where exposure is limited. Figure 12, in Ghosh *et al.*, 1995, shows one such example from an experimentally formed pattern (Figure 3.23). Here, although there are two directions of axial planar trace at high angles to one another, those with the rectilinear trace (and therefore likely to be regarded as the later set) are not parallel but are perpendicular to one another and therefore, when in such close proximity, are not likely to be classified as belonging to the same set. Furthermore, but less obviously and perhaps less convincingly, it can be seen that there is not a consistent match along the axial planar traces between folds of what might appear to be the same set.

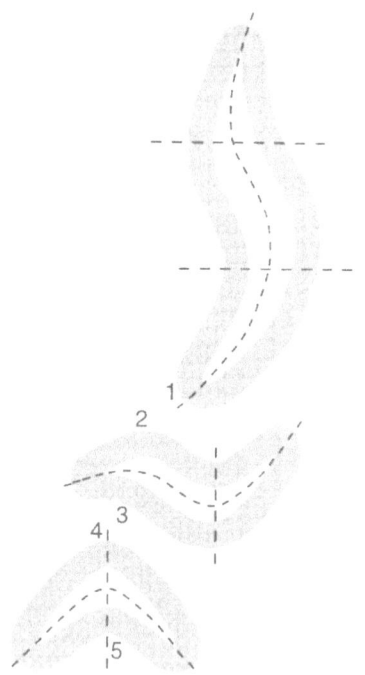

Figure 3.23 Outlines of experimental folds formed by constriction. See text for explanation. Derived from Figure 12, Ghosh, S. K., Khan, D. and Sengupta, S., 1995. Interfering folds in constrictional deformation. *J. Struct. Geol.* **17,** 1361–1373, with kind permission from Elsevier Science Ltd, the Boulevard, Langford Lane, Kidlington OX5 1GB, UK.

The curved axial traces do not run smoothly from one fold to the other with 'synforms' matching 'synforms' etc. In this example, the sinuous axial planar trace of the upper fold outline passes down from a concave **upward** curve (at 1) into the concave **downward** curve of the next fold (at 2), out through a concave **downward** curve (at 3) to the concave **downward** curve (at 4) of the upper part of the lowest fold, and finally, via a concave **downward** curve (at 5), out of the lower part of the lowest fold. Where limited exposure is not a problem then the absence of a consistent overprinting pattern from place to place will prevent any such misunderstanding from arising. Similarly, local departures from the systematic overprinted structure which arise from localized constrictional strain and which do not influence the overall structural pattern will be recognized for what they are and should not affect the structural interpretation.

Ghosh and co-workers have been describing interference patterns considered to be the response to constriction since the 1960s, and studying the development of the structures responsible for them experimentally. Their work has tended to confirm the absence in constrictional structures of the regular relationships shown where structures are the product of overprinting (e.g. Figures 2–5, in Ghosh *et al.*, 1995) and as a result of their recent study, Ghosh, Khan and Sengupta (1995) concluded that; 'Over a relatively large domain the interference pattern produced by general constriction can be recognized by the association of domes and basins with nonplane noncylindrical folds, the occurrence of hairpin bends of hinge lines of open folds, by occurrence of amoeboid outcrop patterns and by absence of a consistent overprinting relation among different sets of folds.'

It is important therefore to bear in mind from the foregoing that one can conclude that folds formed in response to constrictional deformation are unlikely to affect the application of the principles discussed here. Inevitably they will be distinguished from the 'regular' interference structures formed by overprinting of successive sets of folds which will show consistent patterns, whereas those arising from constriction will not. Even if they are not recognized at first, or indeed at all, their presence will represent at most a small hiatus in the overall regularity of the structural pattern (section 8.1.9 and especially Figures 8.20–8.22).

3.2.2 TRANSECTED FOLDS

The use of cleavage and cleavage/foliation intersection to determine attitudes of the fold axial plane and the fold hinge (axis) in well-cleaved rocks is sometimes subject to the limitation imposed by the effect of **transection** (Borradaile, 1978; Powell, 1974). However, in respect of this see Duncan's 1985 discussion of fold and cleavage relationships at Sulphur Creek. Also refer to example (a) in section 10.1.8, and examples (a)–(d) in section 10.1.9. If transecting cleavage is present, i.e. the cleavage is contemporaneous with the folding but lacks a simple axial planar relationship to the fold, or parallelism with the fold hinge (Figure 3.24), then care must be exercised to avoid attributing cleavage and related folds to different structure sets (see for example the problems discussed from the Southern Uplands of Scotland by Stringer and Treagus, 1980). It is, however, suggested that the cleavage will serve as a good first approximation to the attitude of contemporaneous folds. This relationship becomes less of a problem in the case of refolding relationships in that

the changed attitude(s) of the transecting cleavage (like that of other cleavage present) will still reflect the effect of later deformation (Figure 3.25) and the similarity of its behaviour following reorientation will show when plotted stereographically (cf. Figure 2.3). Furthermore, examination of several examples of the particular fold set is likely to show the true relationship between the cleavage and the fold axial plane.

In general the attitude of cleavage, even when it transects, will vary gradually, in a regular fashion, across the terrane, depending on its response to subsequent deformation, in much the same manner as any other foliation (or other structure – see Figure 12.17). Correspondingly it will be susceptible to variations in lithological inhomogeneity and will also be liable to sudden changes of orientation in response to local structural inhomogeneities ('deformation partitioning'). See below. The effect will be even less important where exposure is approximately two-dimensional as in

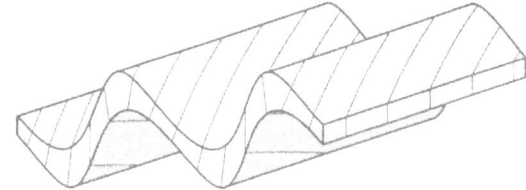

Figure 3.24 Sketch showing the relationship between a fold and transecting cleavage. Note that, although the fold and cleavage are contemporaneous (this is shown by their consistent relationship), the cleavage does not bear a simple axial-planar relationship to the fold.

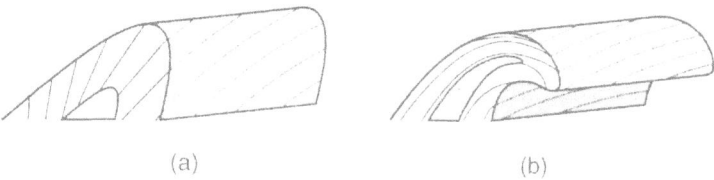

(a)   (b)

Figure 3.25 (a) Fold with transecting cleavage. (b) Fold refolded with consequent folding of the originally planar transected cleavage. Compare the plots of Figure 2.10 showing the distribution of folded linear structures such as poles to cleavage.

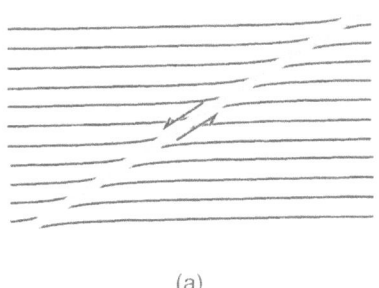
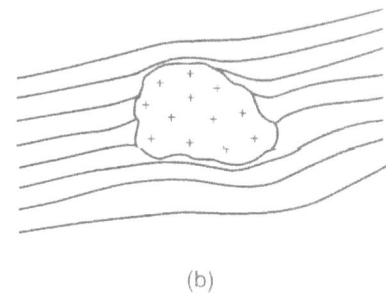

(a)   (b)

Figure 3.26   Effect of local inhomogeneity on foliation attitude. (**a**) Local deflection of foliation by a ductile shear. (**b**) Local deflection of foliation by a later intrusion. The effect is similar to that on a smaller scale associated with the growth of porphyroblast (Figure 3.27).

the case of cross-sections of transected folds of the kind shown in Figure 3.24.

The latest set of cleavages or foliations will be that which is most easily recognized because of its generally consistent attitude across all previous structures (see this relationship in the foliated agmatite of Figure 4.23 and see also Figures 2.10, 2.11 and 5.15).

3.2.3   EFFECTS OF LOCAL INHOMOGENEITY

Locally, the attitude of foliation, like that of other structures, will be sharply affected by structural inhomogeneities such as folds, thrusts, ductile shear zones, etc. (Figure 3.26), or intruded bodies and on a smaller scale, by pods, boudins etc.

On a very small scale, abrupt changes in structural orientation can be caused by deflection around porphyroclasts and porphyroblasts etc. (Figure 3.27).

Large-scale abrupt deflections in the orientation of all structures or breaks in their continuity, probably more easily recognizable by comparing the attitudes of the latest foliations, serve to indicate the presence of post-foliation terrane boundaries (Hopgood and Bowes, 1995), and could provide a useful means of matching separated parts of formerly contiguous segments of terranes (Figure 3.28). The strength of such matching is likely to be

particularly enhanced when used in conjunction with metamorphic isogrades (Figure 3.29). See section 6.1.3 and Chapter 14 (also Hudson, 1987).

For a discussion of the effects of local inhomogeneities the reader is referred to the account by Marques and Cobbold (1995) of the development of natural and experimental non-cylindrical folds formed around rigid inclusions.

Figure 3.27   Deflection of foliation around a local inhomogeneity (porphyroblast or boudin, or in this case, a pod or tectonic 'fish'). This is a small-scale example of local, abrupt change of foliation attitude in response to variation of stress field orientation because of local structural and lithological inhomogeneity. Compare these with pressure shadows. Korssund, Inkoo, southern Finland.

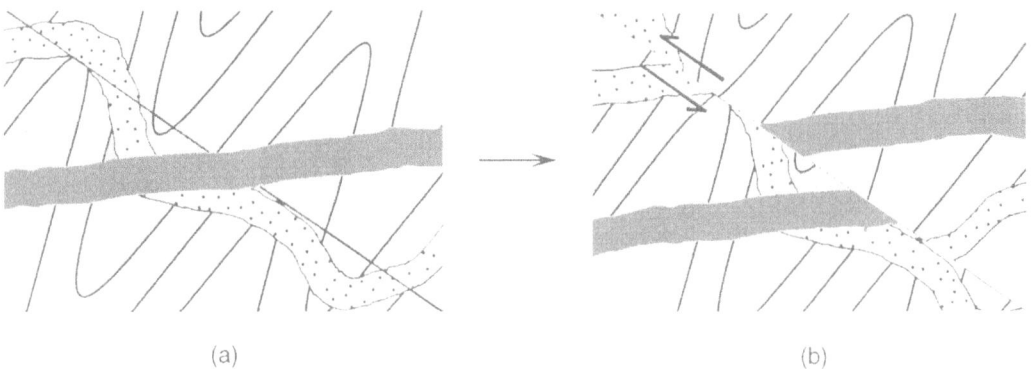

(a)   (b)

Figure 3.28 (**a**) Boundary between terrane segments offset along a major strike-slip fault (**b**) showing the potential for matching of formerly contiguous sections on the basis of structure.

3.2.4 REFOLDING ON PARALLEL AXES

It is not unusual in highly deformed terranes to find that refolding has taken place (often repeatedly) about an axis that is parallel or sub-parallel to an earlier one, so resulting in a very strongly-developed lineated aspect to the rock. However, this strong fabric need not necessarily obscure overprinting relationships completely (Figure 3.30).

However, in those cases where folding has taken place on parallel or sub-parallel axes, it is likely to be more difficult to establish the relationship between the fold sets, and even more so between folds and lineations, especially if these structures formed early in the sequence (Figure 3.31). Nevertheless discrimination of even these structural relationships should be possible in reasonably well-exposed terranes, especially in terms of refolded axial planes. This is because even when two fold sets have nearly parallel axes and axial surfaces, and even if this angular relationship is subsequently obscured by repeated deformation, the individual features of the two sets will often be sufficiently distinctive to allow discrimination between them. This is the case with fold sets F_{br} and F_{bb} described from Jussarö (Hopgood, 1984). Although both sets have parallel axial planes the folds in each case are quite distinctive. Folds of set F_{br} are tight to isoclinal (Figure 7.36c) and those of set F_{bb} are distinguished by their axial planar slip and associated leucocratic veining (Figure 7.36f). See also Chapter 5.

Where individual characteristics are not sufficiently distinctive to be definitive, separation of two fold sets can sometimes be effected when the earlier set has been intersected by features such as tabular veins before the later set has developed (Hopgood 1980, pp. 60–1). This is discussed in detail in the opening section of Chapter 4. On the other hand, the intersection of structures by later tabular features is less likely to provide an easily recognizable basis for distinction in the comparable situation where the two sub-parallel structure sets are **lineations** or **foliations** as in the case of two sets of shear surfaces, or foliations defined by coplanar mineral growth, such as mica. This is because discordance with structures that are only two-dimensional or one-dimensional is less likely to be observed than it would be with folds that are three-dimensional (see section 3.1.3). Furthermore, the attitude of emplacement of later tabular (or linear) features such as veins is even more susceptible to control by the attitudes of pre-existing planar structures, i.e. parallel rather than discordant, so that clear time relationships, between

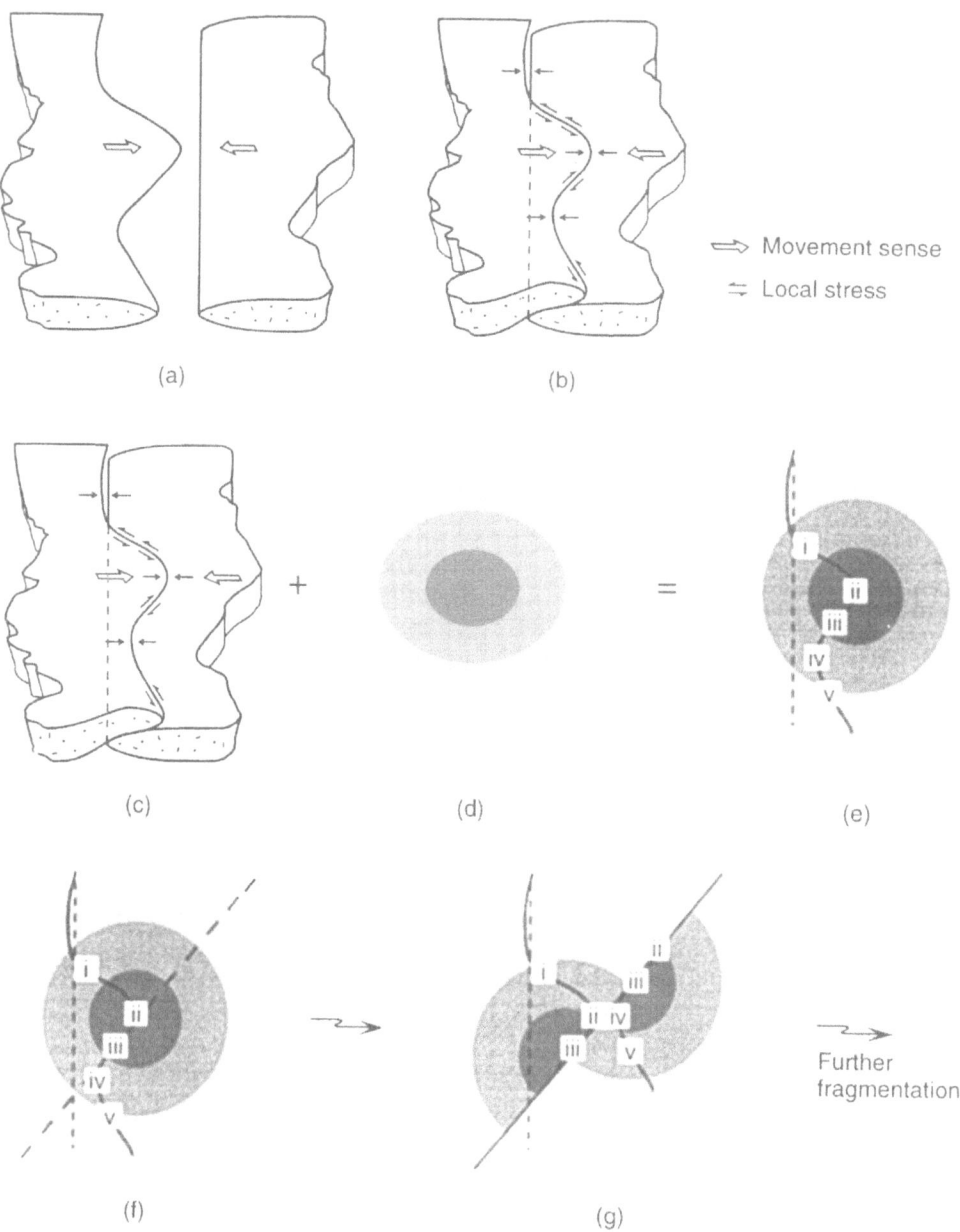

Figure 3.29 (**a**) Converging lithospheric plates showing the changing stress directions arising from variation in obliquity during collision of the irregular plate margins (**b**, **c**). The structure resulting from these, like metamorphic isograds (**d**, **e**), provides a means of matching dismembered tectonostratigraphic terranes (**f**, **g**). As the collision progresses, later stress directions will be superimposed at sites where the directions of previous stresses were different, leading to superimposition of structures – 'structural overprinting'.

(a)　　　　　　　　　　　　　　　(b)

Figure 3.30 Intensely-developed linear structure. (**a**) Leeuwin Block, Ringbolt Bay, Cape Leeuwin, Western Australia. (**b**) Svecofennian migmatites, Southern Finland.

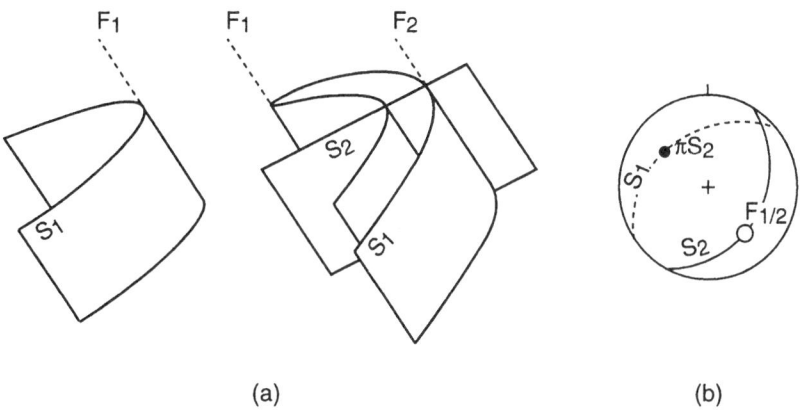

(a)　　　　　　　　　　　　　　　(b)

Figure 3.31 (**a**) Refolding on parallel axes. F_1 and F_2 are parallel to each other and to both the earlier and later fold axial planes. (**b**) Stereo plot of the relationships between the fold elements; earlier folded surface (S_1), second fold axial plane (S_2), common fold axis ($F_{1/2}$) and normal (pole) to S_2 (πS_2).

the two sets of (planar or linear) structures and the tabular veins separating them would not be clearly evident, especially after folding.

Refolding on parallel axes can arise if the orientation of the later, overprinting stress field, while not precisely the same as that of the earlier one, is nearly so. Then the existing fabric can exert a strong control on the orientation of later folds causing the later set of folds to develop on the same hinges as the earlier set (Figure 3.32).

According to Odonne and Vialon (1987) this can happen even when the angle between the compression directions causing the two fold sets is as much as 50° or even more. The earlier hinge directions rotate ('hinge migration') and are 'adopted' during the later deformation.

3.2.5 NON-TECTONIC FOLDS

One of the difficulties which must be overcome in deciphering the structure of high-

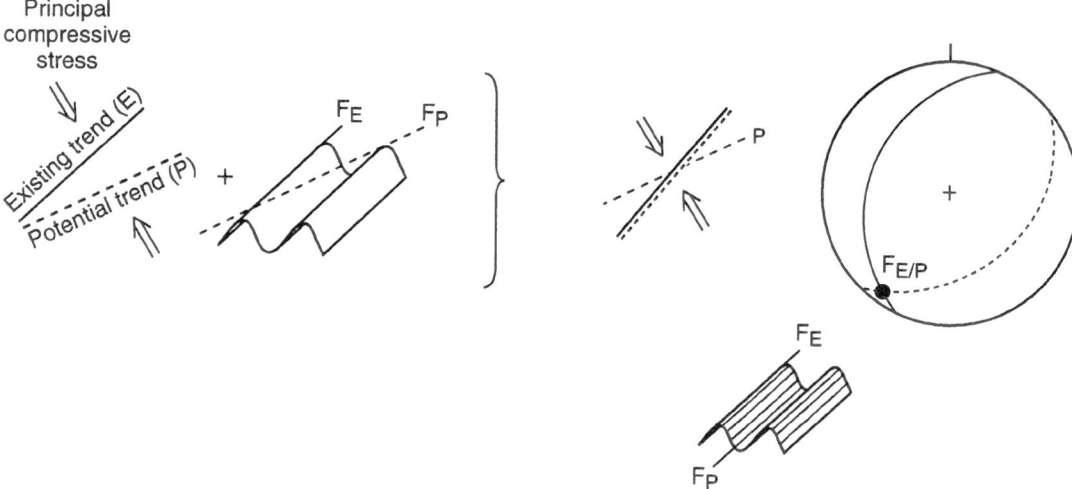

Figure 3.32 Control of the direction of a later hinge development parallel to an existing trend (E) instead of parallel to the potential trend (P), in response to superimposed sub-perpendicular compressive stress (cf. Odonne and Vialon, 1987). The stereo plot shows the ultimate attitude of superimposed existing and potential trends (F_E and F_P).

grade metamorphic rocks such as migmatites is the problem of determining whether or not the structures observed are essentially the outcome of deformation of rocks during tectonism or whether they originated prior to metamorphism and tectonism. It is always possible that the structures observed formed during igneous emplacement as flow folds in tabular bodies such as dykes or sills (Figure 3.33) or as drag folds in lava flows, or in sediments as a result of deformation prior to lithification and subsequent metamorphism.

The identification of non-tectonic structures (stemming from the effects of soft-sediment deformation) depends essentially on the fact that structural patterns will not be consistent over the wider 'area' of the study, i.e. from one exposure to another. Clearly, use of this criterion to identify non-tectonic structures could mean that there will be a transition between structures that are apparently tectonic (i.e. because observation of their consistent pattern was within only a very limited area) and those that are clearly non-tectonic, depending on the

Figure 3.33 Hinges of igneous flow folds in a thin dacite vein intruded into the Lewisian complex, Isle of Iona, Inner Hebrides, Scotland.

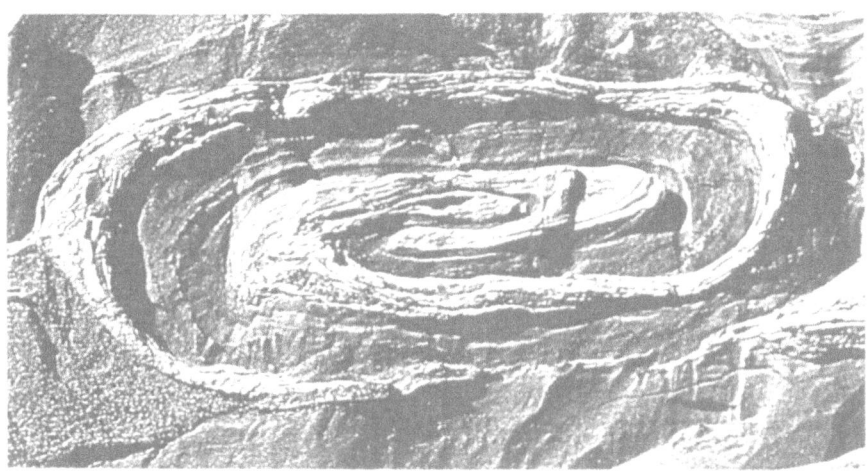

Figure 3.34 Rollup structure in unconsolidated sand and gravel. (From the cover photograph of *Geology*, **14**, No. 4, April 1986. Reproduced by permission of Scott Paterson).

scale of observation. For example, a local penecontemporaneous slumped mass of sediments could very well, within a restricted segment of rock a metre or two across, contain the same structural pattern, produced by the local stress field generated during the slumping. However, neither this stress field nor the structures it produced, would be constant, in terms of that particular orientation, a few metres away, outside the slumped mass. Similar reasoning would apply in the case of other penecontemporaneous sedimentary structures such as convoluted ball, or 'rollup', structures (Figure 3.34). Therefore on the broader scale such structures would not be penetrative, i.e. consistent in orientation throughout the segment, so that their true nature is easily recognized. On the other hand an imposed stress field of tectonic origin will, apart from minor local variation, be consistent with only gradual change, over a large crustal segment (Figure 12.17, and cf. Hopgood and Bowes 1972, Figure 2). However, one should remain aware of the effects already discussed, of even minor examples of sharp changes in stress orientation around local inhomogeneities such as the hornblen-

dite pod of Figure 3.27, and the examples shown in Figure 3.26.

Other features such as small-scale local unconformities between slump folds and the overlying laminae (Figure 3.35) are also clear indicators of soft sediment deformation. The distinction between such unconformities and tectonic dislocations (ductile shear etc.) is usually clear and in any case can be confirmed, if there is any doubt, by thin section examination of the contact.

The presence of blocks and smaller clasts amongst the folds is also clear evidence of soft sediment deformation (Figures 3.35 and 3.36).

The discrimination of soft-sediment folding from tectonic folding by analytical means can be achieved in a number of ways. An example of the effective use of shape factor analysis involves compaction-sensitive strain markers such as accretionary lapilli. The successful application of this technique in Proterozoic volcanogenic rocks in western Australia has been described by Boulter (1983) where the superimposed relationship of the compaction fabric to post-folding pre-consolidation clastic dykes is clearly demonstrable.

Figure 3.35 Soft sediment folds with unconformities truncating folds immediately to the left (above in the sedimentary succession) of the slump folds at the centre and above the hammer head. It is interesting to note the similarity between the shape of the fold at the centre of this photograph initiated in a sedimentary environment and that of the ductile 'scar' folds formed deep in the crust that are shown in Figures 10.15 and 10.35. Ardwell Flags, Scotland. Photograph courtesy of B. J. Bluck.

Discussion of other criteria useful in the distinction between pre-lithification folds and tectonic structures can be found in structural geology textbooks. Hobbs, Means and Williams (1976) list several such and discuss their merits in Section 3.6 (p. 156):

(a) Soft sediment structures may show:
(i) the presence of undeformed trace fossils;
(ii) opposite facing isoclinal folds (Figure 3.37);
(iii) truncation of folds (Figure 3.35) by

(a)

(b)

Figure 3.36 Soft sediment folds. (**a**) Irregular isoclinal and asymmetrical folds. Devonian sandstone, Scotland. (**b**) Planar-limbed, sharp-hinged fold and (**c**) irregular-limbed, round hinged fold associated with irregular clasts above pencil. Karelia, Finland.

(c)

overlying beds (as discussed above);
(iv) absence of joints and veins unrelated to folds;
(v) deformation confined to a single bed, and
(vi) 'chaotic' folds (Figure 3.38), although Hobbs, Means and Williams (1976) say this is true of some tectonic structures also, perhaps referring to constrictional structures or perhaps subscribing to the view that some tectonic folding is 'wild' (see section 1.1 on 'wild folds' and 'wild structures').

(b) Tectonic structures show:
(vii) the bending of metamorphic foliations and mineral grains around folds;
(viii) fossils deformed in a systematic manner, and

(ix) the presence of axial planar metamorphic foliation rather than fracture cleavage (although preferred orientation of mineral growth has been described from penecontemporaneous folds (Williams *et al.*, 1969; Moore and Geigle, 1972).

The certainty of any diagnosis will of course be increased if several of these criteria are recognizable together, although each one of (i), (vii) and (viii) is considered by Hobbs, Means and Williams (1976) to be definitive on its own.

Folds, thrusts and faults in unconsolidated sediments have been observed in areas affected by glaciation where overriding glaciers and ice sheets have provided the imposed stress (Figure 3.39) and Funder *et al.*, 1984, (Figures 6 and 7).

Figure 3.37 Opposite-facing, intrafolial isoclinal folds formed during soft sediment deformation. Port Gower, Scotland.

Figure 3.39 Folds developed by deformation of unconsolidated sediments beneath moving ice. Nushagak Bay, Alaska.

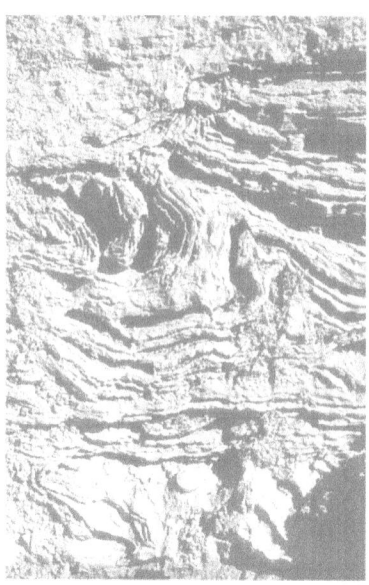

In some instances where the ice itself has folded, the fact that the folds exhibit structures characteristic of those formed in rocks during tectonism, with axial planar cleavage, tension gashes and crumpling in the hinge zone (Figure 2.16) implies that comparable large-scale structures can form prior to metamorphism in consolidated layered sediments also. Therefore, such structures developed in sediments might, after metamorphism, be indistinguishable from tectonic structures. Nevertheless, the absence of a consistent, widely-distributed structural pattern would still provide a means of distinguishing these structures.

Figure 3.38 'Chaotic' folds. Folds formed by soft sediment deformation of fluvio-glacial muds. Qeqetok, Southwest Greenland.

THE ROLE OF STRUCTURES INTERPOSED IN THE SUCCESSION

4

4.1 THE SIGNIFICANCE OF GEOMETRICALLY SIMPLE STRUCTURES IMPOSED ON THE DEVELOPING SUCCESSION

Earlier it was stated (section 1.1.2) that the analysis of complex structural relationships in migmatites is paradoxically often made easier by some of those very factors which contribute to this complexity.

From time to time during the developmental history of migmatites the number of sets of planar structures in the succession is increased – by the introduction of cleavages and joints during deformation, and by the emplacement of tabular minor intrusions during igneous activity. In addition, syntectonic metamorphic mineral growth produces new foliations and lineations in the succession. Ultrametamorphism can lead to anatectic partial melting with the production of neosomes emplaced either concordantly or discordantly as thin tabular or irregular veins and pockets which often result in the formation of agmatites. While all of these imposed structures complicate the succession, and in some cases tend to obscure the pre-existing structure locally, each serves not only as a potential strain marker but also as a potential **interval marker** and, where the structure is in some way distinctive, this marker can be a useful means of subdividing the succession.

4.2 BEHAVIOUR OF SIMPLE (PLANAR) STRUCTURES DURING DEFORMATION

There is, however, another reason why these structures can be of particular value to the structural geologist. Some of them (veins, cleavages, elongate minerals etc.) impose planar or linear geometrical features on the already complex structure and these newly added structural elements, although they further complicate the succession locally, can in fact be used to help to resolve the overall structural pattern. This is because, as planar (or linear) structures, their geometry is simple compared with that of the existing folds, and therefore when they respond by folding during subsequent deformation they produce relatively simple geometrical forms which are more easily deciphered than the complex forms of the refolded earlier fold structures (Figure 4.1).

This is an important although not immediately obvious aspect of migmatites, viz. that contributions to their structural complexity, stemming from the introduction from time to time during their developmental history of new ('imposed') elements, can often be a positive advantage in structural analysis. As noted above, these elements may include not only more or less regular features such as tabular (geometrically planar) igneous bodies such as dykes, or cleavage and joints and linear structures such as mineral lineation, but also irregular bodies such as pockets of

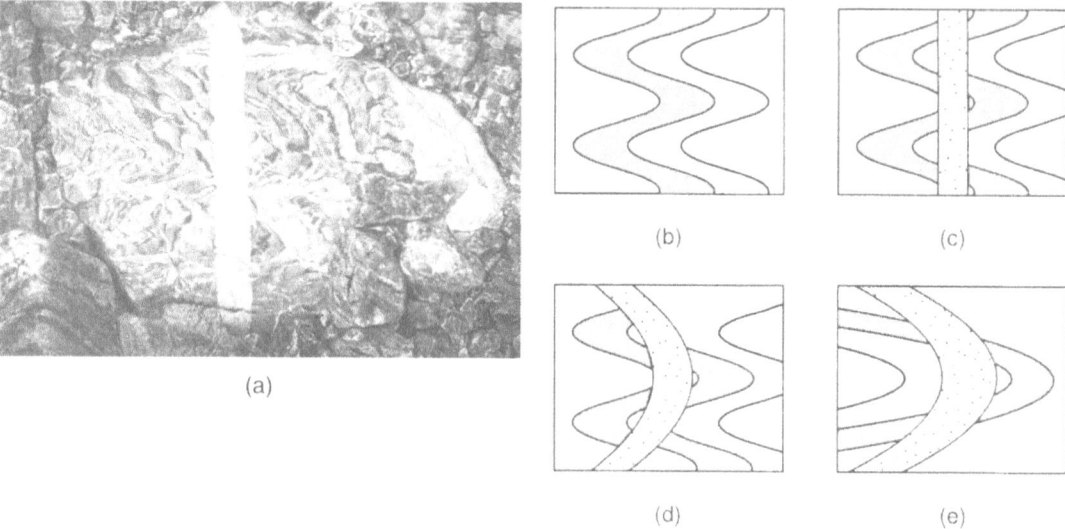

(a)

(b)

(c)

(d)

(e)

Figure 4.1 (**a**) Folded Moine metasediments cut by tabular leucocratic vein (submerged boulder). Sketches **b–e** show the development of a relatively simple fold form from a comparable tabular intrusive vein in folded rock. (**b**) Folded host rock prior to the intrusion of the vein. (**c**) Intrusion of the tabular vein. (**d**) Progressive folding of the vein accompanied by coaxial refolding of the host rock. The host rock itself shows no obvious separate record of the later (post-intrusion) folding which is revealed only by the fold in the vein. (**e**) Part of (**c**) enlarged. Compare Ramsay 1967, Figure 7.2, p. 344. Highlands, Scotland.

quartzofeldspathic material (neosomes) arising from partial melting (section 4.3.3(c)) or other igneous bodies. This is especially important when the later deformation has the potential to form only gentle, open folds (expressed only in the introduced planar structures) because, without the visible evidence of the modified introduced structure, such slight modification of the pre-existing complex fold pattern would almost certainly be unrecognizable (see Figure 10.3, section 10.1.2). These intermediate, imposed structures need not necessarily be penetrative and when developed only locally they assume even greater importance as the **only** potential means of detecting the existence of sets of weakly expressed folds, particularly when these have axes or axial planes parallel to other fold sets (Figure 4.2).

Furthermore, if such geometrically simple structures introduced into the complex at a particular time in the deformational history

subsequently have imparted to them even later simple secondary structures like cleavage, their value as strain indicators is further enhanced (Figure 4.27 in section 4.3.2).

Intervening structures of this kind can also be particularly valuable as indicators of the presence of more than one set of later structures in parallel orientation such as later foliation parallel to an earlier set (Figure 4.3a). Without the intervening structure which it affects, the superimposed secondary foliation would normally be indistinguishable from the earlier set.

Consider the structural relationships explained by Figure 4.3. This shows a leucocratic vein containing more or less equant porphyroblasts which was emplaced through the existing foliation, cutting it at a high angle. Subsequently deformation and syntectonic metamorphism caused the development of a new foliation parallel to the first. It also formed folds in the vein (presumably by slip

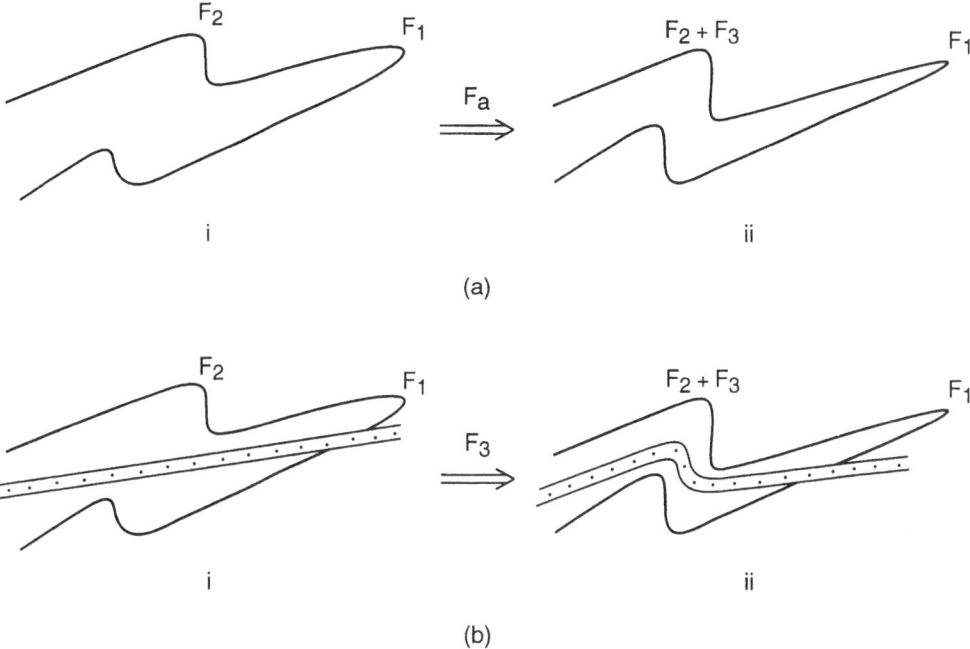

Figure 4.2 An isoclinal fold (F_1) refolded (F_2) and refolded again (F_3) on a later axial plane parallel to the earlier to form $F_2 + F_3$. The second refold (F_3) is revealed only by the deformation of a vein (an 'imposed' structure) intruded between the two fold events. (**ai**) First refold (F_2). (**aii**) Second refold (F_3). (**bi**) Tabular vein intruded after F_2. (**bii**) Second refold (F_3) on the same axial plane as F_2, recognized and distinguished from F_2 only because of the fold in the vein which is the *only* pure F_3 fold in the structure. The refold in F_1 is a composite ($F_2 + F_3$) fold.

parallel to the foliation), with axial planes parallel to the foliation, and 'ellipsoidal' modification of the porphyroblasts with preferred orientation of their long axes coplanar to both the earlier and later foliations. This demonstrates that the foliation observed in the host rock must be **composite**, the product of at least **two** foliation-forming events. Without the vein, with its folds and the aligned porphyroblasts, there would be nothing to suggest that the foliation in the host gneiss is the product of more than a single deformational event.

Secondary (planar) structures have greater value when they are at a high angle to the margin of the host intrusions. This is because at least one of the two planar structures (intrusion and superimposed foliation) is likely to have an orientation favouring deformation in

response to a later stress system whatever its particular orientation. Together the two structures provide the potential to respond to later deformation three-dimensionally. This situation is geometrically comparable to that where overprinting takes place between folds rather than between simply planar or linear structures, a condition discussed in section 3.1.3. See also Figure 4.27 in section 4.3.2. For example, if a structurally homogeneous tabular intrusive body had been subjected to compression normal to its margins, or to slip parallel to them, no obvious macroscopic record of this deformation would be preserved. Therefore, where the host rocks are already highly folded, as in the case under consideration here, the deformation arising from this later compression, or slip, would be unrecognizable in the field. The effects of **compression** could be

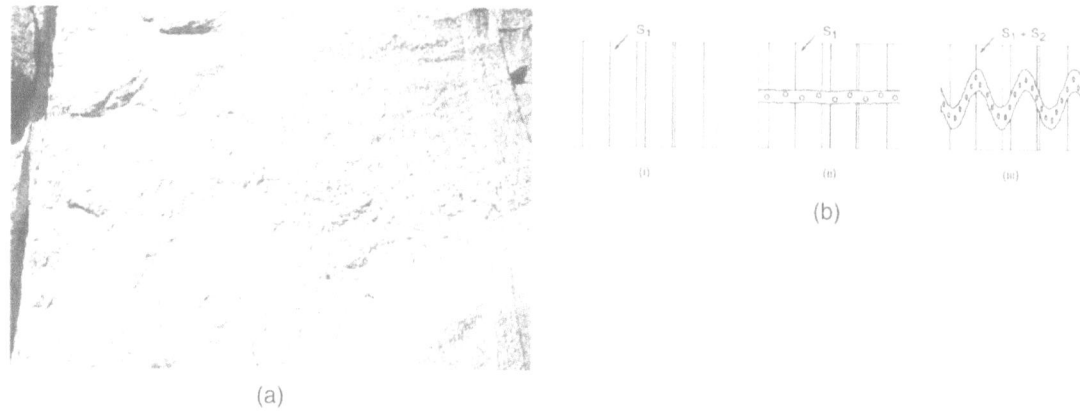

(b)

(a)

Figure 4.3   (**a**) Irregular 'horizontal' leucocratic vein with vertically aligned porphyroblasts in vertically foliated tonalitic gneiss. (**b**) Sketches showing the derivation of this structure from (**i**) early gneissic foliation S_1, (**ii**) intrusion of tabular vein containing porphyroblasts, (**iii**) imposition of later foliation (S_2) parallel to S_1 with deformation of porphyroblasts from 'spherical' to ellipsoidal shape, with long axes parallel to S_1/S_2. The deformation of the vein and the ellipsoidal porphyroblasts demonstrate the composite nature of the foliation in the host rock. Svecofennian migmatites, Barosund Ferry, southern Finland.

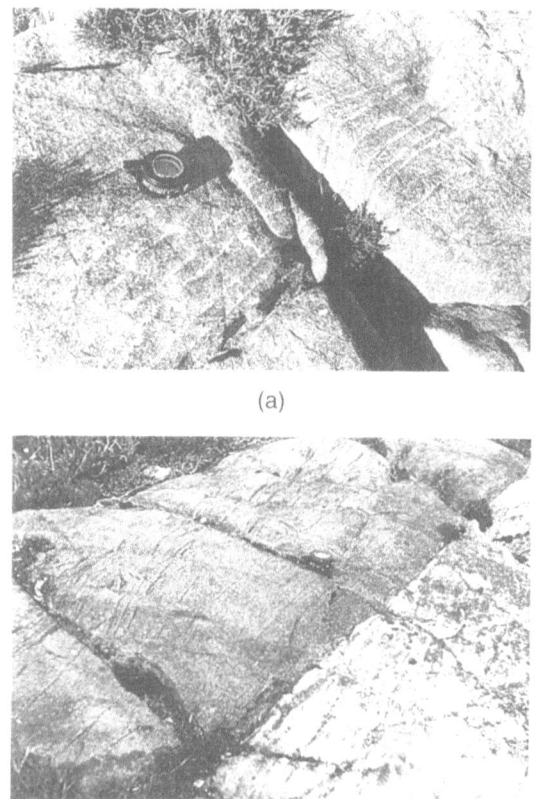

(a)

(b)

gauged only if the original width of the intrusion was known, and the effects of **slip** would not be recognized in the absence of clearly discernible strain indicators. However, had the intrusion possessed planar structure at a high angle to its margin this would have been likely to buckle under compression, or fold passively in response to slip (see Figures 4.29–4.33 in section 4.3.2).

The presence of secondary structures of this kind in minor intrusions has the additional value in allowing discrimination of the host intrusion from other intrusions where there is more than one set of similar-looking bodies in the overall sequence. The specific secondary structure (e.g. crenulation cleavage, fracture cleavage or platy mineral growth) in one intrusion will distinguish it from the others lacking the structure.

Other structures like quartzofeldspathic veins, shears, and less commonly, cooling joints in igneous intrusions, especially if the

Figure 4.4   (**a**) Primary igneous layering in a metabasic intrusion. (**b**) Primary flow foliation preserved in a discordant metabasic intrusion. Lewisian complex, Rona, Inner Hebrides, Scotland.

intrusions are tabular, (Figure 4.35 in section 4.3.3) and even intangible geometrical features such as axial planar traces will similarly provide means of demonstrating later deformation (Figure 4.26 at the end of section 4.3.1), and these are discussed later. The same is the case for primary foliation (crystallization layering, depositional layering, and flow and shear foliation) in larger igneous masses (Figure 4.4). Because quartzofeldspathic veins, both cross-cutting and concordant, form an integral part of the composition of migmatites, the interpretation of their deformational history in terms of relationships of veins to host rocks is particularly important. Concordant as well as discordant bodies can be used in this way, provided these have affected pre-existing structures (e.g. cleavage or mineral growth which is axial planar to earlier folds), or are associated with agmatites affecting pre-existing structures (Hopgood and Bowes, 1978).

4.3 ILLUSTRATIONS OF THE IMPORTANCE OF SIMPLE INTERPOSED STRUCTURES IN RESOLVING (STRUCTURAL) SUCCESSIONS

4.3.1 STRUCTURES EMPLACED IN (INTRUDED INTO) THE HOST ROCK

(a) Planar structures

The planar contacts of tabular bodies intruded into folded rocks clearly demonstrate the overprinted (i.e. cross-cutting) relationships of the bodies (Figure 4.5).

Such later planar structures are particularly valuable where subsequent folds (i.e. those formed later than the introduced tabular body) are very open and could otherwise remain undetected because the foliation was already complexly folded (Figures 4.2 and 4.8). Correspondingly, later deformation can be recognized in rocks such as agmatites with structure that is or appears to be irregular, when they have hosted tabular intrusion prior to the later deformation (Figure 4.6).

Figure 4.5 Margin of a metabasic body intruded into folded gneiss. Its planar contact with the folded host gneiss clearly demonstrates its discordant (post-folding) nature. Lewisian complex, Rona, Inner Hebrides, Scotland. (cf. Figure 4.34.)

The determination of the relative age of folded foliation in rocks that have been highly deformed is considered by some to pose serious problems (Park, 1969). Even so it is remarkable how often very early planar structure, e.g. original lithological layering such as bedding, can still be recognized in spite of polyphase metamorphism and multiple deformation. This is illustrated by Figure 10.22 in section 10.1.8 and Figure 11.9 in Chapter 11, in which an early layering (a lithological banding S_s) has been repeatedly deformed and has not only been cut by a strong syntectonic metamorphic foliation (S_a), itself folded, but is also intensely transposed more or less parallel to S_a.

Linear structures are also useful indicators of subsequent folding (Ramsay 1967, pp. 461–90) but the effects on them of later deformation are not so readily discernible because the deformed intersection of one-dimensional structures on the outcrop is less obvious than that of two-dimensional structures (Figure 4.7). However, in some circumstances it might be necessary to use them in the absence of any other means of proving the effects of later, comparatively slight deformation in complexly folded rocks (see also the discussion on

(a)   (b)

Figure 4.6   (**a**) Agmatite cut by tabular metamorphosed basic intrusion. (**b**) Detail showing the planar form of the contact. The intrusion provides the potential to identify later deformation, even if this is represented by only slight warping of the contact, an effect that would not be recognizable if superimposed on the irregular structure of the agmatite. Lewisian complex, Rona, Inner Hebrides, Scotland.

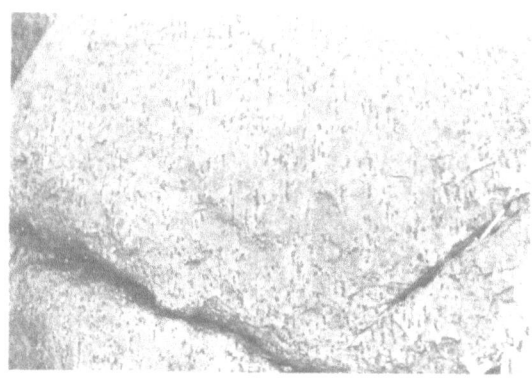

◄ **Figure 4.7**   Garnet aggregates whose statistically linear trace could be used to demonstrate later deformation in a manner comparable to the trace of foliation. Leeuwin Block, Western Australia.

the 'Relative value of folds compared to foliations and lineations as indicators of overprinting relationships', section 3.1.3). The importance of imposed linear structures is considered further in 4.3.1(b) below.

Tabular minor intrusions of contrasting colour and texture to the host rocks, especially discordant ones such as thin dykes, sills and veins, provide ideal reference structures in these situations. In quartzofeldspathic gneisses, basic intrusive bodies are especially valuable as they stand out more clearly from the light-coloured host rocks (e.g. Figures 4.8a, 4.8b) so that subsequent simple folds will be clearly

discernible, whereas granitic veins are likely to show more clearly in darker host rocks (Figure 4.8c).

Note that interposed structures of this kind are probably much more common than cursory inspection might at first suggest and their overprinted relationships could pass unnoticed without careful observation, particularly in cases where the intrusive structures are very small (Figure 4.11), or where discordant relationships are not striking (Figure 4.9).

Often closer examination shows that apparently concordant basic sheets (or for that

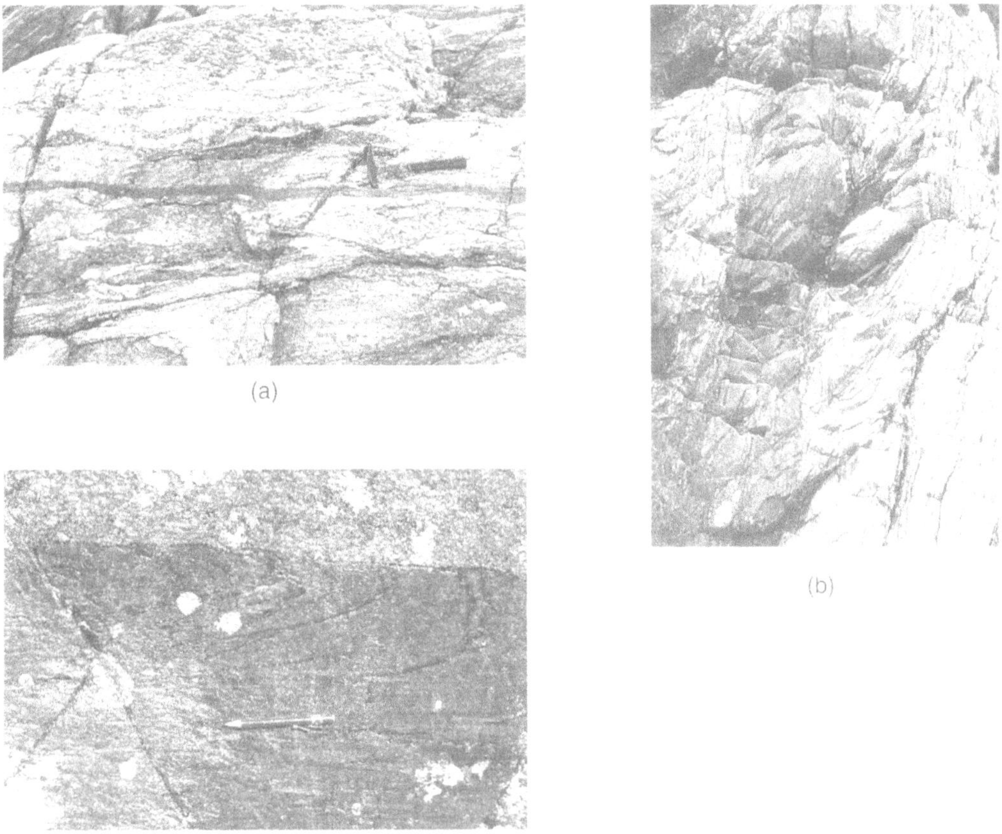

(a)

(b)

(c)

Figure 4.8 Three examples of planar-side intrusions in folded host rocks. (**a**) Minor basic intrusions discordant to isoclinal folds and clearly later than the folds. (**b**) Discordant basic intrusion more or less axial planar to fold hinges. Lewisian complex, Rona, Inner Hebrides, Scotland. (**c**) Discordant granitic vein parallel to the axial plane of an isoclinal fold and replacing one limb of the fold. Svecofennian migmatites, Sundharun, Jussarö area, southern Finland.

Figure 4.9 Small boudins cut and enclosed by a thin basic intrusive vein concordant to the dominant foliation which is isoclinally folded. The enclosure of the boudins by the vein could easily pass unnoticed and in consequence so could the time relationships between the earlier boudin-development and later intrusion. Note that the basic vein material encloses the three upper boudins which lie on the inside of the isoclinally folded intrusion. The pencil is on the fold axial planar trace. Svecofennian complex, Kaijalohja, Hyypiänmäki, Finland. ▶

matter, quartzofeldspathic or other veins) are in fact slightly discordant to foliation. Where this foliation happens to be in the limbs of isoclinal folds, whose hinges could be found, the discordance of the host foliation at the hinge, where it is much more obvious, will confirm them as introduced structures that are post-isoclinal folding (Figure 4.10).

Although they are more difficult to see, and might easily pass unnoticed, even very narrow intrusive veinlets and their relationships to the host rock structure should not be ignored as they can be useful indicators of subsequent deformation (Figure 4.11).

These structures are particularly useful for showing relationships such as that between the time of foliation development and the intrusion. An example of the relationship between foliation, folds and a later structure (foliation/lineation) in a discordant dyke which is fortuitously (and usefully) at a high angle to the pre-existing foliation and fold axial planar trace, is shown in Figures 4.12a and 4.12b.

Tabular intrusive bodies can occasionally play a role in determining more subtle structural relationships. Consider the igneous body

of Figure 4.13. This has been emplaced into foliated rock (Svecofennian migmatite) and the intrusion exhibits a foliation that is not parallel to the margins of the body (and therefore cannot be attributed to, say, relict flow foliation) but is parallel to the foliation of the host rock. This indicates that some at least of the foliation development in the host gneiss **post-dates** the intrusion, i.e. some was formed **at the same time** as the foliation imposed on the dyke. The host foliation must therefore be **composite** (compare Figure 4.2). This apparent 'passing through', or extension into, the intrusive bodies of the host rock foliation is not uncommon in migmatites and can often be discerned (if only faintly), even in very coarse-grained pegmatoid veins. It could be explained by later shear in a direction parallel to the foliation in the host rock and parallel to the margins of the intrusive body (if shear had been oblique to the margins, the shape of the intrusive body would of course have been modified accordingly, either causing the contacts to become non-planar or causing rotation with change of thickness of the body. (See Hopgood, 1966, Figure 6).

Figure 4.10 Post-fold vein (below pencil) sub-parallel to the limb of an intrafolial isoclinal fold. Leeuwin Block migmatites, Western Australia.

Figure 4.11 Narrow intrusive veinlet (below pencil stub) discordant to small folds in amphibolite and therefore later than this folding. The effects of even slight post-intrusive deformation would be revealed by the regularity of the structural pattern superimposed on the vein. Lewisian complex, Rona, Inner Hebrides, Scotland.

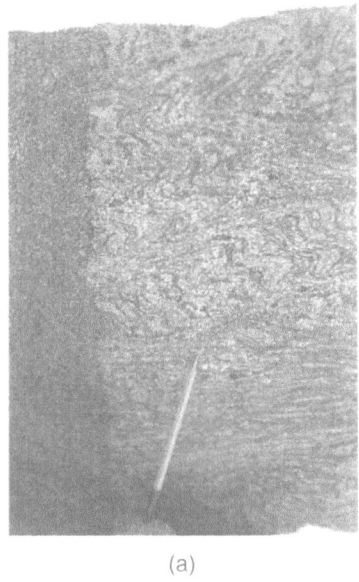
(a)

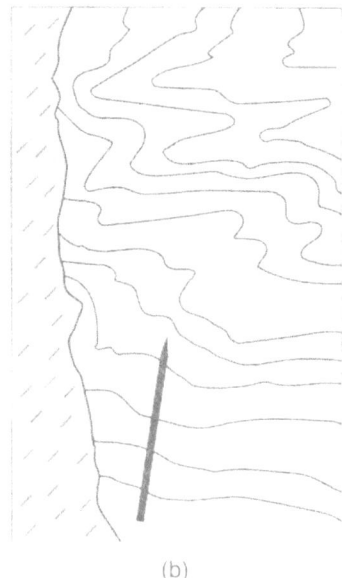
(b)

Figure 4.12 (**a**) Foliation imposed on an intrusion emplaced after the folding of the foliation in the gneiss. The foliation which is expressed as very fine intermittent linear traces inclined at approximately 45° to the contact in the intrusion can be discerned only with difficulty among the structural complexity of the host rock. (**b**) Sketch of the structural relationships in (**a**). Leeuwin Block migmatites, Cowarumup, Western Australia.

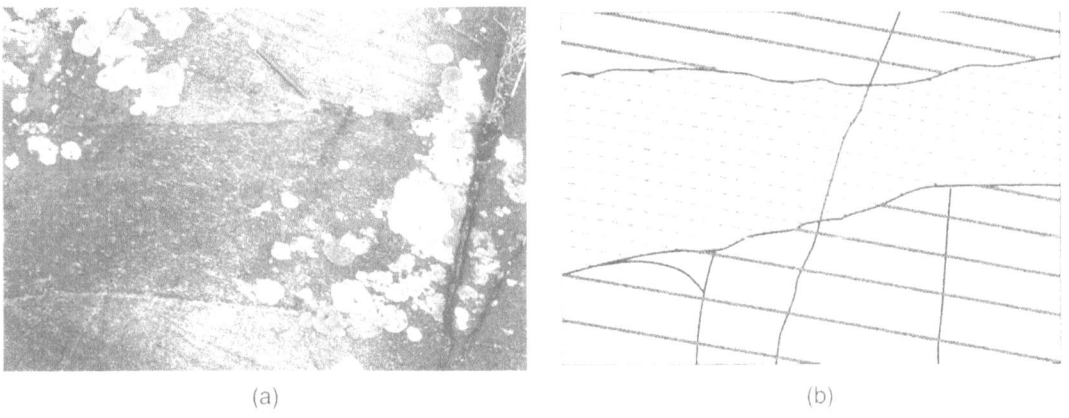
(a)   (b)

Figure 4.13 (**a**) Metamorphosed minor basic intrusion discordant to the lithological layering (foliation) in the host gneiss and showing a foliation parallel to the host foliation. (**b**) The relationships between the two foliations, clear at the 'lower' contact indicates that the attitude of the foliation which developed in the basic body was controlled by that of the host foliation to which it is parallel. This demonstrates the fact that the foliation in the host rock is composite, in part developed prior to the intrusion and in part developed afterwards. Svecofennian migmatites, park above South Harbour, Helsinki, Finland. Compare Figure 4.2.

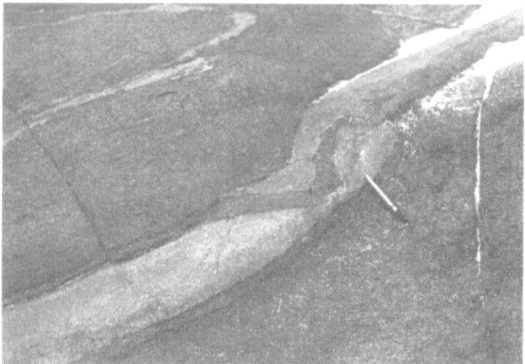

Figure 4.14 Leucocratic vein emplaced parallel to the hinge of an open asymmetrical fold (partly 'wedged' off at pencil) in foliated amphibolite. Leeuwin Block migmatites, Cape Leeuwin, Western Australia.

Figure 4.15 Leucocratic vein developed parallel to one limb of an asymmetrical fold in gneiss. Lewisian complex, Rona, Scotland.

As was noted earlier, the generalization regarding the value of darker basic intrusive rocks as intermediate structures does not always hold of course, because in some instances the host rock is dark. In that case a light (leucocratic) intrusion shows more clearly. Particularly useful indicators are quartzofeldspathic veins parallel to fold axial planes or to one limb of a fold (Figures 4.14 and 4.15).

In the case of concordant intruded bodies, especially where the host rocks are discretely layered so that the sharp boundaries of the body contrast less strongly with the layering, the intrusive relationships are less obvious. Nevertheless the true nature of such bodies can be recognized in those cases where enclaves are present (Figure 4.16). Furthermore, the age of the intrusion relative to the fabric in the host rock can sometimes be determined from the attitude of the fabric in the xenolith, for example where the fabric in the enclave is rotated with respect to that in the host rock the fabric must have developed prior to the intrusion (Figures 4.16b and 4.16c).

In those cases where the country rock is not obviously foliated, or is almost isotropic (as in the case of parts of some plutons), minor

intrusions may provide the only means of recognizing (and measuring) subsequent deformation on the outcrop. In such cases, the relationships between the pre-existing (weak) fabric in the country rock and the later deformation shown by the structure in the minor intrusion, could be difficult to recognize (Figure 4.17).

Intervening planar geometrical structures are particularly important in distinguishing between earlier and later folds with parallel (or nearly parallel) axial planes (Figure 4.18). In the absence of the intervening structures the existence of the later fold would not be recognized (see also Figure 3.31 in section 3.2.4, Figure 7.8 in section 7.1.1, Figure 11.13 in section 11.1.2, and Figure 12.4 in section 12.2).

As well as demonstrating overprinting relationships, 'complications' of this kind in the succession provide good indicators of the time of the events when the igneous bodies involved enable the events to be dated isotopically.

Consider the structural relationships shown in Figure 4.19a and 4.19b. Here, the structure

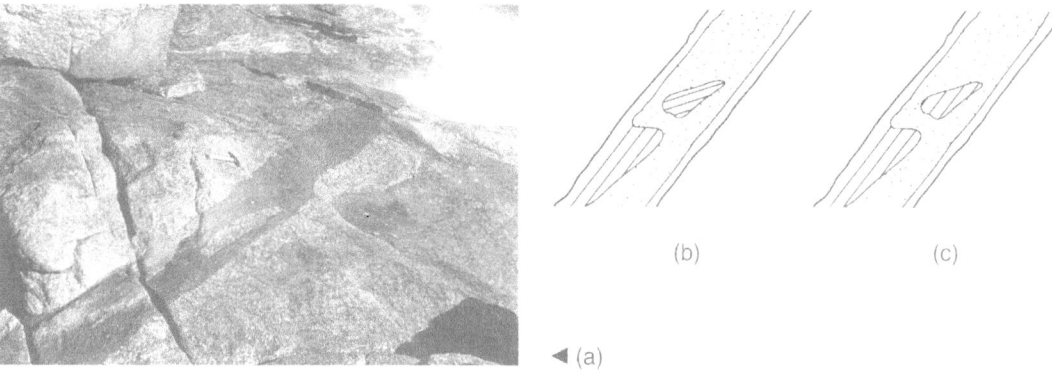

Figure 4.16 (**a**) Enclave of coarsely foliated gneiss in a basic intrusive sheet; (**b**) and (**c**) Examples of possible relationships between such enclaves and their host intrusions. In (**b**) the enclave and its foliation are rotated, therefore the vein would have been emplaced after the foliation was developed. In (**c**) the foliation in the rotated xenolith is parallel to the host foliation and therefore the vein must have been emplaced before the latest foliation development in the host rock. Leeuwin Block migmatites, Western Australia.

Figure 4.17 Weakly foliated granitic gneiss cut by narrow (grey) intrusive veins which could act as potential strain markers. Leeuwin Block migmatites, Western Australia.

suggests that the gneiss has been affected by at least two sets of folds below the dark horizontal intrusion, viz. tight upright folds (F_t) on the right and open upright folds (F_o) on the left. There is a suggestion also of a third, inclined asymmetrical fold set (F_a) to the left of F_t with an inclined axial plane (S_a) and this shows more clearly in Figure 4.19b.

The tabular basic intrusion is markedly discordant to the recognizable structure in the gneiss, having been emplaced **through** the folds noted above, and this discordance between the structure and the margins of intrusion shows clearly in the figures. It can also be seen in the figures that the intrusion is distinctly folded on upright axial planes by consistently regular folds of a greater wavelength than those already described. This means that after the emplacement of the intrusion both the host and intrusion were deformed by the folds of at least one other set, viz. open warps (F_w), therefore there are at least four (3 plus F_a) fold sets. These relationships are summarized in Figure 4.20a.

However, without the tabular intrusion, the structure would appear as shown in Figure 4.20b and it would be very difficult indeed, if not impossible, to detect the effect of the later warping on the already complex fold pattern (cf. Figure 4.2).

(b) Linear structures

Linear traces can also be used to show the effects of later deformation (see Figure 4.7 again) although their value is limited because they are one-dimensional rather than two-dimensional structures. This means that the chance of their being favourably orientated

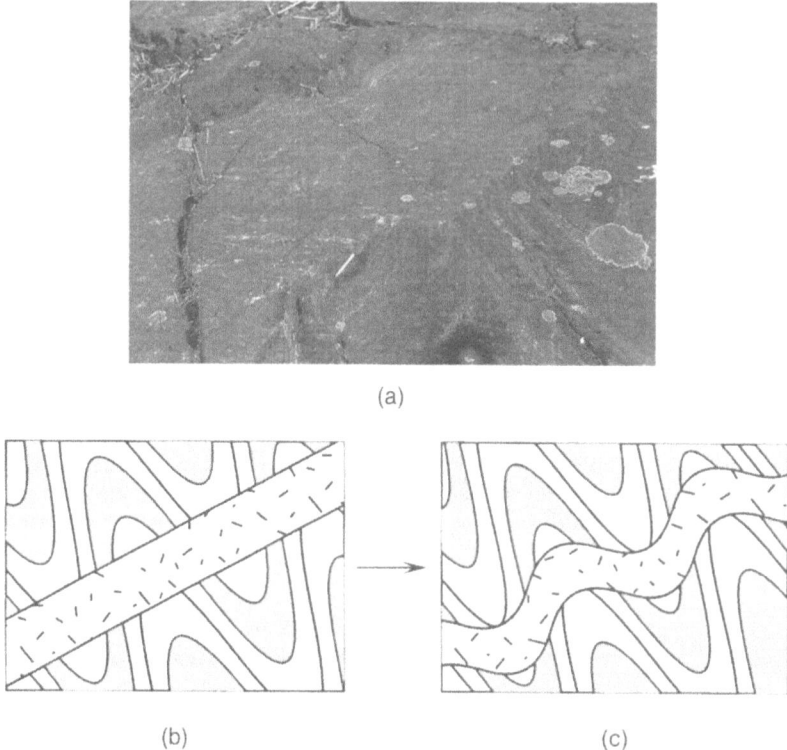

(a)

(b)   (c)

Figure 4.18 (**a**) 1 m wide leucocratic vein discordant to folded gneiss and subsequently deformed by folding on axial planes parallel to those of the folds in the host rock. (**b**), (**c**) Sketches showing the sequence of events causing (**a**). Svecofennian migmatites, Skåldö Ferry, southern Finland. Compare Figure 4.1.

(a)   (b)

Figure 4.19 (**a**) Gently folded, sub-horizontal basic minor intrusion discordant to tight upright folds in amphibolite facies quartzofeldspathic gneiss. (**b**) Detail of the structure in the host gneiss below a gentle fold in the intrusion. To the left of the tight upright fold in the centre (F_t) there can be seen an asymmetrical fold (F_a) with an inclined axial plane dipping steeply to the right parallel to the pencil. Lewisian complex, east coast, Rona, Inner Hebrides, Scotland.

(a)

(b)

with respect to applied stress directions is correspondingly reduced (see section 3.1.3 for a discussion of the relative importance of one-, two- and three-dimensional structures in strain analysis). Furthermore, because they are one-dimensional they are less likely to intersect the exposure surface and therefore are less likely to be observed.

Linear structures are probably most valuable as indicators of strain history where their relationships with respect to folds can be identified (Figure 4.21). They are useful where more than one set of lineations can be observed together on the exposure and the relationships between the two sets is identifiable. Such a case is where one set (associated say, with the hinges of an earlier fold set) is seen to have been folded by a fold associated with another (later) set of lineations (Figure 4.22).

Other, less obvious examples of linear and planar structures which play a comparable role in demonstrating the effects of later deformation are discussed below and are illustrated in the following figures. While some of these may not be as prominent as those previously discussed they are valuable nevertheless, especially when there are no other appropriate

Figure 4.20 (**a**) Recognition of refolding on parallel axial planes where an approximately horizontal planar tabular intrusion has been emplaced late in the deformational sequence. Prior to the emplacement of the intrusion the foliation of the host rock had been folded at least twice, by tight folds (now with a vertical axial plane – S_t) and by steep open folds (axial plane S_o). The sub-horizontal thin tabular amphibolite body has cut all structures, as shown clearly by the discordance of the contact with the limb of the open fold (F_o). As can be seen from the open warps in the originally planar amphibolite body, both host and intrusion have subsequently been gently folded on axial planes (S_w) which, because they too are upright, are parallel to S_t. (**b**) Structure of Figure 4.19 with the tabular basic intrusion 'removed' showing how it would appear without the basic intrusion. In the absence of the folds in the intrusion there is no obvious evidence for more than three sets of folds, S_t, F_a and F_o. Lewisian complex, east coast, Rona, Scotland.

Figure 4.21 Linear structure associated with folds. Linear garnet aggregates aligned parallel to isoclinal fold hinges (faintly discernible hinge closures indicated by convergence of the gross foliation at centre and bottom right edge) in deeply weathered gneiss. Leeuwin Block migmatites, Cape Leeuwin, southern West Australia.

structural features, in which case they are of course indispensable.

The relationship between foliation development and agmatization of folded rocks is shown in Figure 4.23. In this example foliated gneiss has been folded and later, during partial melting and the development of a leucocratic neosome, it has been subjected to fragmentation. Movement causing separation of blocks has resulted in the development of an agmatite. Here it can be seen that the linear trace of another foliation cuts both palaeosome blocks and neosome and, because this trace is rectilinear (and therefore not showing any effect of later deformation), the foliation is clearly the latest structure visible. The rectilinear trace provides a potential strain marker capable of demonstrating subtle warping of the agmatite, warping so slight that it would otherwise be undetectable in the complexity of the agmatite structure.

Figure 4.24 shows a foliated palaeosome block in an agmatite comprising blocks of tonalitic gneiss and amphibolite in coarse leucocratic neosome. The earlier foliated gneiss and neosome have both been affected by a foliation which can be seen to overprint the palaeosome block. The neosome is crossed

by a coarse linear trace parallel to the pencil and there is also evidence of this foliation affecting the edge of the block as shown by the faint linear trace in its marginal zone. The relationships between the structures are sufficiently clear to allow the determination of

(a)

(b)

Figure 4.22   Three examples showing two sets of linear structures. (**a**) Earlier, steeply-plunging lineations (parallel to the upright pencil) in the limbs of isoclinal fold hinges which plunge gently to the left. The isoclinal hinges are associated with the sub-horizontal lineation parallel to the pencil on the right. The variation in plunge of the earlier lineations (lower left) results from their curvature across the isoclinal hinges. Leeuwin block migmatites, southern Western Australia. (**b**) Slab showing two interesting sets of lineations. Note that the finer, horizontal traces parallel to the pencil cross the coarser plunging lineations without deviation, indicating that the former are later. From Letovice, Czech Republic. (**c**) Coarse, plunging linear structure (mullions) below, and finer, more steeply-plunging linear traces (above) which curve across the mullions indicating that because they were deformed by them, they predate the mullions. Letovice, Czech Republic.

(c)

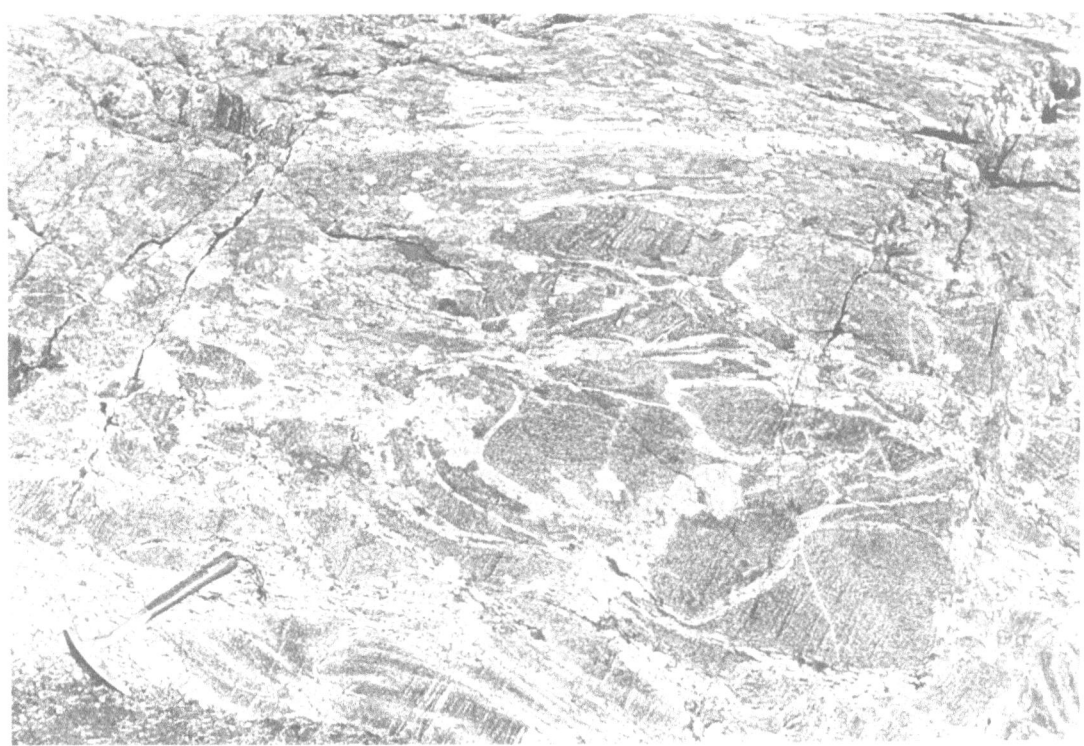

Figure 4.23 Foliated agmatite. New foliation (S_n), which trends across the exposure approximately parallel to the irregular NE-trending fractures, was imposed after (1) folding of the gneiss and (2) fragmentation during neosome formation. This provides a potential strain marker. Deformation of the later planar foliation (which would be shown for example by curvature of its linear trace) would enable subsequent deformation to be detected relatively easily, whereas the same deformation might be undetectable amongst the complexity of the folded, agmatitized earlier foliation. Lewisian complex, Isle of Barra, Outer Hebrides. See also Figure 4.6.

the sequence of events responsible for their formation, which is explained in the figure and caption.

Figure 4.25a shows a tabular basic minor intrusion intruded into quartzofeldspathic gneiss. The numbers in the following discussion refer to structures shown in Figure 4.25b. The linear trace (5) on the basic intrusive body (4) seems to be syntectonic with open folds (3) in the foliated (1) and isoclinally folded (2) host rock. This is because the trace appears to be parallel to the axial plane of the folds, thus suggesting that it represents the trace of an axial planar foliation. However, the fact that the linear trace also shows on the intrusive vein demonstrates that, because it is later than the vein, it must of course be later than the folds which pre-date the intrusion which cuts them. In fact the linear trace appears to be the latest in the sequence of events shown in the column of Figure 4.25b.

In addition to the planar structures discussed above, **intangible** simple geometrical features such as fold axial planes (or their linear traces) defined solely by the positions of successive hinges of folded layers can also serve as indicators of overprinting relationships (Figure 4.26).

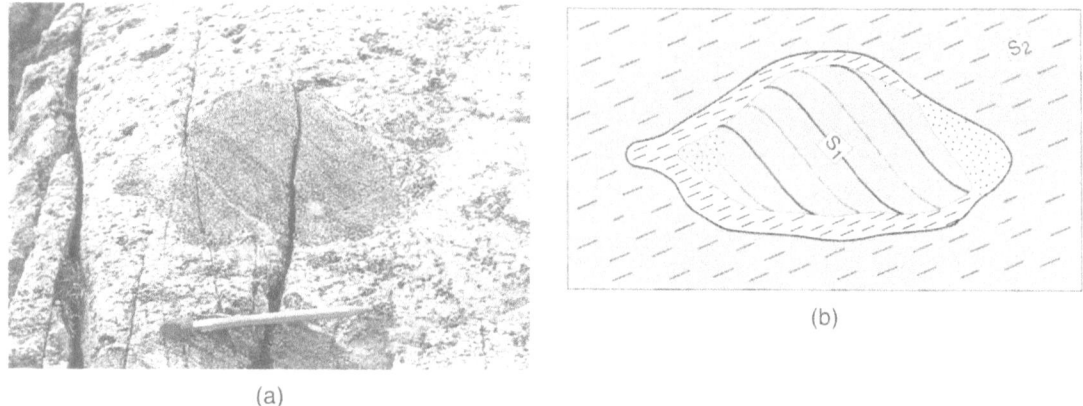

(a)

(b)

Figure 4.24 (a) New, post-rotation foliation in a palaeosome block, sketched in (b). The block lies in neosome in which coarse relict foliation can be detected by the alignment of smaller elongate fragments. Rotation of the palaeosome block containing the earlier foliation (S_1) took place after agmatite-formation and this was followed by the development of a second foliation (S_2) in the agmatite. This second foliation, inclined to the first, is discernible in the rim of the block, where it can be seen to be parallel to that in the neosome. The relict 'S_1' foliation in the neosome, labelled S_2 in (b), is therefore composite, comprising ($S_1 + S_2$) in parallel orientation. Without the intermediate event (rotation of the block) the effects of the second foliation-forming event would be undetectable. See also example 8 in Section 3.1.2. Lewisian complex, Rona, Inner Hebrides, Scotland.

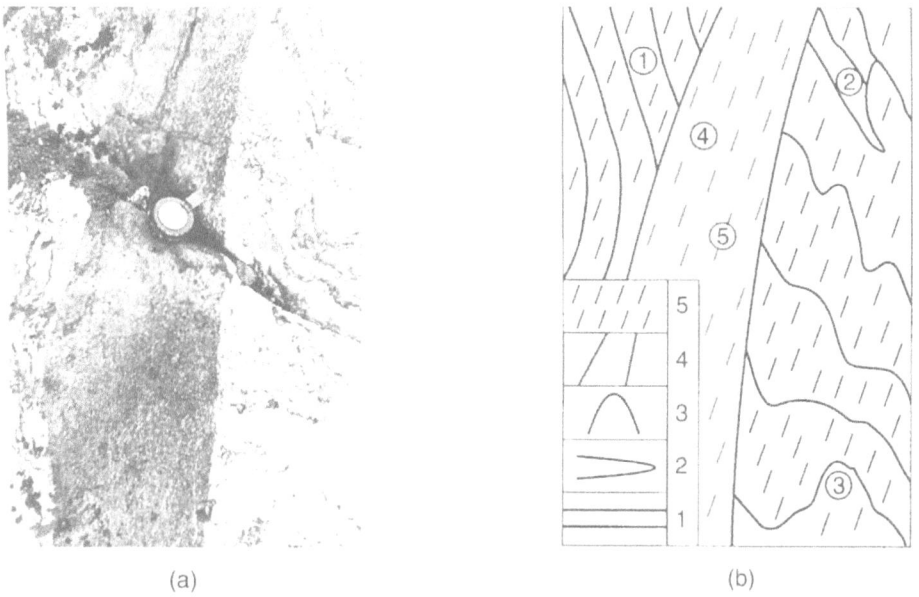

(a)

(b)

Figure 4.25 (a) Folded gneiss cut by a thin, lineated (?foliated) tabular basic body. (b) Explanation and succession of the structures shown in (a). Lewisian complex, Rona, Inner Hebrides, Scotland.

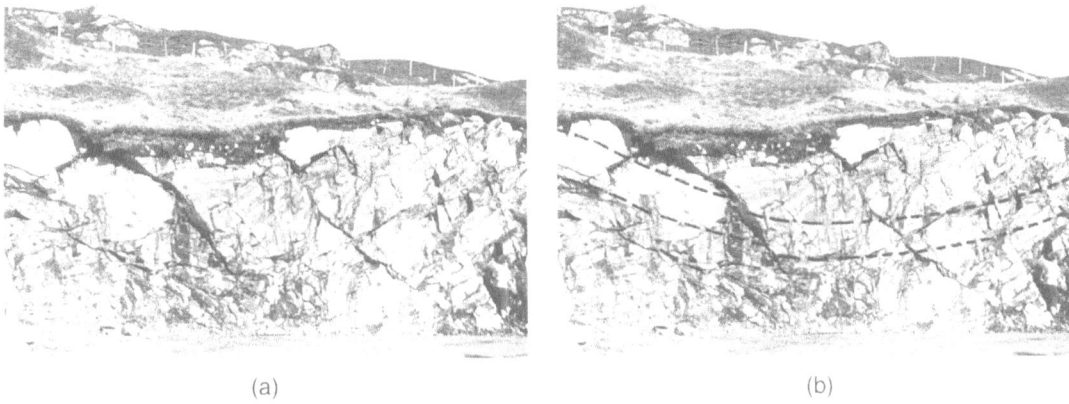

(a)   (b)

Figure 4.26 (a) Recumbent folds with curved axial traces in gneiss. Although there is no tangible structural surface involved, the curvature of the traces of points joining successive fold hinges allows demonstration of the fact that the recumbent folds have been deformed by warping on later steep fold axial planes. (b) Explanation of (a). From Figure 15, Hopgood, 1980. Reproduced with the permission of the Royal Society of Edinburgh. Lewisian complex, Bernera Bridge, Harris, Outer Hebrides, Scotland. See also Figure 11.7.

4.3.2 COMPOSITE INTERPOSED STRUCTURES – SUPERIMPOSED SECONDARY STRUCTURES

In terranes affected by metamorphism and deformation, tabular intrusions of the type just discussed, which have been interposed in the succession, may have simple secondary structures such as cleavage or foliation imposed on them as a result of tectonism. In that case, as composite structures they are effectively three-dimensional strain markers. This condition is by no means uncommon (Figure 4.27) and where it arises it constitutes something of a bonus to the structural geologist, especially when the imposed new structure is inclined at a high angle to the margins of the tabular body. See also the discussion in section 3.1.3, and Figure 4.3.

The new foliation can have other advantages besides those of providing evidence of another tectonothermal event and a means of relating this to other structures in terms of overprinting relationships. If the new structure is distinctive in character (texturally, structurally or mineralogically, or all three

combined) it can (as noted earlier) provide the means of distinguishing one particular set of tabular intrusions from others that are similar-looking, even when all are foliated, provided the others lack this **particular** new foliation (Figure 4.28).

A planar intrusive vein or dyke when emplaced at a **high angle** to the dominant foliation will provide a useful indication of strain where the stress orientation was such that the effect on the foliation is not obvious. This is an extension of the condition just described (and shown in Figure 4.27) where foliation inclined at a high angle to the margin of a planar intrusion is used as a strain indicator. In the case of compression normal to the foliation, and in the absence of, say, initially spherical strain markers, there would be no means of recognizing that the foliation had been attenuated in response to the compression, whereas there **is** the likelihood on the other hand that a discordant intruded vein (Figure 4.29) might have buckled, so demonstrating shortening parallel to the vein and thinning of the foliation perpendicular to this direction.

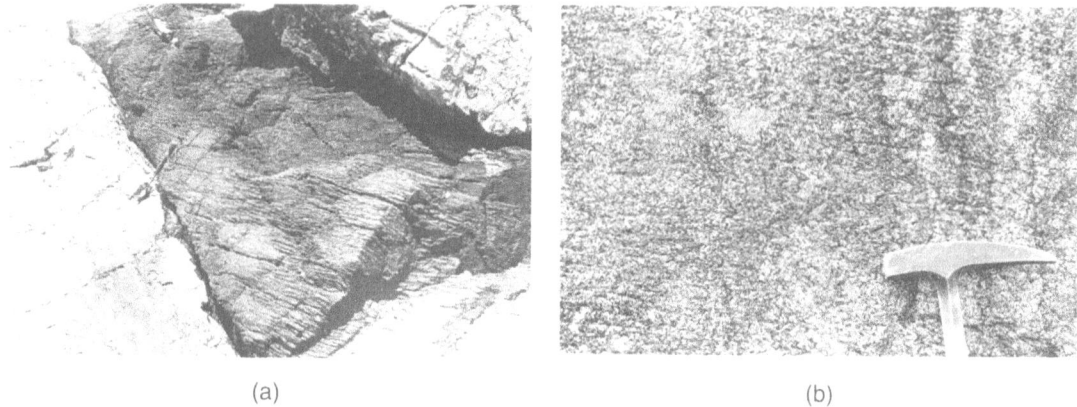

(a)   (b)

Figure 4.27   (**a**) A metamorphosed tabular basic intrusion strongly cleaved by crenulation cleavage developed at a high angle to its margins. This is an example of an interposed structure (the intrusion) exhibiting a secondary imposed structure (cleavage). (**b**) Crenulation cleavage detail. Amphibolite, Lewisian complex, Rona, Inner Hebrides, Scotland.

Figure 4.28   Uncleaved, but foliated (parallel to the margins) tabular amphibolite body cutting folded amphibolite facies migmatites. The absence of cleavage offers a simple and distinctive means of discriminating this amphibolite from others displaying cleavage which are older, having been emplaced prior to the cleavage-forming event. Lewisian complex, Rona, Inner Hebrides, Scotland.

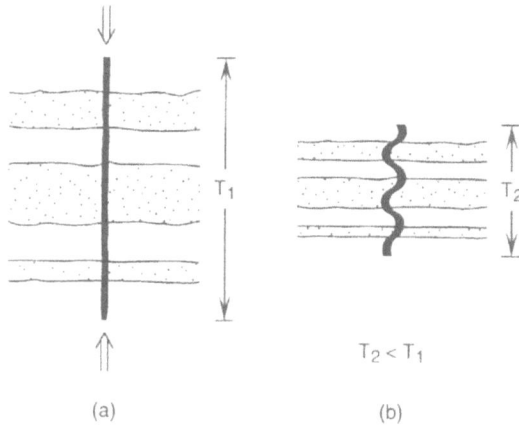

(a)   (b)

Figure 4.29   (**a**) Sketch of a tabular vein emplaced at a high angle to foliation prior to deformation. (**b**) Effect of compression perpendicular to the foliation demonstrated by buckles in the previously tabular vein.

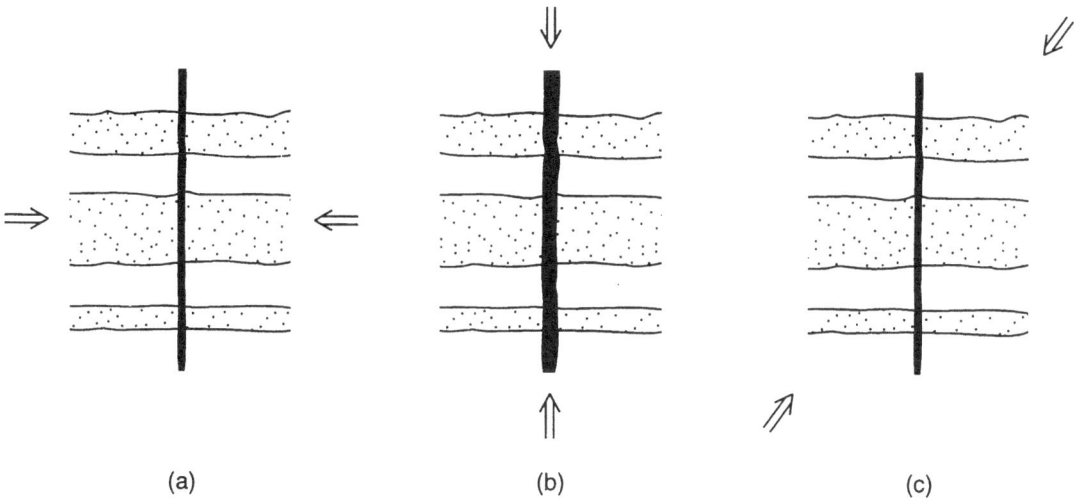

Figure 4.30 (a) Sketch of a thin tabular vein intruded perpendicular to the foliation prior to deformation. (b) Thickening of the vein as a result of compression perpendicular to the foliation. (c) Compression inclined to both vein and foliation. In none of these cases is the effect of the compression likely to be detectable without the presence of a strain marker to show the original geometry.

However, in the absence of buckling and in the case where the tabular intrusive body is superficially homogeneous, the response to compression (Figure 4.30) perpendicular to its margins, leading to thinning of the body (a), or parallel to the margin, leading to thickening of the body (b), or in some direction lying between (a) and (b), will not be apparent to the observer in the field without recourse, say, to detailed thin section examination. Nor will the effect on the host rock be apparent, whether it has previously been folded or not. Although the form of pre-existing folds will be modified in ways dependent on the direction of compression, without the presence of some suitable strain marker in the host rocks the effect of this deformation would not be noticed.

On the other hand, planar or linear structure (foliation or lineation imparted, say, during folding after the emplacement of the intrusive body) would represent a useful strain marker which, provided it was not parallel to the margins of the body, would record the later compressions (Figure 4.31).

Correspondingly, the effects of post-intrusive ductile shear, especially if it was parallel to the margins of the body, could pass unnoticed if neither the host nor the intrusive rock contained strain markers such as megacrysts (Figures 4.31(a)(i) and 4.31(b)(i)). In contrast, foliation (or lineation) imposed on the intrusive rock would allow recognition of the later shear (Figures 4.31(a)(ii) and 4.31(b)(ii)).

In the absence of such later foliation or cleavage, the presence of marginal apophyses from the intrusive body could similarly serve to demonstrate the stages in later, post-intrusive, strain history (Figure 4.32). These would have the potential to record **all** deformation **immediately** following the emplacement of the body whereas a later imposed foliation would respond only to events **later** than its development, and this development could have been at some significant interval of time after the intrusion. Therefore, without the apophyses, there would be no means of distinguishing any events that might have taken place **between** the **igneous** emplace-

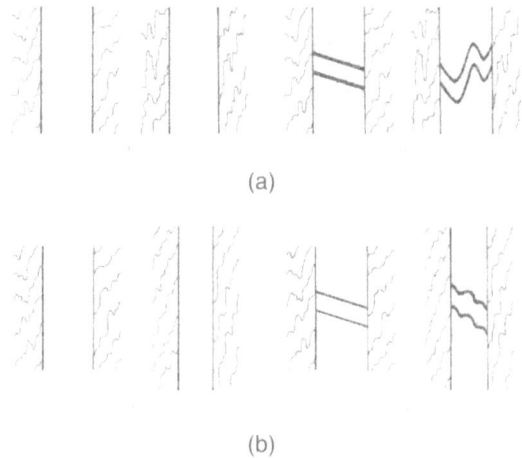

(a)

(b)

Figure 4.31 Sketches showing segments (stippled) of a tabular intrusive body affected by (**a**) slip parallel to its margins, and (**b**) compression normal to its margins, in complexly folded rocks. Where the body is homogeneous, (**ai**) and (**bi**), there is no obvious evidence of deformation within the body, but where the body contains initially planar structures at an angle to the margins (**aii**) and (**bii**) their distortion demonstrates the effects of later deformation and may allow the determination of the nature and amount of that deformation. From Figure 13, Hopgood, 1980. Reproduced with the permission of the Royal Society of Edinburgh.

Figure 4.32 Sketch of a quartzofeldspathic pegmatite vein with apophyses. The shape and angular relationship between the apophyses and the margin of the pegmatite vein imply sinistral shear stress at the time of emplacement, whereas curvature of the apophyses indicates that subsequently, post-intrusive dextral shearing took place parallel to the vein margin. Drawn from a photograph. Lewisian complex, Leenish, Barra, Outer Hebrides, Scotland (Hopgood, 1971a).

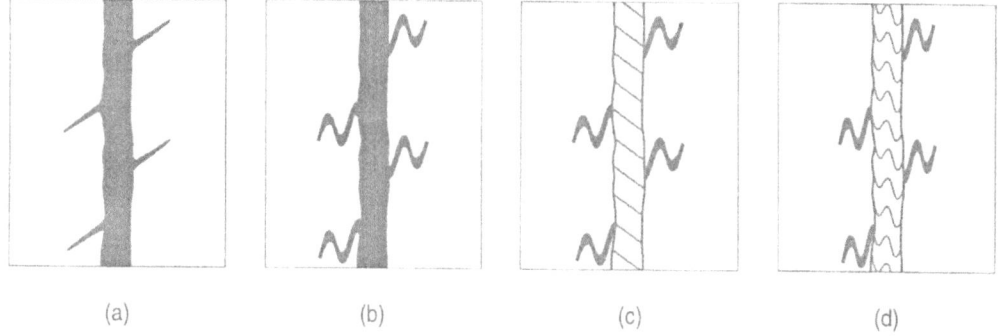

(a)   (b)   (c)   (d)

Figure 4.33   (**a**) Sketch of an intruded vein with apophyses. (**b**) Buckles in apophyses caused by compression perpendicular to vein margins. (**c**) Cleavage developed in the vein after compression (cf. Figure 4.12). (**d**) Buckled cleavage and apophyses resulting from later compression perpendicular to the vein margin. The folds in the apophyses are composite, comprising the effects of compression both before and after the cleavage was formed.

ment and the development of the imposed foliation (Figure 4.33).

In all cases discussed above which involve the deformation of imposed secondary structures, their response to that deformation can be used in the structural analysis. Here one can include apophyses, or indeed primary (flow) foliation in igneous bodies (Figure 4.4) and even cross-bedding, although this is likely to occur only in thicker igneous bodies (Figure 4.34) which would be less prone to respond to deformation than would those that are thin. These structures can be used even where the margins of the intrusive body that hosts the structures are not exposed, provided of course the nature of the structures and their relationship to the intruded body can be, or have already been, established elsewhere on the outcrop.

4.3.3 LESS COMMONLY OBSERVED INTRODUCED STRUCTURES

(a) Cooling fractures

Besides the examples discussed so far there are other, less common structures imparted to intrusive rocks which, in some circumstances, might be the only means of determining some parts of the deformational history. Planar shrinkage fractures such as those formed during the cooling of igneous bodies emplaced in the succession are one such example (Figure 4.35), another is joints that have originated tectonically.

In some ways analogous to spaced fracture cleavage, such fractures provide open, **low friction** surfaces, which therefore are likely to become folded in preference to the coherent host rock foliation or the welded margins of an intrusive body during subsequent deformation (Hopgood and Bowes, 1980). This is especially so if compression likely to cause flexure is involved and the direction of maximum principal compressive stress is more or less perpendicular to the foliation and/or igneous contact. In that case the foliation and contact

Figure 4.34 Primary igneous layering including current-bedding in an ultramafic body in the Lewisian complex. This structure provides a second potential planar strain marker inclined to the other planar marker, the dominant foliation. Lewisian complex, Achiltibuie, Northwest Highlands, Scotland. Compare Figure 4.4.

will not respond by folding whereas the open cooling joints lying more or less perpendicular to the foliation and contacts, and therefore lying at the more favourable angle, will respond to compression by flexure and flexural slip folding. Figure 4.35 shows upright folds in fractures in a metamorphosed composite intrusion (amphibolite) in the Lewisian complex on the Island of Inishtrahull, Eire (Hopgood and Bowes, 1980). The fractures can be seen to pass without interruption from the coarsely crystalline, foliated meta-igneous rock on the right and centre of the photograph into the fine-grained dark amphibolite on the left. As well as evidence for later modification of their geometry by flow, the folds show features such as apical gapes (centre). The latter are consistent with flexural-slip folding which has exploited the open joints in preference to both the foliation and the parallel contact between the two rock types as well as the foliation in the host gneisses (beyond the field of view). It appears also, that slip parallel to the joint fractures which occurred during folding caused the stepped offset of the contact between the two bodies of meta-igneous rocks (Figure 4.35b).

 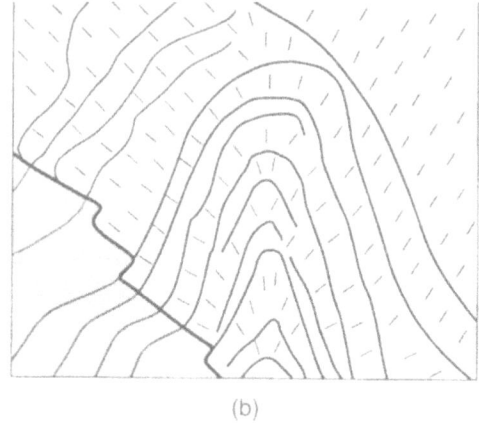

(a)   (b)

Figure 4.35   (**a**) Folded cooling joints in a metamorphosed basic composite intrusion. The contact between different parts of the intrusion, which is shown by the lithological differences near the pencil to the left, dips to the right. (**b**) Sketch showing the relationship of the contact between the two rock types to the cooling fractures and fanned foliation. Amphibolite, Lewisian complex, Inishtrahull, Eire.

(b)   Tension gash zones and shears

As noted earlier, introduction or 'establishment' of a structure in the host rock includes not only the physical emplacement of a new, three-dimensional rock body such as a tabular intrusion but also the development in place of cleavage, planar mineral orientation or lineation of various kinds ranging dimensionally in scale from mineral lineation to rods and mullions or fold hinges. There are other structures that can play a similar role in identifying the effects of 'post-emplacement' deformation. These include other non-penetrative structures: tabular zones of structures such as tension gash zones and shear zones which, although less common than pervasive (penetrative) structures such as cleavage, are nevertheless useful and indeed their deformation from their originally tabular form might be the sole evidence for subsequent deformation in the host rock (Figures 4.36a and 4.36b).

(c)   Agmatites

The neosomes of agmatites constitute an integral component of migmatites and as such

provide the potential for a group of markers of another type that, besides contributing to the succession, could also be used to subdivide it. They may be emplaced more than once in the sequence, and neosome sets can be distinguished one from the other by reference to their relationships to sets of other structures such as folds, as well as to other sets of neosomes in the succession. These relationships are often strikingly shown, particularly in agmatites (Hopgood and Bowes, 1978). Not only is the **distinction between** neosome sets possible but also, as with any other structural features, the **position** of a neosome in the deformational sequence can be determined by its discordance both to other neosomes and to structures such as folds, foliations and lineations imposed on the succession.

Consider the relationships observed at different localities between two neosomes (N_e, N_x), palaeosome (P_p), and fold structures (F_p, F_l) represented schematically in Figure 4.37 (see also Figure 4.38, discussed below, which shows an example of a neosome enclosing a palaeosome block containing folds.) One of the neosomes (N_x) contains a palaeo-

(a)   (b)

Figure 4.36   (**a**) Tabular zone of *en echelon* tension gashes. An example of an introduced simple (planar) potential strain marker. Moine, Kyle of Lochalsh, Scotland. (**b**) Thin shear zone (under hammer) parallel to the axial plane of horizontal tight folds. A further example of an introduced simple (planar) potential strain marker. Lewisian complex, Torridon, Northwest Scotland.

some (P_p) within which there are structures F_p (Figure 4.37a). Since F_p is contained within the palaeosome of N_x, F_p must therefore be earlier than neosome N_x. The relationship between N_x and F_p is thus: $F_p \ldots + \ldots N_x$.

There is also evidence to show that the palaeosome (P_p) pre-dates an early neosome (N_e) and evidence that N_e is earlier than N_x (Figure 4.37b) and also that it was deformed by folds (F_l) which are later than F_p (Figure 4.37c). The relationships between P_p, N_e, N_x, F_p and F_l are thus: $F_p \ldots + \ldots P_p N_e \ldots + \ldots (F_l/N_x)$.

From further observation F_l can be seen to have developed **before** the emplacement of neosome N_x so that on the basis of all the evidence available, the succession is: $F_p + P_p N_e + F_l + N_x$.

The succession just discussed is comparable to that of the structure shown in Figure 4.38a where there is also a succession of folds, palaeosomes and neosomes. The development of this structure as shown in Figure 4.38b is: $HG + F_i + P_1 N_1 + F_{ii} + P_2 N_2$. Figure 4.38a shows dark stringers of folded (F_i) palaeosome (P_1) in blocks of light grey neosome (N_1),

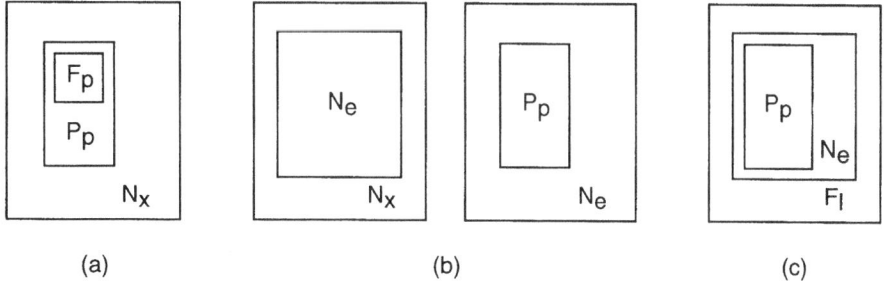

(a)   (b)   (c)

Figure 4.37   Representations of the relationships observed at different localities between folds (F_p, F_i), palaeosome (P_p) and neosomes (N_x, N_e) with the innermost rectangles being earliest in each case. Agmatite (**a**) comprises palaeosome P_p containing fold F_p. The two agmatites of (**b**) comprise neosome N_e containing palaeosome P_p, and neosome N_x containing N_e as a *palaeosome*. In (**c**) fold F_l affects the neosome N_e which contains fold P_p as the palaeosome.

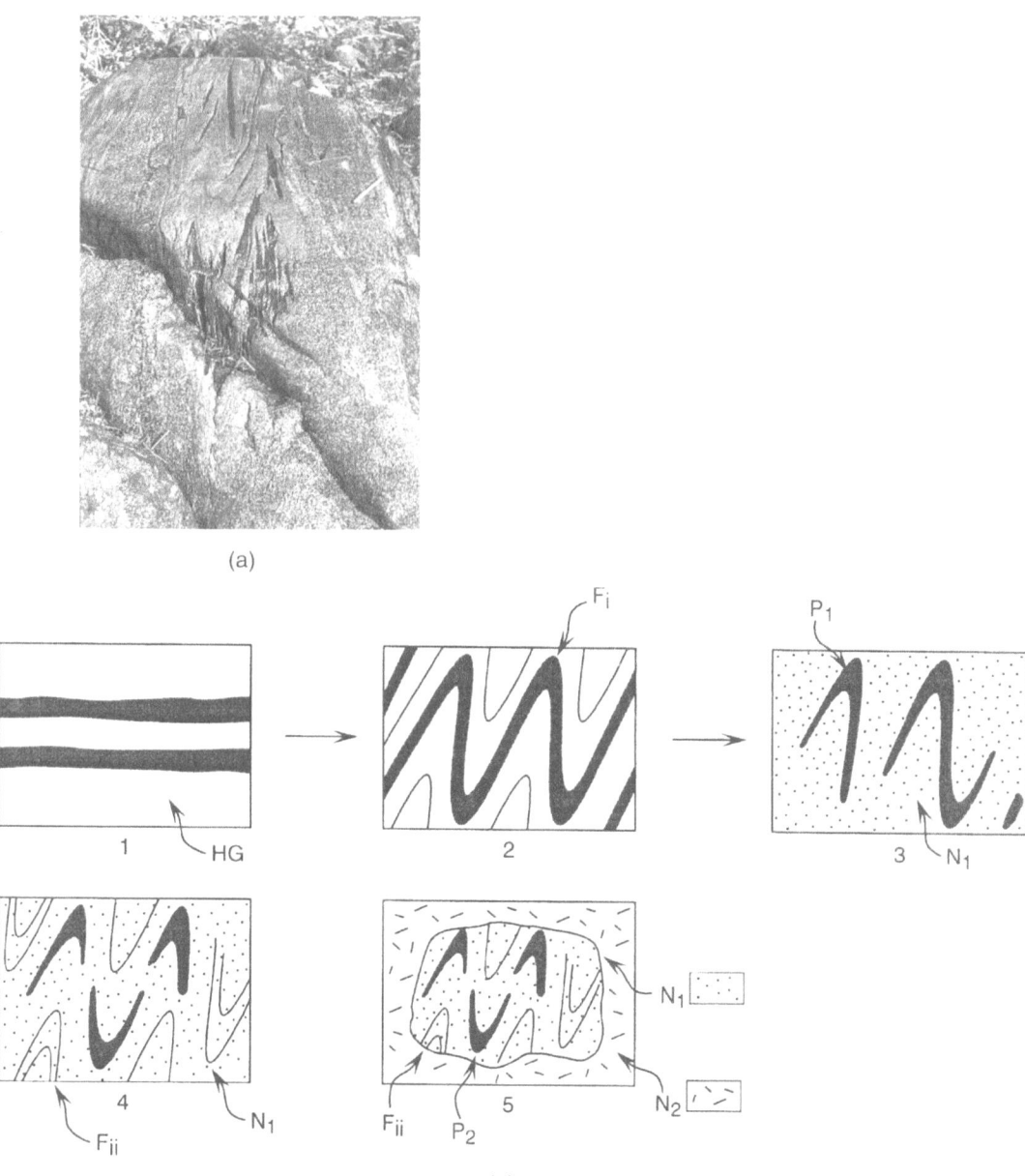

(a)

(b)

Figure 4.38   (a) Dark stringers of folded (F_i) palaeosome (P_1) in blocks of light grey neosome (N_1) itself folded (F_{ii}) and constituting a second palaeosome (P_2) in coarse 'granular' new neosome (N_2) below the pencil. (b) Sequence of events (1–4) leading to the structure of (a), beginning with (1) the host gneiss (HG) with interfoliated basic layers. (2) Initial folds, F_i. (3) Development of first neosome N_1 leaving relict blocks of F_i folds as the first palaeosome, P_1. (4) Development of later folds in neosome N_1, F_{ii}. (5) Development of second neosome N_2 with relict blocks of the pre-existing rock (interfoliated gneiss and basic layers together with folded N_1 neosome) preserved as the second palaeosome, P_2. $P_2 = [HG + N_1 + P_1]$. Compare Figures 5.37, 5.38 and 13.11.

itself folded to form F_{ii}, and constituting a second palaeosome (P_2) in coarse 'granular' new neosome (N_2) underlying the pencil. Figure 4.38b shows the sequence of events (1–4) leading to the derivation of the observed structure of (a) from the host gneiss (HG) with interfoliated basic layers prior to deformation. (1) Formation of the initial folds, F_i. (2) The anatetic development of the first neosome N_1 leaving relict blocks of F_i folds as the first palaeosome, P_1. (3) The formation of later folds (F_{ii}) in the neosome N_1. (4) Development of a second neosome (N_2) leaving relict blocks of the pre-existing rock (interfoliated gneiss and basic layers together with folded N_1 neosome) preserved as the second palaeosome, P_2. Note that P_2 is a composite palaeosome comprising [HG + N_1 + P_1].

The potential to determine relationships between neosomes and structures in such cases has the merit of allowing distinction between neosomes that appear to be lithologically the same (Figure 4.38 and Hopgood, 1984, Figure 3g), and this is an important consideration in the study of migmatites where there may be several generations of neosome within the overall succession.

Neosome-forming events throughout the developmental history of migmatites can thus be placed in relative order and their petrogenesis considered in relation to their isotopic age. Isotopic dating of the partial melt of neosomes forms one of the bases for converting relative chronology to absolute chronology (Hopgood *et al.*, 1983; Bowes *et al.* 1995 p. 113). Conversely, characteristic neosomes may be used to separate otherwise similar fold sets. See also the discussion in section 5.2.10 ('Agmatites and fold identity'); section 13.3.2 ('Using agmatites to identify structural relationships'); Figure 5.19 and the discussion on the uncertainty of identity of structures and successions in section 5.2.6 ('Identification (identity) of folds (structures)'); section 4.1 ('Significance of geometrically simple structures imposed on the developing succession') and Figures 4.1, 4.2 and 4.18 to 4.20.

FOLD (STRUCTURE) SUCCESSIONS AND DEFORMATIONAL SEQUENCES

5.1 SUCCESSIONS AND SEQUENCES

Interference patterns in highly deformed rocks are often enigmatic, sometimes spectacularly so (back cover and Figures 10.4 and 10.38), and it is not surprising that observers are sometimes overawed by first impressions of the structure. No doubt it is this initial sense of overwhelming complexity, particularly when experienced by an observer with little experience of such complex structures, that has prompted the use of expressions such as 'wild folding' (Berthelsen *et al.*, 1962) referred to earlier. Whether or not there is justification for conclusions like those of Williams (1985), that the problems in attempting to resolve such structural complexity outweigh the advantages, depends on the purpose of the study rather than on the nature of the structure. Undoubtedly there will be difficulties in undertaking such a study, but daunting first impressions should be ignored and a determined effort made to adopt a systematic analytical approach to resolving the structure.

It is the intention here to demonstrate ways of surmounting the problems inherent in the structural analysis of complexly deformed migmatites (see also section 2.2) and to show how it is possible, in spite of any such first impressions of intractability, to determine their structural succession.

5.1.1 STRUCTURAL SUCCESSIONS AND WHAT CAUSES THEM

A **succession** of **structures** or structural succession is comparable to a stratigraphical succession (a series of rock units following one another, e.g. in the case of sedimentary rocks, units lying one above the other) in that it comprises a series of structures developed one after the other in response to a **sequence** of deformational **events**. In other words, a structural succession is the response of a unit that is composite (commonly layered, or foliated) to multiple, or 'polyphase', deformation, e.g. folding (see also section 2.2.4).

Following from definitions in the North American Stratigraphic Code, p. 861 (North American Commission on Stratigraphic Nomenclature, 1983), a **structural succession** can be considered to comprise the ordered series of units in a **structural complex**.[1]

---

[1] Strictly, in the present case, it more closely resembles a succession of **lithodemic** units, defined in the North American Stratigraphic Code (North American Commission on Stratigraphic Nomenclature, 1983, p. 859) as follows:

'A lithodemic[6] unit is a defined body of predominantly intrusive, highly deformed, and/or highly metamorphosed rock, distinguished and delimited on the basis of rock characteristics. In contrast to lithostratigraphic units, a lithodemic unit generally does not conform to the Law of Superposition. Its contacts with other rock units may be sedimentary, extrusive, intrusive, tectonic, or metamorphic (Figure 3)'. The concept of 'succession' still applies even though the order will *not* be one of a *vertical* succession of horizontally disposed units (i.e. like a 'layer cake'). "[6]From the Greek demas, -os: "living body, frame"."

In considering a consecutive train of objects or events, e.g. a **series** such as a **succession** of structures or a **sequence** of events, it is important to bear in mind the fact that a concept of this kind, certainly in geology at least, does not necessarily imply either (i) a known beginning or a known end or (ii) a knowledge of whether the series is complete within the known (or unknown) end members.

In other words the series may be:

(i) open-ended i.e. extend beyond either its apparent beginning or its apparent end, and/or
(ii) internally incomplete i.e. **incomplete** within the known end members.

Depending on both the intensity and permanence of the imprint produced by individual events, neither of the conditions (i) or (ii) above is **ever** likely to be known and the series is always subject to revision as more data are obtained.

For example, in considering the case where evidence has been obtained for nine successive events, two possibilities are that,

(A) these might comprise the sequence 1–n where n = 9, **or** more likely,
(B) they could be part of the sequence 1–n where n is unknown but here (Table 5.1) might comprise, let us say, at least 83 events where '9' happens to be the 83rd event, '8' is actually the 73rd, '7' is in fact the 66th event, . . . etc. thus:

Table 5.1

A. Sequence 1–n (n = 9)	B. Sequence 1–n (n = ?)
9	83
8	73
7	66
6	62
5	61
4	53
3	18
2	7
1	2

A 'sub-succession' (or sub-sequence) comprises the known part of a succession (or sequence) 1–N which exists, but of which only the sub-succession (sub-sequence) is **known** to exist. In special cases the sub-succession could equal the whole succession 1–N.

As noted in section 2.2.4 the expression multiple, or 'polyphase' deformation is used here in a non-genetic sense, i.e. in the sense of processes causing successively-formed structures which are themselves deformed in turn by those structures produced even later in the sequence. Multiple deformation results in a succession of structural sets (fold sets) which can be identified and correlated consistently throughout the terrane under consideration. It is important to bear in mind that neither the nature of the deformation nor the genesis of the structure is inherent in this definition so that the determination of the structure sets themselves is independent of genetic criteria.

As we have just seen, the structural succession established at some particular place on the basis of refolding, or overprinting, of fold (structure) sets may not necessarily be complete. In fact, in the initial stages of the study at least, as will be seen later (section 8.1.7), even if it is not certainly **known** to be incomplete, it must at least be **suspected** to be so. It will be a succession only in the sense that the structure (fold) sets are in the correct order and will therefore always be subject to continual revision in terms of the number of fold sets as the study progresses. The successions in different parts of the same terrane may well differ in terms of which sets are present, and also in terms of the number of sets present, but not in terms of the relative order of the sets (see the discussion on the 'Absence of structures from the succession', section 8.1.8).

5.1.2 DEFORMATIONAL SEQUENCES

Commonly the geometry of the structures in the succession implies a deformational sequence which is consistent with the concept of an orogenic episode which develops to reach a climax and whose intensity then gradually wanes, with later folds becoming progressively less strongly developed and more 'brittle' in aspect. See, for example, the illustrations of structures typical of a developing succession (Figures 7.13 to 7.20). In effect this suggests the action (probably during gradually varying pressure and temperature) of a series of 'deformational pulses' caused by stress fields having differing orientations with respect to the successively formed (fold) structures. The 'pulses' could stem from one or more factors, ranging from surges in the external applied stress (such as 'rapid' variations in the rate of movement caused by changes in the driving force, e.g. lithospheric plate movement) or more probably the naturally uneven rate of movement ('lurches') comparable to the rapid release of strain energy by fault movements (strike-slip) in pulses arising, (i) from changes in the obliquity of collision between the irregular boundaries of steadily moving plates (see Figure 3.29), (ii) from internal causes such as the sudden overriding of frictional resistance to movement etc. (cf. Ramsay, 1967, p. 518), (iii) from crossing orogenic belts separated by a **long** time interval and (iv) to successive deformations in one orogenic cycle (very common).

In contrast to the above (while not affecting the ultimate structural picture) is the consideration that the overall external stress field remains more or less constant in orientation but the internal response to this varies directionally (cf. Figure 3.29). This might very well be the response to a given set of stress directions of differently orientated surfaces subject to deformation and to internal variations in friction and viscosity within the deforming rock body. For example, spatial variation in deformation rate (due to friction or changing viscosity with changing temperature, water content etc.) in a developing fold, say, could result in considerable curvature of the structure. Departure of the hinge (axial) direction of the fold from its original attitude perpendicular to the maximum compressive stress direction might (in extreme cases) be such that some parts of the structure become orientated so that their hinge and limb (zones) are more or less parallel to the direction of compressive stress. Hence it could be possible to have relatively small-scale cross-folds imposed on a relatively large-scale fold in response to the same stress field (Figure 5.1).

However, these cross-folds have nevertheless formed in sequence (i.e. they are successive) from whatever cause, so that the deformational sequence results in a succession comprising an **original** hinge (H_o) which in this case is relatively large-scale, paralleled in part, and crossed in part by penecontemporaneous (but consistently later) smaller-scale **cross**-hinges (H_c). This appears to be similar to

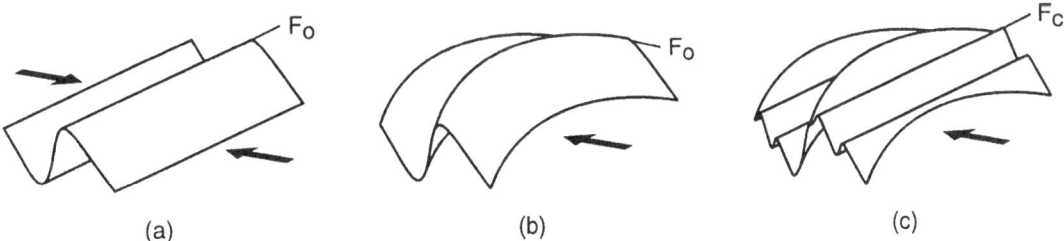

(a)   (b)   (c)

Figure 5.1 Stages in the development of overprinting in response to the same stress during progressive deformation. (**a**) An early fold, the 'original' fold (F_o) with hinges (H_o) formed by simple horizontal compression. (**b**) Curvature of the initially rectilinear fold hinges (H_o), e.g. because of frictional variation across the structure. (**c**) Overprinting with the development of cross-folds (F_c) with rectilinear hinges (H_c) superimposed on the curved F_o hinges.

the arrangement of folds in the Jura Fold Arc in France and Switzerland (Figure 2.7). Circumstances might be such that syntectonic mineral growth (e.g. L_o and L_c parallel to the fold hinges H_o and H_c) associated with the deformational sequence resulted in the presence in the rocks of structures such as the following:

(i) H_o with L_o and H_c with L_c, if metamorphism and recrystallization took place over an extended period of time, i.e. throughout the folding of F_o and F_c, giving the succession, $H_o + L_o$ followed by $H_c + L_c$, or

(ii) H_o with L_o and H_c but *not* L_c because mineral growth was restricted to the former part of the development of the composite fold structure ($F_o + F_c$), giving the succession, $H_o + L_o$ followed by H_c, or

(iii) H_c with L_c but H_o without L_o because mineral growth was restricted to the latter part of the development of the composite fold structure ($F_o + F_c$), giving the succession, H_o followed by $H_c + L_c$.

5.1.3 FUNDAMENTAL ASSUMPTIONS REGARDING STRESS AND DEFORMATION

As discussed in section 2.3, 'Assumptions inherent in the approach used', it is generally assumed that:

1. the stresses acting in the orogen have been more or less penetrative;
2. with the exceptions noted above in section 5.1.2, the stresses have been transmitted with orientations that have been more or less constant over a 'limited segment' of the orogen, except in proximity to the earth's surface or other interface and where local inhomogeneities caused by such factors as changes in rock type have affected the direction in which they have acted, e.g. around granitoid plutons, and basic and ultrabasic masses;
3. if there have been changes (i.e. spatial variations) in contemporaneous stress directions then the stress directions are broadly considered to have varied slowly but more or less regularly throughout the orogen so that the orientation of a given set of structures would be seen to change only gradually from one part of the orogenic belt to another (Hopgood and Bowes, 1972 and see Figure 12.17).

Inhomogeneities of the type just referred to could be responsible for variation in fold style. Similarly, variation in crustal depth and consequently variation in pressure, temperature and pore fluids would affect the degree of plasticity and response of the rocks to applied stress for a given strain rate so that these factors can also affect the style of the fold structures.

If the structure of the terrane reflects the effects of more than one orogenic episode it is quite likely (even probable, but not a proven accepted fact) that structures of earlier episodes may be indistinguishable from one another (as a series of relict, rootless intrafolial folds, say) because of the cumulative effects of prolonged or intense later deformation (Table 5.1). Their relationships may thus be irresolvable in terms of geochronology. On the other hand, the presence of igneous bodies emplaced at different times in the sequence and which could be dated isotopically might well enable this difficulty to be overcome (Hopgood *et al.*, 1983).

5.2 IDENTIFICATION, RECOGNITION AND CLASSIFICATION OF STRUCTURES

5.2.1 IMPORTANCE OF STRUCTURAL IDENTIFICATION AND RECOGNITION

With the recognition of interference patterns (evidence that the rocks have been affected by multiple folding), the need for structural analysis is established. Consequently this leads to the need to observe and record the data necessary to set up a structural succession. In collecting this data there are two

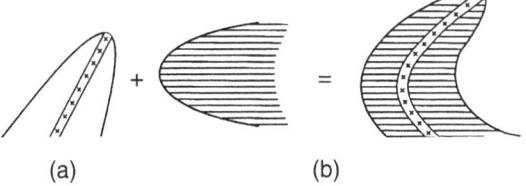

(a)   (b)

Figure 5.2 (a) Uncleaved fold with an axial planar vein refolded by a fold with well-developed axial planar cleavage to produce the composite structure (**b**). The presence of the axial planar vein and absence of cleavage of the earlier fold (**a**) distinguishes it from the later fold (**b**) which is cleaved and lacks the axial planar vein.

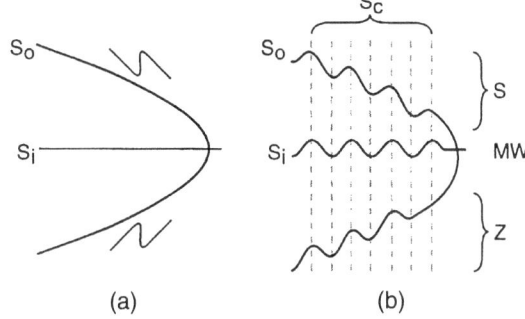

(a)   (b)

Figure 5.3 Sketches to show how folds of the same set can have distinctly different profiles. (a) Initial fold (F_i) in foliation S_o with axial planar cleavage (S_i) showing sites of potential 'S' and 'Z' folds. (**b**) Smaller cross-folds (F_c) with 'S', 'Z', 'M' and 'W' profiles superimposed on both the F_i-folded foliation (S_o) and the axial planar cleavage of the larger F_i fold. o = original, i = initial, c = cross. Compare Figure 2.14.

aspects of the field work the observer will need to concentrate on from the outset. One is the **observation** of refolded (overprinted) relationships between structures; the other is the **recording** of detailed descriptive data for individual folds. The latter provides the basis for identifying the structures so they can then be assigned to specific sets, i.e. to groups of coeval folds – those having the same refold relationships to other fold sets. The collection of this descriptive data is particularly important in those cases where folds can be demonstrated to have overprinted relationships with each other. The refold relationships mean that folds having whatever specific (i.e. characteristic) features that have been observed in a particular case can be distinguished unambiguously (Figure 5.2), i.e. refold relationships can be seen between folds which can be identified on the basis of observed features.

Note: When using fold geometry, e.g. fold profile, as a first step in the identification of a fold it is important to remember that relative size of the earlier and later structures has a significant bearing on their geometry. In the case of superimposition (overprinting) of relatively small structures on a larger fold, the profile of the later, imposed folds will vary, depending on their position in the earlier, larger structure. In this respect, see also the cases discussed in section 10.1.8, 'Fold/cleav-

age relationships', section 10.1.9, 'Structures associated with one or two folds' and section 10.1.10, 'Multiple fold relationships'. See also Figure 10.25, and the discussion on fold profile, and Figures 2.13 and 2.14.

In Figure 5.3 the axial plane of F_c which folds S_o is S_c. The intersection of S_c with F_i axial surface (S_i) produces symmetrical folds ('M' and 'W') folds. Its intersection with S_o produces 'Z' folds on one limb of F_i and 'S' folds on the other so that F_c does not have a unique profile in this setting.

Where the scale of both F_i and F_c is the same then the relationship between S_c and S_i and S_o is further complicated, especially where F_i is isoclinally refolded. Even if both sets are accompanied by axial planar cleavage, isoclinal refolding can lead to parallelism of S_i and S_c cleavage in parts of the composite structure, such as close to the hinge of F_i in Figure 5.4, and in such cases relationships between the two fold sets may be very difficult to establish.

Consider the isoclinal folds F_i (without axial planar cleavage) and F_c (cleaved) of Figure 5.4b. In this case F_c folds near the F_i hinge

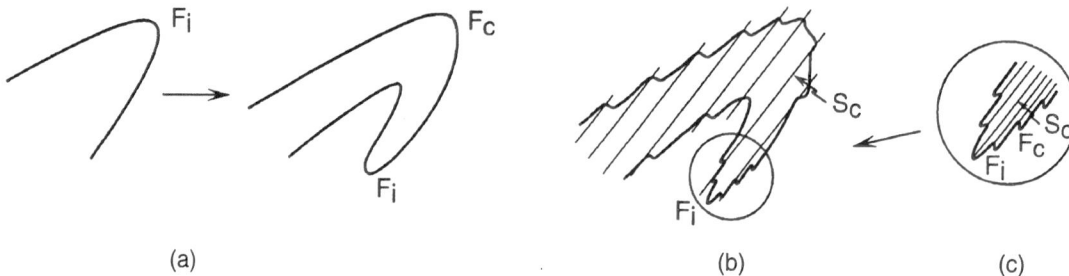

Figure 5.4 (**a**) Simple refolding (F$_c$) of an uncleaved fold (F$_i$). (**b**) Refolding of F$_i$ where F$_c$ is associated with axial planar cleavage development (S$_c$). The imposition of the cleavage and small-scale F$_c$ hinges in the hinge of F$_i$ (**c**) makes F$_i$ (originally without cleavage) look very like F$_c$, particularly because the symmetry of the small-scale F$_c$ folds here accords with that of small-scale F$_i$ folds. Where such conditions are suspected, distinguishing features must be sought. Compare Figures 10.28 and 6.29.

(Figures 5.4b and 5.4c) have opposite symmetry sense and these accord with the symmetry of F$_i$ drag folds with which they could be confused. If F$_c$ is associated with an axial planar cleavage this will also be imposed on the F$_i$ hinges and where these are parallel to F$_c$ hinges (c) they will not be easily distinguishable (if at all) from F$_c$ so the true status of the major fold (as F$_i$ and **not** F$_c$) might not be appreciated (c). This illustrates the significance (and limitations) of vergence of relatively small structures as a means of identifying major structures (section 10.1.3).

At the outset, when recording descriptive structural information about folds, it is usually advisable to concentrate first on collecting data from those folds that show clear overprinted relationships. Thereafter the study can be extended to the examination of individual folds that show particularly distinctive, or clearly defined, structural features, and especially features that can be matched to those in folds with demonstrable overprinted relationships (section 6.3). This examination should be undertaken in spite of any initial impression that, just because the complexity of the interference patterns in some exposures is not immediately resolvable, the structure is indecipherable (see for example Figures 10.1 and 11.2). The existence of the complex interference patterns causing difficulties in this particular area of the study should be noted at the time, and the locality returned to later for further examination when more has been learned about the structural relationships from other exposures.

In the case of isolated folds it is unlikely at the outset that they could be assigned to their correct sets and improbable that even their approximate position in the structural succession could be fixed except in very general terms, e.g. as 'later' or 'earlier' structures (section 5.2.8). Furthermore, it might well transpire that even by the end of the study some folds have still not been placed unambiguously for the various reasons discussed in section 2.3, 'Assumptions inherent in the approach used', (e.g. Assumption 5 regarding the non-diagnostic character of orientation, geometry and style).

It is only exceptionally that isotopic methods are sufficiently precise to enable the age of a particular structure to be determined, say with respect to a 'dateable' discordant body (Figure 5.5). This is because the time span necessary for a structural succession to become established may be comparatively short in terms of isotopic 'disturbance' (Hopgood *et al.*, 1983) and may lie within the acceptable range of experimental error for the particular isotopic technique.

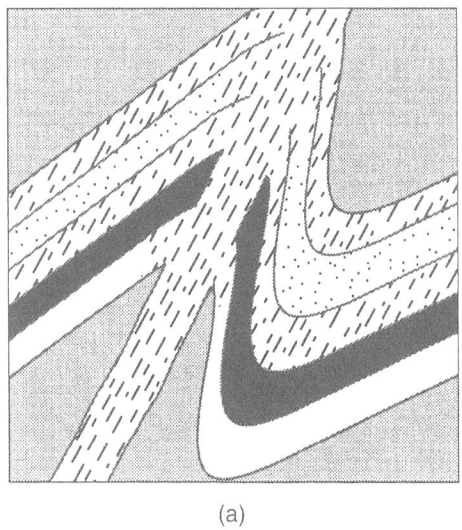

(a)   (b)

Figure 5.5 (a) Sketch of the type of association between a fold and a neosome vein that is likely to be isotopically dateable. The intimate relationship of the vein and its apophyses to the folded layers and the parallelism of the preferred orientation of the mineral growth to the fold axial plane suggests that it was contemporaneous (or penecontemporaneous) with the folding, i.e. that it is essentially syntectonic (b) Fold in migmatite with a syntectonic axial planar leucocratic vein of the type shown in (a). Note there is no evidence of discordance between the vein and the host rock. This is shown by the absence of clearly defined margins to the vein and also by the continuity between leucocratic layers in the host rock and the apophyses from the vein.

Ultimate confirmation of the identity of the structures of a particular set depends on their overprinting relationships with respect to the identified sets following **and** preceding them (section 3.1.1, 'Principle of overprinting').

'Identification' is used here in the biological sense and is based on a number of attributes (commonly all of those observed), the principal attribute in the case of structures (and the only unambiguous one) being the overprinting (or refold) relationships which a structure (fold) holds with others earlier and later in the succession (Figure 5.6). It has a basis similar to that used in taxonomy (classification based on specific criteria or attributes) and numerical taxonomy (classification based on specific numbers of criteria or attributes).[1]

It could be said that the 'ultimate' attribute of a structure is its overprinting relationship to other structures. However, while a single structure of any set may have few criteria (perhaps only one criterion) a structural **succession**, on the other hand, with its greater number of attributes more closely fits the conditions necessary for using numerical taxonomy as a basis for its identification. This

[1] See for instance Robert R. Sokal (1966, pp. 108) (and Sneath and Sokal, 1973) on the distinction between 'classification' and 'identification'. 'This is the difference between the terms 'classification' and 'identification'. When a set of unordered objects has been grouped on the basis of like properties biologists call this 'classification'. Once a classification has been established the allocation of additional unidentified objects to the correct class is generally known as 'identification'. Thus a person using a key to the known wild flowers of Yellowstone National Park 'identifies' a given specimen as a goldenrod. Some mathematicians and philosophers would also call this second process classification, but I shall strictly distinguish between the two. Here I am principally concerned with classification in the biologists' sense'.

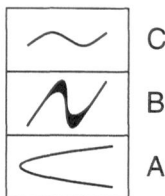

Figure 5.6 Sketch showing a succession of three folds, A, B and C. The identity of fold B depends on its relationship to folds A and C. B folds always deform A folds and are themselves always deformed by C folds.

is very important in correlation (see, for example, section 5.2.2, 'Classification of structures'; section 5.2.6, 'Identification (identity) of folds (structures)'; section 7.1, 'General features of fold successions', and section 12.3, 'Identification of structures'.

However, it is the combined (total) attributes which together comprise the characteristics, or style, of the structure and serve to identify it in the first instance.

5.2.2 CLASSIFICATION OF STRUCTURES

The classification of an object, like its identity, will depend on its attributes. Consider the heads shown in Figure 5.7. These all have some features in common and their classification will depend on using different numbers of attributes (features). For example it is likely that numbers 6, 7 and 8 would have the power of speech and could be classified accordingly, while numbers 4, 5, 6, 7 and 8 could be classified as a group that are likely to have the potential to hear. Numbers 4 to 8 could be classified in terms of having the ability to see.

On the other hand, however, while it can be said that each of numbers 6, 7 and 8 have in common the seven features; ears, hair, nose, mouth, eyes, eyebrows and eyelashes (and on that basis might be said to be identical) each is in fact readily and unambiguously distinguishable from the others by virtue of other

features which they possess, such as eye colour, hair colour, hair length, nose shape etc. The various attributes possessed by each of five of the heads shown in Figure 5.7 are listed in Table 5.2.

In summary, individuals 4, 5, 6, 7, 8 possess eyes, a nose and a mouth and individuals 5, 6, 7, 8 possess eyes, a nose, a mouth and ears while three (6, 7, 8) possess eyes, a nose, a mouth, ears, eyebrows and hair. These are their distinguishing features (identity) in taxonomical and numerical taxonomical terms.

The attributes discussed are not the only means of distinguishing these individuals. If considered further it could be seen that even if their eye and hair colour etc. were the same they would still be distinguishable because we could use other more subtle means of making the distinction. Such distinction might even be possible in the case of identical twins. As the number of distinguishing features (i.e. different attributes) decreases, observational power becomes an increasingly important aspect of discriminating between individuals. This discrimination entails identifying, or recognizing all the features (attributes), facial and otherwise, of the individuals and integrating these. This process of integration to define the character of the individual is usually performed unconsciously and when one is taxed with explaining the differences between individuals, or describing them uniquely in terms of a given set of parameters, it may prove difficult if not impossible to do so, even though one has no difficulty in distinguishing each individual unambiguously.

In making this distinction or identification, use is being made of the equivalent of what is known in structural geology as the '**style**' of the structure (section 6.2). Just as the features contributing to a person's appearance include every attribute, the style of a structure such as a fold also includes every aspect – its associated cleavage, its geometry, its hinge and limb features, related veining, and so on.

5.2.3 IMPORTANCE OF EARLY PROVISIONAL CLASSIFICATION OF STRUCTURES

Clearly, classifying structures is not an entirely simple, straightforward process and classification is likely to present some problems, at least in the initial stages of the study when only some of the structural features have been identified. Nevertheless it is good practice to endeavour to classify structures provisionally in as precise a manner as possible and as early as possible in the course of the study. Any such classification is of course always subject to confirmation or review as the study progresses.

While new evidence might subsequently show some of the groups formed by the initial subdivision of structures to be invalid, this condition, (this 'over classification') is much to be preferred to that where structures initially grouped together as penecontemporaneous are later found to belong to more than one set. Resolving the latter case would entail working back through the whole study area, re-classifying and subdividing structures on the basis of the new evidence which shows that the group of folds belongs to more than one set, and hence having to revise the entire structural succession on the basis of the new subdivisions. In effect this means having to do much if not most of the work again, whereas rectifying the former condition simply entails abandoning the invalid subdivision and amalgamating the groups concerned (Figure 5.8).

In case (ai) of Figure 5.8 the grouping was found to be incorrect and subdivision was needed, but only following the discovery of new field evidence. In the initial stages of the study the asymmetrical fold had always been found associated with igneous vein material (grey in Figure 5.8) so that the igneous vein was interpreted as being contemporaneous with the asymmetrical folding as well as with the syntectonic leucocratic limb-parallel vein (dotted). However, further investigation elsewhere showed that igneous vein material of

Figure 5.7 Drawings of human heads in various stages of development, from embryo to old age. Each is distinctive and recognizable in terms of its identity, in spite of the fact that, for example, the three individuals in the lower part of the drawing each possess the same comparable features viz. two eyes, a nose, a mouth, two ears etc. (Haeckel, 1910).

Table 5.2 Attributes possessed by the five human heads shown in Figure 5.7.

Feature	Individual				
	4	5	6	7	8
Nose	X	X	X	X	X
Mouth	X	X	X	X	X
Eyes	X	X	X	X	X
Ears		X	X	X	X
Eyebrows			X	X	X
Eyelashes			X	X	X
Hair			X	X	X

the same composition is in fact **discordant** to leucocratic veins of the same composition as the limb-parallel vein material related to the fold so that reinterpretation of the vein relationships was necessary (c). Therefore, although the leucocratic vein material **is** syntectonic with the asymmetrical folding, the grey igneous veins are **later** than this folding (aii). (cf. the neosome example of Figure 5.37).

In case (b) amalgamation is necessary because the initial subdivision is not supported by subsequent field evidence. The asymmetrical fold was provisionally separated from the one associated with the limb-parallel

vein material until, after further observations, the consistent relationship between the vein and fold came to be accepted in the absence of evidence to the contrary. This conclusion ultimately received confirmation with the discovery of evidence for the intimate association of the vein material and the folding (cf. Figure 5.8).

It is important to note the distinction between cases (a) and (b) and the implications of the two different approaches. In (a) data have been grouped under the same heading at an early stage in the study but new evidence found later showed that the group constitutes several different structures, whereas in case (b)

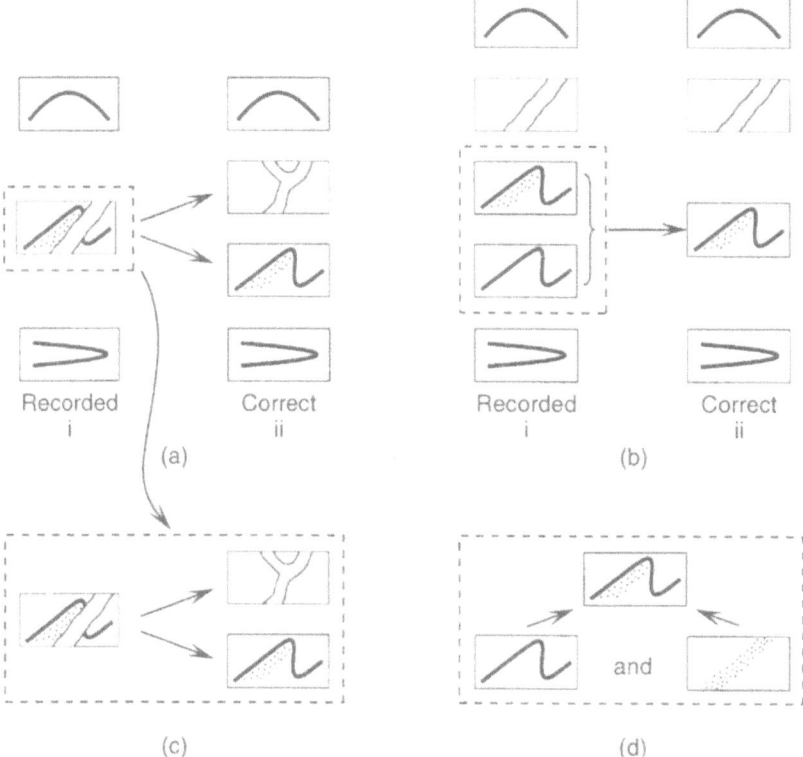

Figure 5.8　Two cases where further examination leads to modification of initial observations. (**ai**) Initially recorded observations. (**aii**) Corrected data. (**bi**) Initially recorded observations. (**bii**) Corrected data. (**c**) Further observation shows that subdivision of (**ai**) is necessary and this leads to the corrected data of (**aii**). (**d**) Further observation shows that combination of (**bi**) is necessary, leading to the corrected data of (**bii**).

the data were retained in several separate categories which could, if the subdivision proved to be unwarranted, be readily combined at the end of the study. It is not possible to subdivide the combined data of case (a) without repeating much of the whole study whereas the retention of the separate categories in case (b), or their subsequent combination if required, does not entail significant further work.

Therefore in the early stages of the study, structures with clear overprinting relationships can be placed provisionally in the correct relative position in the succession. Each of those whose relationships are less clear can be restricted to a limited number of possible positions in the succession until any doubt about their correct position can be resolved unambiguously. Examples of this situation are discussed in sections 10.1.7, 13.2 and 13.3.

In contrast to the above cases, certain structures (for reasons such as 'intensity' of development, characteristic geometry, associated mineralogy etc.) will be so distinctive that they can be identified unambiguously. Such structures are suitable for use as datum or **key** structures and these are discussed later (section 6.3).

5.2.4 ILLUSTRATION OF PROVISIONAL SUBDIVISION

As an illustration of the importance of provisional subdivision of structures ('over-classification'), consider a simple case of a fold with an associated axial planar hornblende growth (Figure 5.9a). Although all the hornblendes are coplanar and parallel to the fold axial plane, observation suggests that not all the hornblendes are the same and that there might be two distinct populations of hornblende crystals, each exhibiting a different habit; (i) slender, acicular and (ii) blunt, prismatic. In such a case, and as a matter of course, the attitudes of the long axes of the crystals of each habit should be recorded separately in two groups, so that each group can be presented stereographically as a separate plot.

Plotted together, the data produce a scatter diagram with points distributed along a great circle representing the fold axial plane (Figure 5.9b).

On the other hand, separate plots of the long axes of the crystals show that the blunt prisms lie along the same great circle (Figure

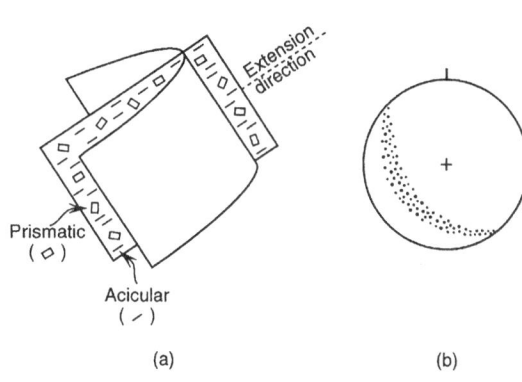

Figure 5.9 (a) Sketch showing the orientation of the hornblende crystals as being constrained parallel to the fold axial plane. (b) Stereo-plot of the attitudes of all hornblende C-axes, showing grouping within a great circle girdle maximum, i.e. within a plane (which is coplanar with the fold axial plane).

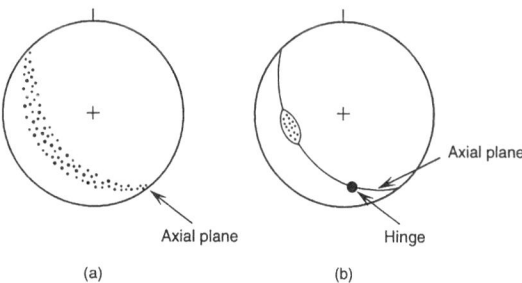

Figure 5.10 (a) Stereo-plot of the prismatic hornblende C-axes showing grouping in a great circle maximum, i.e. with a coplanar arrangement. (b) Plot of the acicular hornblende C-axes showing clustering approximating to a point maximum, i.e. with a linear arrangement, at a high angle to the fold axial direction.

5.10a) whereas the acicular crystals cluster to form a point maximum lying on the great circle (Figure 5.10b).

Together these relationships provide some clues regarding the developmental history of the fold. If the acicular hornblendes are aligned parallel to the extension direction of the structure (Figure 5.9a) the attitudes of the two populations of hornblendes could imply that there was early growth of hornblende prisms within the fold axial plane followed by overgrowth of the acicular hornblendes parallel to the extension direction as the form of the fold was modified by slip parallel to its axial plane.

Plotting the data separately, as in this case, has provided additional critical information which would not have been obtained had the attitudes of all the hornblendes been recorded and plotted together. If, on the other hand, the situation had been such that separate plotting had failed to produce distinct patterns on the scatter diagram, then the data could simply have been combined on a single diagram subsequently, with no loss of information and without the loss of significantly more time.

Distinction between mesoscopic fold structures belonging to different sets does not necessarily involve plotting procedures and can often be achieved simply by inspection in the field. In such cases it is dependent on observational ability, something that tends to become enhanced with increasing experience of the structure in the area of study. An illustration of this is provided by two fold sets in the Lewisian complex in the Inner Hebrides of Scotland. On the island of Rona asymmetrical folds with similar geometry and axial trend were distinguished and ascribed to the 4th and 5th sets (Hopgood and Bowes, 1972). Although both sets of folds are steeply inclined as well as asymmetrical, they are readily disting-uishable, not only because their axial planes are inclined to one another significantly, but also because their axial planes happen to dip consistently in the opposite sense to one another.

Note: It should be borne in mind that the opposite dip sense of the axial planes, while a useful characteristic in this instance, is not a fundamental distinguishing characteristic of the two fold sets. It is accidental, arising purely because of the use of the horizontal as a datum. For example, supposing the total structure had been tilted such that the axial planes of both fold sets (while still retaining their relative angular separation) dipped in the same sense, they would still be distinguishable because of this angular difference, but not in terms of their dip **sense**. (See also a further reference to this in section 12.3). Thus it is the consistent difference in **angular relationship** between the attitudes of the two sets of axial planes that is significant.

5.2.5 IDENTIFICATION AND CORRELATION OF STRUCTURES

The correlation of structures is an essential part of the procedure of establishing structural successions, but obviously before structures can be correlated they must be identifiable. The **recognition** of structures and their **correlation** are thus interdependent and the degree of success in correlating structures will be closely related to the number of distinctive features possessed by individual structures. The larger this number is, the greater will be the certainty of recognition of the structures. Therefore in order to correlate structures most effectively it is important to recognize as many distinctive features as possible which contribute to their character or **style** (section 6.2). These include – syntectonic mineralization (particularly if it is in some way distinctive, such as when it is particularly strongly developed or unusual in type and/or preferred orientation); attitude of the structure (more especially in the later sets); characteristic association of (leucocratic or other) veining (e.g. axial planar) or, as is often the case, when it is parallel or sub parallel to one limb of the fold (Figure 5.11), or extreme attenuation or shearing-out of one or both limbs.

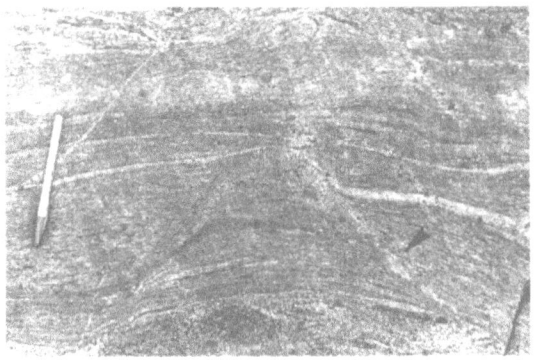

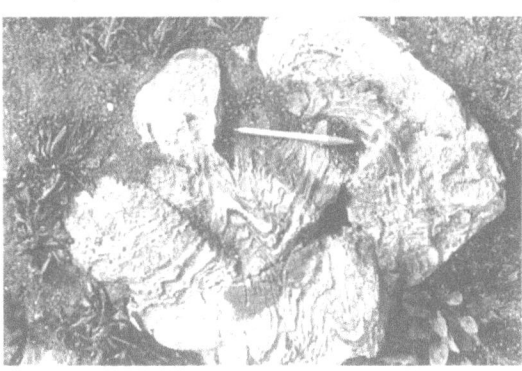

Figure 5.11 Fold with development of potential 'veining' by diffuse leucocratic neosome (arrow) parallel to the limbs. Svecofennian migmatites, Jussarö region, southern Finland.

Figure 5.12 Complex fold pattern in a fold hinge. Svecofennian migmatites, Jussarö region, southern Finnish Archipelago.

Geometrical features (such as the relative thickness of fold limbs or their thickness with respect to the hinge, crumpling of hinge zones, round as opposed to angular hinges, etc.) may sometimes be characteristic **locally** (Figure 5.12).

Also it is important to make use of the temporal significance of interposed reference features. Structures such as shears, syntectonic veins (Figure 5.5), discordant tabular structures such as veins (Figure 2.1) or minor intrusions (especially basic), are all potentially useful, particularly if they are amenable to isotopic dating and so can be used to find the age of the fold as well as place it in the succession. Figure 5.13 shows three different examples of age relationships between folds and post-fold intrusions and, in one case (Figure 5.13b), a shear.

Within the limitations imposed on the techniques by the circumstances, it may be possible to some extent to confirm, at least provisionally, the identity of folds based on style (mineralization and other characteristic features, section 6.2) by plotting attitudes in equal area projection. For the reasons given earlier (Chapter 2, and Figure 2.11) regarding the relative size, complexity etc. of particular sets, those of corresponding sets are likely to exhibit the same distributional behaviour and thus produce comparable patterns on the scatter diagram (Figure 5.15). As a general rule the elements of earlier folds will have a wider and apparently less regular scatter than the elements of later folds. And of course, axial orientation will be less regular than axial planar orientation in any case, and will tend to give less regular plots (Figure 5.14).

Usually, however, stereographic projection in representing the attitudes of structural data from highly complex terranes fills a more useful role in summarizing or demonstrating diagrammatically the relationships between folds of different sets rather than providing a simple technique for structural analysis (Figure 5.15). Exceptions are where stereographic analysis is used to separate only two or three sets, or to distinguish geometrical relationships between folds and lineations, or to show age relationships between fold sets that are difficult to see by direct observation on the outcrop, e.g. the identification of large, very open folds, or in some cases the relationship between folds where wavelength is similar to those of other sets. Stereographic projection can also be very useful for establishing axial and axial planar attitudes by plotting limb attitudes, particularly of very open folds (Figure 5.16b).

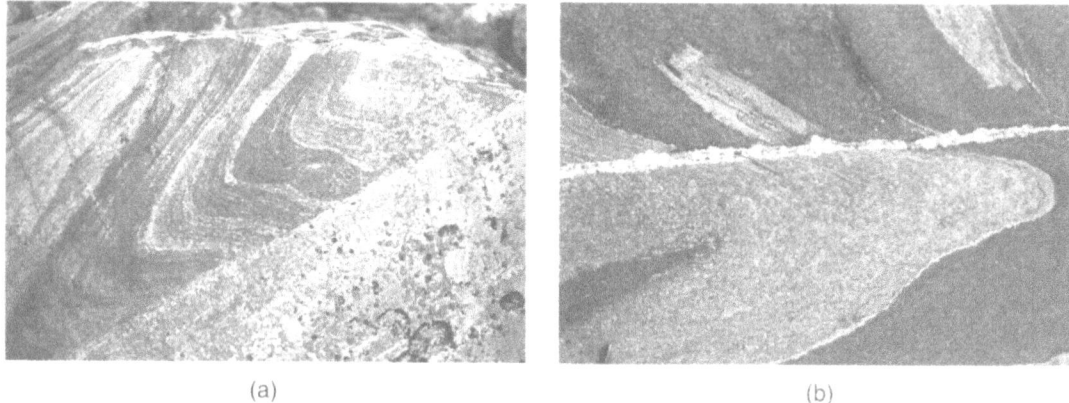

(a)

(b)

Figure 5.13   Three different examples of folds cut by (later), potentially dateable, leucocratic veins. (**a**) Thin, slightly irregular, tabular vein cutting a tight fold (top of exposure). The vein, being slightly curved, pre-dates the tabular leucocratic intrusion with its margin parallel to the fold axial plane. Because of its rectilinear edge, the intrusion is clearly later than any significant post-fold deformation. Pre-Ketilidian migmatites, Fiskenaesset region, Southwest Greenland. (**b**) Post-fold, and post-sinistral shear, thin tabular vein parallel to the limb of the fold. Belemoride migmatites, Soviet Karelia. (**c**) Thin, irregular vein, cutting a fold approximately parallel to its axial plane and, together with the axial plane, curved by a post-intrusive broad warp with an axial trend approximately perpendicular to the vein. Svecofennian migmatites, Karelia, Finland.

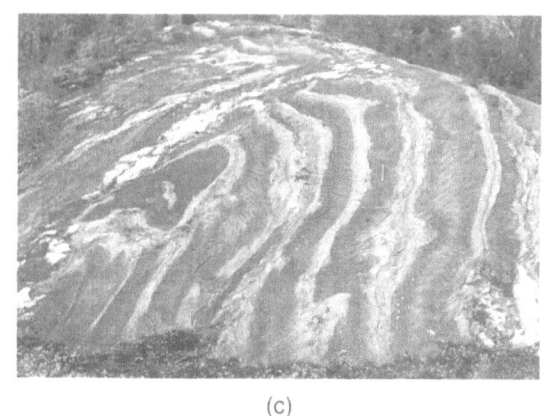

(c)

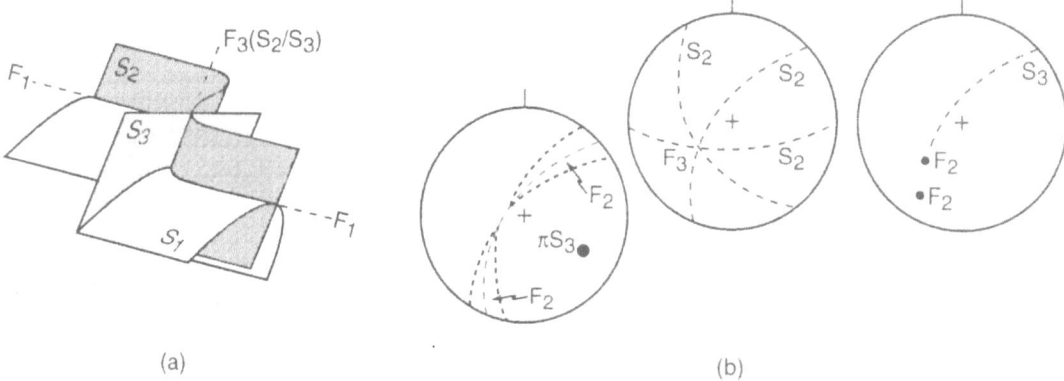

(a)

(b)

Figure 5.14   Sketches to show how reorientation by overprinting widens the range of fold element attitudes; the third fold axis (F_3) has a single plunge direction (at the intersection of S_2 with S_3) but the second axis (F_2) has a much wider distribution (at the intersection of S_1 with S_3) while F_1 can have any attitude defined by the intersection of S_1 with S_2. (**a**) Refolded surface (S_1) showing F_1 (at S_1/S_2), with F_3 at the intersection of the first (S_2) and second (S_3) fold axial planes. (**b**) Stereo plots of the structural elements of (**a**). The second fold axis (F_2) is dispersed in a great circle zone (=S_3 with pole πS_3). S_2 and S_3 intersect in F_3. See also Figure 5.15.

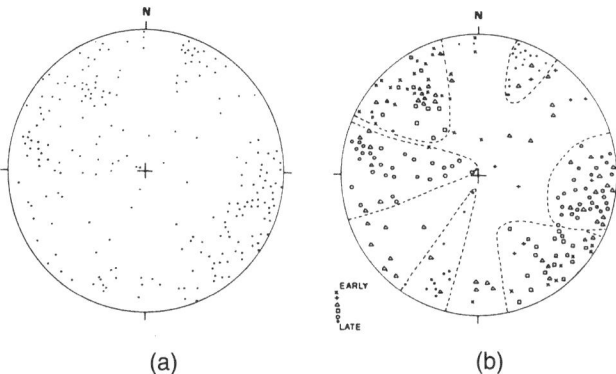

(a)   (b)

Figure 5.15 (a) Equal area scatter diagram of the axial attitudes of folds. Folds of comparable sets are indistinguishable. (b) When fold sets have been identified the clustering of folds into groups related to comparable sets (enclosed by dashed lines) is recognizable. Lewisian complex, Outer Hebrides, Scotland. From Figure 6, Hopgood, 1980. Reproduced with the permission of the Royal Society of Edinburgh.

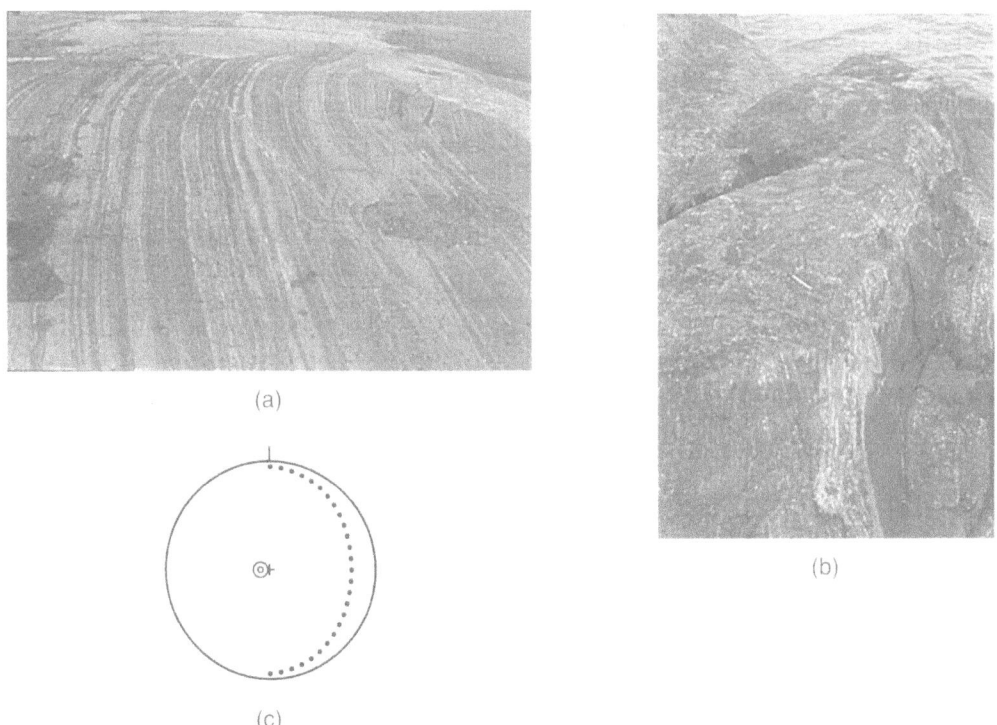

(a)

(c)

(b)

Figure 5.16 (a) Profile, on an essentially horizontal surface, of a very open fold (a 'warp') in steeply foliated tonalitic gneiss. Svecofennian migmatites, Kulosaari, Helsinki. Compare this with (b) the appearance of a comparable fold in three-dimensions. Enskär, southern Finnish Archipelago. The three-dimensional fold is not so easily recognizable as the profile is (see section 10.1.2). (c) Equal area plot of poles to foliation showing the attitude of the axis (the vertical pole of the great circle) of the fold in (a).

5.2.6   IDENTIFICATION (IDENTITY) OF FOLDS (STRUCTURES)

Identification of individual folds and specific fold sets, in effect the establishment of the existence of such fold sets, depends primarily on the potential to recognize the component folds of a set consistently, or at least to recognize a significantly large proportion of these fold structures consistently. Recognition of the individual folds is dependent to some extent on their relationship to other identified structures (as will be seen, this relationship is certainly very important in confirming their identity) and sometimes to a large extent on their characteristics, or style (section 6.2). The likelihood that identical folds or component members of a partial set will be correctly identified from their style alone clearly depends on the number of specific features they possess: the greater the number of these attributes, then clearly the greater will be the likelihood of correct identification.

Identification of structures is important, because it is an essential prerequisite for relating them to specific sets. It is especially important (and often difficult) in the case of folds (such as open folds) later in a complex succession when the only means of detecting their presence is on the basis of how they relate to, i.e. reorientate, earlier (already identified) folds. In contrast to superimposed folds that are moderately tight (close to isoclinal), and therefore easily recognizable by direct observation on the outcrop (Figure 5.17), late very open structures and warps are not always obvious. This is particularly so where the structure is already complex, and their effect on pre-existing structures is not easily discernible (see the discussion on 'Open folds', section 10.1.2). Such is the case where the existence of very late structures is revealed (only with difficulty) by the relatively slight axial planar curvature that accompanies the changing attitudes of the overprinted folds. While slight curvature of earlier structures may be relatively easy to detect on perfectly

planar exposures such ideal conditions are seldom encountered in practice. This condition is illustrated by Figure 5.18. Although the axial planar curvature of the later fold in the figure does not stem from a fold that is particularly open, the effect shown is comparable to that just described.

Correct identification of the folds of an earlier set (by recognizing the significance of their changing orientation) will provide the potential for the later structures to be recognized (Figure 5.19). On the other hand, failure to identify them correctly could almost certainly lead to at least two errors. Not only would there be the incorrect establishment of more than one fold set where only one exists (see section 5.2.3), and the error in that part of the succession which this implies, but also there would be the failure to recognize the later, overprinting structure (Figure 5.19b) and this would lead to a further mistake in the fold succession (Figure 5.19c).

Structures (folds) can be correlated in much the same way as lithological units are in lithostratigraphy (Table 5.3). For example, of three units, e.g. structures A, B and C at some particular locality, B and C can be correlated with B and C of the group of structures B, C, D and E say at a second locality while structure C can be correlated with C of the pair of structures C and F say at a third locality.

Structural correlation based on structural characteristics differs from that of stratigraphical correlation in that structural style replaces stratigraphical parameters such as lithology and palaeontology (sections 5.2.2 and 6.2). However, bear in mind that correlation based on style alone is always subject to confirmation on the basis of overprinted relationships. In its broadest sense, style, a concept whose meaning is commonly misunderstood, can be said to embody **all** the features of a particular structure, not solely the geometry of the structure – features such as limb thickness, hinge shape, axial planar features, associated mineral growth (and therefore the pressure and temperature conditions during growth) and

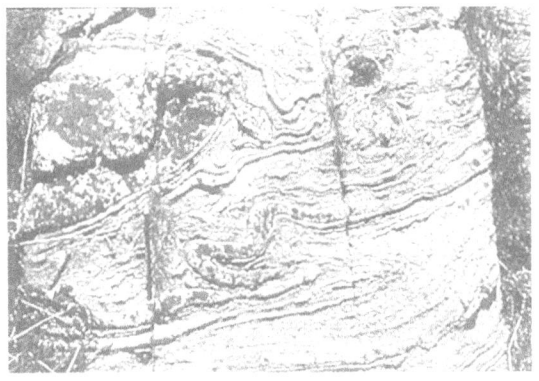

Figure 5.17 Photograph showing clearly distin-guishable refolded (overprinted) relationships between two fold sets at the centre of the picture, a horizontal tight-isoclinal set and a steeply-inclined (upright) close set. Because the hinges and axial planar traces of the isoclinal folds curve around the axial planar trace of the close folds, the isoclinal folds are earlier than the close upright folds. Even later deformation is implied by the curvature of the axial planar trace of the close folds. Other folds are visible in the upper part of the photograph. Sveco-fennian carbonate rocks, Åland Islands, Finland.

Figure 5.18 Convex surface showing early iso-clinal folds (centre of photograph) with axial planar traces gently curved by later deformation. Between the pencil and the arrows the later, overprinting folds are obvious and easily recognized as the cause of the curvature of the axial planar trace of the isoclinal folds. However, if only that part of the exposure shown to the left of centre (left of the arrows) had been visible then the true cause of the curvature would have been less obvious and might easily have been attributed to the convexity of the exposure. Svecofennian migmatites, Åland Islands, Finland.

orientation. This means that the characteristic features of any fold belonging to a set result-ing from a particular deformational 'phase', when considered together are distinctive, and may very well be unique because of the number and interrelationships of all the para-meters involved.

The fold sets of different relative ages in any exposure together constitute the local **partial succession** for that exposure. The par-tial successions from different exposures when correlated and integrated make up the 'full' or 'complete' succession, strictly the total **recognized** succession for the region being studied (see also Chapter 12, and particularly Figure 12.31).

Furthermore, the style of all these fold sets in their particular relative order comprises the overall character (**total style**) of this specific succession which enables it to be distinguished (in a manner analogous to the distinction of a particular stratigraphical succession) from any other superficially similar succession else-where, even though it may have the same number of fold sets.

5.2.7 NEOSOME IDENTITY AND CORRELATION

In addition to using style and the presence of discordant igneous bodies emplaced at inter-vals in the succession, correlation of succes-sions in high-grade deformed rocks like migmatites may be further strengthened by taking into consideration the neosomes intro-duced at different times in the deformational history of the rocks. These, like discordant igneous bodies, provide another means of subdividing the structural succession in terms of relationships between structural and dis-tinctive lithological features, in this case the relationship between the structures in the suc-cession and the partial melt (neosome). See also section 4.3.3(c).

The converse is also true, viz. that these bodies of newly-formed melt emplaced from time to time at deep crustal levels as an inte-

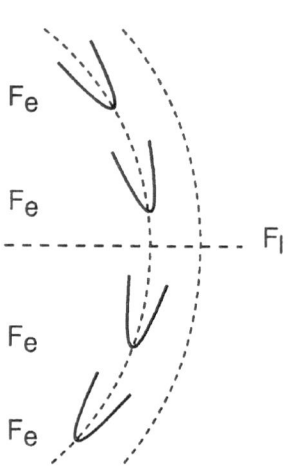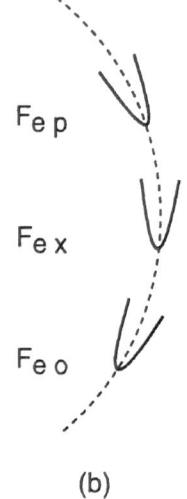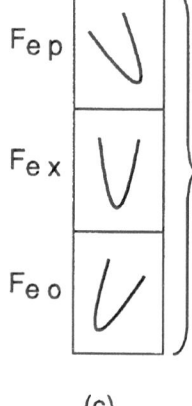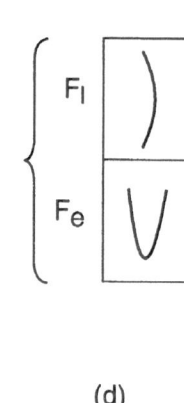

(a)   (b)   (c)   (d)

Figure 5.19   (a) Earlier folds (F$_e$) folded by a later fold (F$_l$). (b) Failure to identify folds belonging to the earlier set (F$_e$) as F$_{ep}$, F$_{ex}$ and F$_{eo}$. (c) **Incorrect**, and (d) **Correct** successions recorded for cases (a) and (b).

gral component of migmatites may be distinguished from one another by their relationships to the deformational structures in the succession. As shown earlier (Figures 4.23, 4.24 and 4.38, and sections 4.3.3(c), 5.2.11 and 13.3.2) neosome relationships are often very clear, particularly in agmatites (Hopgood and Bowes, 1978). For example, the time of a particular neosome in the deformational sequence can be determined by its discordance to particular structures in the palaeosome of an agmatite. Consider the case where two neosomes, one coarse and the other fine-grained, both post-date folds in an agmatite (Figure 5.20). The question of whether both neosomes are coeval, representing coarse and fine phases of the same event, is resolved when examination shows that although both neosomes post-date the folds, they are discordant to one another. Therefore the agmatite is composite, an agmatite within an agmatite, and includes two palaeosomes. The amphibolite folds comprise the palaeosome (P$_1$) of the earlier agmatite which has a fine-grained neosome, Nx. The fine neosome (Nx) together with the

folds comprise the second palaeosome (P$_2$) of the agmatite and this is overprinted by the second, coarser phase neosome (Ny). The succession is composed of three sets of structures, the first being the amphibolite folds (palaeosome P$_1$). These are followed by the neosome Nx and finally by neosome Ny.

This example shows that on the basis of their relationships to structures in the succession as well as to each other, neosomes that appear to be lithologically identical can be distinguished from one another. Neosome-forming events throughout the developmental history of migmatites can thus be placed in

Table 5.3   Correlation between structures A–F at Localities 1–3

Locality 1	Locality 2	Locality 3
		F
	E	
	–	–
	D	–
C	C	C
B	B	–
A	–	–

(a)

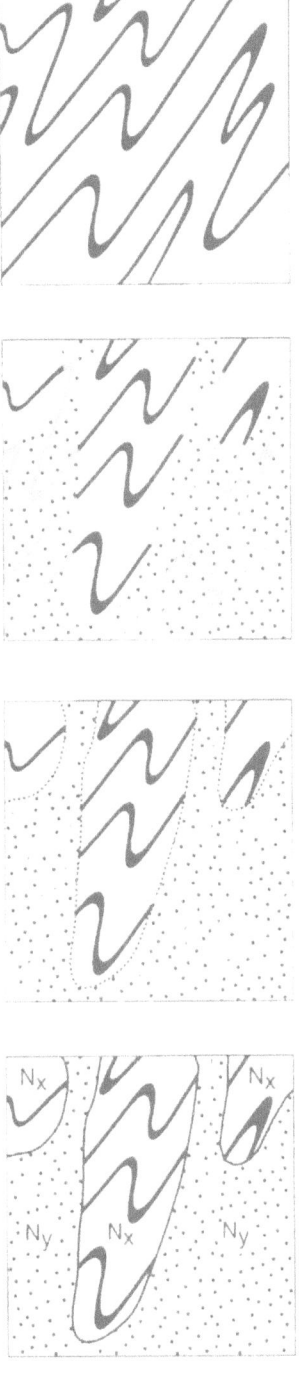

Figure 5.20 Discrimination between neosome sets on the basis of their relationships to fold sets and agmatites. (**a**) Photograph of a composite agmatite. This shows neosome (N_y, beneath the pencil) enclosing palaeosome blocks (P_2, at the centre). Palaeosome P_2 is made up of pale grey fine-grained neosome (N_x) enclosing an earlier palaeosome (P_1) comprising the hinges of earlier (black) amphibolite folds. (**b**) Sequence of events leading to the development of (**a**). This begins at the top with the palaeosome (P_1) comprising black amphibolite folds in white neosome (N_x). Next, the onset of partial melting leads to the production of the second (dotted) neosome (N_y). This progresses, with differential disappearance of P_1 and N_x to leave a composite agmatite made up of N_y enclosing blocks of the palaeosome P_2 which comprises N_x enclosing the amphibolite folds, P_1. Svecofennian complex, Skåldö, southern Finland. (cf. Figures 4.23, 4.24, 5.37, 5.38, 13.11).

(b)

their correct relative order and their petrogenesis considered in relation to their isotopic
age (Hopgood *et al.*, 1983).

Although the presence of distinctive neosome bodies and agmatites adds further com-
plication to the succession, it nevertheless
strengthens the certainty of the correlation
of the succession from place to place as a
consequence. This is because the increase in
complexity also adds to the number of characteristic features (attributes) in the succession.
(See also the comments in section 1.1.2,
'Migmatites – what they imply to the structural geologist', and Chapter 4, 'The role of
structures interposed in the succession'.)

5.2.8   IDENTIFICATION OF ISOLATED FOLDS

Because structural identity and identification
are important concepts in correlation and in
the determination of successions, it is worth
continuing the discussion by examining their
significance in some further detail.

In reality of course, when it comes to the
identification of individuals from any number
(n) of isolated folds exposed in outcrops over
a limited area, use is made of other features
besides fold geometry. Otherwise the determination of discrete fold sets would be very
difficult indeed.

Consider the prospects of recognizing and
determining the relationships between folds
in terms of discrete sets in the following
different circumstances when the only clear
attributes shown by the folds are:

(i)   Geometry (shape) only.
(ii)   Geometry and axial planar cleavage in
       some cases.
(iii) Geometry, cleavage and a vein parallel
       to one limb in some cases.
(iv)   Geometry, cleavage with parallel mineral
       growth and a vein parallel to one limb.
(v)   Geometry, cleavage, mineralization, vein
       and diffuse axial planar neosome.

(vi)   Geometry, cleavage plus shear parallel
       to one or other limb.
(vii) Geometry, cleavage plus shear parallel
       to both limbs.
(viii) Geometry, cleavage plus shear parallel
       to the axial plane.
(ix)   Various combinations of the above.

Three folds such as those in Figure 5.21
observed in isolation in the field might belong
to the same set or to different sets because
the basis for discrimination between them
(viz. geometry alone as in case (i) above) is not
sufficiently firm. The slight differences in
shape between them might simply be the
result of local differences in response to the
same stress because of differences in layer
thickness, lithology etc.

However, folds such as those in Figure 5.22,
with similar geometry plus similar axial planar
cleavage ((ii) above) **are** likely (although by no
means certain) to belong to the same set.

Folds such as those in Figure 5.23, with
similar geometry, comparable axial planar
cleavage and veining parallel to one particular
limb ((iii) above), are reasonably certain to
belong to the same set.

Folds like those in Figure 5.24 with similar
profiles, a vein in one particular limb and
spaced axial planar cleavage with parallel
mineral growths ((iv) above) are even more
likely to belong to the same set. In other
words it could be said that the number of
similar attributes possessed by all three structures (similar geometry, comparable cleavage,
similar axial planar mineral growth and
veining parallel to one limb) is now such
that the degree of certainty that folds such as
these belong to the same set is reasonably high.

Folds like those of Figure 5.25, which all
have similar geometry, a vein in one particular
limb, spaced axial planar cleavage with parallel mineral growth and diffuse axial planar
neosome (v) above) are very likely to belong
to the same set. In other words the degree of
certainty that they are related to one another is
very high indeed.

Figure 5.21 Folds F_x, F_y and F_z displaying similar profiles.

Figure 5.22 Folds F_x, F_y and F_z with similar geometry plus axial planar cleavage possibly belong to the same set.

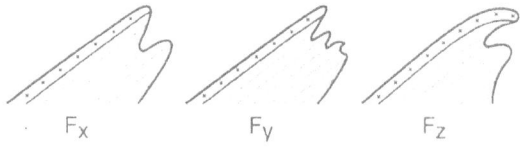

Figure 5.23 Folds F_x, F_y and F_z with similar profiles, spaced axial planar cleavage and a vein in one limb are likely to belong to the same set.

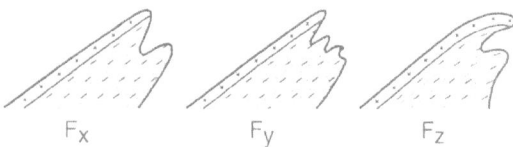

Figure 5.24 Folds F_x, F_y and F_z with similar geometry, a vein in one limb and spaced axial planar cleavage with parallel mineral growth.

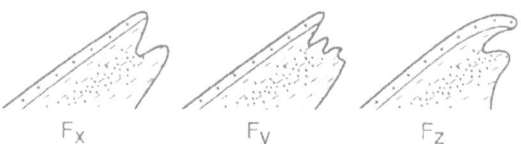

Figure 5.25 Folds F_x, F_y and F_z with similar geometry and possessing a vein in one limb, spaced axial planar cleavage with parallel mineral growth and diffuse axial planar neosome.

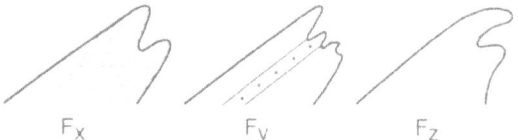

Figure 5.26 Three folds F_x, F_y and F_z with comparable geometry but with **different** associated structural features; viz., axial planar cleavage, axial planar vein and axial plane-parallel mineral growth.

In all the five preceding cases of three folds having very similar, but not identical geometry (Figures 5.21–5.25), their association with an increasing number of additional features increasingly enhances the certainty of their relationship. In other words as the number of comparable attributes possessed by each structure increases so too does the likelihood that all three belong to the same set.

Before proceeding further with this discussion it should be appreciated that the converse

condition holds, i.e. if different folds are consistently associated with different structural, mineralogical and igneous attributes then it is likely that they belong to **different** sets, each set characterized by the different group of attributes.

Consider again the three similar-looking folds exposed in three different places (Figures 5.21–5.25). But in this case suppose that, although the folds have comparable geometry, those at different locations do not all possess the same features (as was the case of those shown in Figures 5.21–5.25) but instead consistently show associated features in three different combinations (Figure 5.26). They can be put into three groups, each group having one of the three attributes of cleavage, axial planar veining, or syntectonic mineral growth. This suggests that they belong to three separate fold sets, F_x, F_y and F_z say, each corresponding to one of the three groups.

Therefore, although a collection of folds may all have the same basic geometry, the fact that they consistently fall into different groups each of which possesses specific attributes, suggests that they belong to separate fold sets, even if each group differs from the others by only a single attribute.

From the first group of folds (Figure 5.21), whose relationship depends entirely on similarity of geometry, through groups with similar geometry plus comparable cleavage etc. to the group with several structural features, the probability that all the folds of the group belong to the same set increases progressively in discrete steps. This is because the features in each case become progressively distinctive, viz. from geometry which can be closely similar for different deformational events and whose distinctive features can be blurred or obliterated by later deformation, to cleavage which could develop similarly at more than one time, to veining which is indicative of particular pressure and temperature conditions (although not necessarily unique conditions), to a distinctive mineralogical association which could be characteristic of specific pressure and temperature conditions and therefore likely to be unique.

When to the various possible combinations of the discriminating features (say two or three) there is added an increase in the possible **number** of structural attributes associated with the folds, the strength of the identity, i.e. the probability or certainty of their belonging to a particular set, further increases. This increase in certainty is directly proportional to the number of features that contribute to the

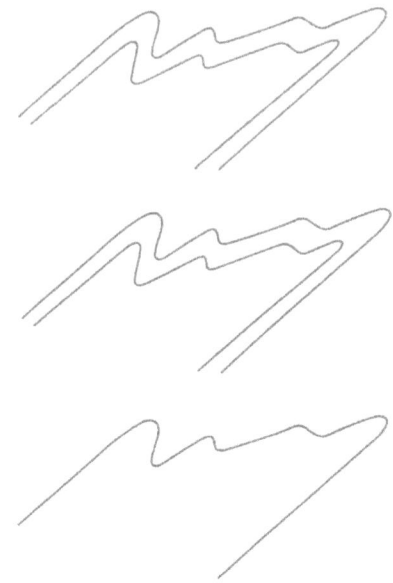

Figure 5.27   Three folds with comparable, though not identical, profiles.

characteristics, or style of the fold so that where this number (1–n) is low the degree of certainty is low, as follows:

1 or 2 attributes	There is a **low** degree
1 + 2 + 3 attributes	of certainty.
1 + 2 + 3 + 4 attributes	
1 + 2 + 3 + 4 + 5 attributes	

$1 + 2 + 3 + 4 + 5 + \ldots$ etc. $\ldots\ldots\ldots\ldots + n$ attributes. There is a **high** degree of certainty. Also, conversely, the probability that folds possessing all n attributes but **not** belonging to the same set is extremely remote indeed. And similarly this degree of certainty would apply even where folds that possess these attributes have some other, though comparable, geometry such as that shown in Figure 5.27.

The degree of certainty (probability) P with which isolated folds can be identified as belonging to a particular set α will bear some relationship to the number N say, of structural features related to the folds. The same kind of relationship will also exist in the case of folds belonging to set β. If, however, consistent refold relationships exist between folds of the two sets α and β, the same degree of probability P of correct identification will depend on the association with each set of a number of structural features which can be smaller than N, i.e. with the addition of another attribute common to the two sets (viz. their consistent refolded relationships), N will be reduced by some number x for the same degree of probability P of correct identification. For example, supposing a fold set α possesses N characteristic structural features. A number of folds are observed to have fewer than the N (N–x, say only two or three) structural features associated with this set. Nevertheless it is still considered likely that they belong to set α. The folds of this first group consistently have the same temporal (refold) relationships to another group of folds which likewise have only two or three (n–x) structural features which this time are characteristics of a different set, β. Correspondingly they are considered to belong to set β. In spite of the fact that none of the folds possess all structural features characteristic of either set α or set β, because they have the same mutual refold relationships, the probability P that each of these folds has been correctly identified as belonging to set α or set β is very much higher than it would be if each of the folds with its two or three associated structural features was isolated, without refold relationships. For the correct identification of the isolated folds (i.e. folds lacking overprinted relationships) to have the same degree of probability P, the number of associated structural features must be N.

Hence the degree of certainty (probability) with which folds possessing a particular number of associated structural features can be identified as belonging to specific sets is greatly enhanced in those cases where refold (overprinted) relationships exist.

In practical terms this means that in the case of isolated individual folds reliable identification of the component members of specific fold sets can be made only when there are several (say 4 or 5) distinct structural features associated with the folds. These features comprise the **style** of the folds (section 6.2). On the other hand, only two or three structural features are necessary for reliable identification when refold relationships exist between folds of two (or more) sets.

1. Consider the case of three folds F_x, F_y and F_z which look similar. Although each looks somewhat like the other, they all have slightly differing geometry and there is an insufficient number of associated structural and mineralogical features to ensure that they can be assigned to the same set with any degree of confidence. Each of them refolds a fourth fold F_p (Figure 5.28), but even though they bear the same refold relationships to F_p, i.e. they are later than F_p, this is still not enough to warrant placing them in the same set (see section 8.1.5, 'Determining the structural succession: principles').

2. Supposing on the other hand that, besides refolding F_p, each of the folds F_x, F_y and F_z also

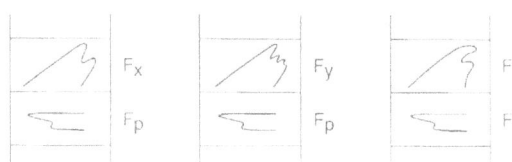

Figure 5.28 Three folds F_x, F_y and F_z each refolding fold F_p.

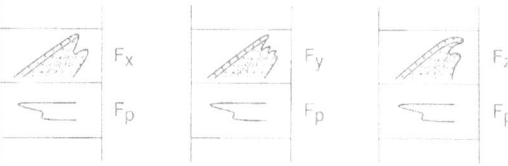

Figure 5.29 Three folds F_x, F_y and F_z with similar characteristics (geometry, veining, cleavage and neosome) and all refolding F_p.

has the same kind of veining, cleavage and neosome associated with it (Figure 5.29). In this case there must be a high degree of probability that F_x, F_y and F_z belong to the same set.

3. Supposing the folds F_x, F_y and F_z of case 1 not only refold F_p but are also themselves refolded in every case by fold F_m (Figure 5.30). In this case there is a reasonable possibility, but no certainty, that F_x, F_y and F_z belong to the same set (refer again to section 8.1.5).

4. If, in the fourth instance, F_x, F_y and F_z also have the same veining, cleavage and neosome associated with them, as well as possessing the same refold relationships with F_p and F_m, then they are near **certain**, in terms of reasonable interpretation, to belong to the same set (Figure 5.31).

Extending this argument to include local partial successions comprising, say, two or three fold sets determined at some exposure, it is reasonable to conclude that if the same succession can be found consistently at other exposures over a wider area of outcrop, the likelihood that the identification of its correlative components and their order of succession is fortuitous is extremely remote.

The preceding discussion highlights the indivisible interrelationship between identity, or identification, and correlation. There is a reciprocal interdependence between the two.

(1) Each is mutually supportive of the other, i.e. in the first instance identification of identical folds is necessary before correlation can be made between them to establish them as components of a set.

(2) Consistent correlation between folds of individual sets reinforces, i.e. confirms, the original identification.

The observer is therefore constantly reassessing relationships (often unconsciously) – weighing up, comparing, accepting and rejecting, throughout the process of establishing a structural succession.

For example, if a group of folds (G_1), such as those of Figure 5.21, appears likely to belong to a specific set (Z_1) and:

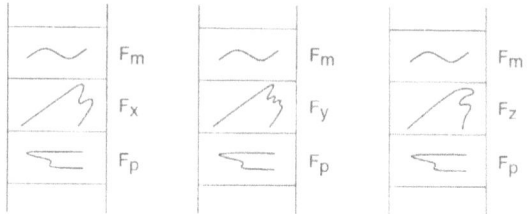

Figure 5.30 Three folds, F_x, F_y and F_z with the same geometry refolding F_p and refolded by F_m.

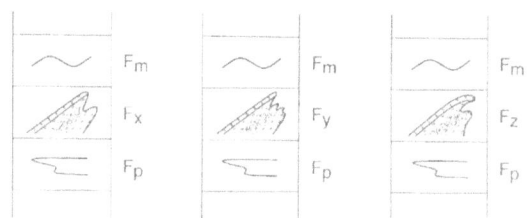

Figure 5.31 Three folds Fx, Fy and Fz with the same associated attributes (geometry, veining, cleavage and neosome) fold F_p and are refolded by F_m.

(i) consistently refolds another group of folds (G_2), e.g. F_p of Figure 5.30, that in turn are likely to belong to another specific set (Z_2), and

(ii) is refolded by yet another group of folds (G_3) which seem likely to belong to yet another specific set (Z_3), e.g. F_m in Figure 5.30 (refer back to the beginning of this discussion for reasons for correlation) then it is more than reasonably certain that not only do G_1 folds all belong to the same set (Z_1) but also that G_2 folds are all members of another set (Z_2) and similarly G_3 are part of yet another set (Z_3).

5.2.9 CERTAINTY OF IDENTIFICATION OF ISOLATED STRUCTURES – REALITY OF THE 'SUCCESSION' OR 'COINCIDENCE'

Among points of concern relevant to the concepts of identity and succession is the belief

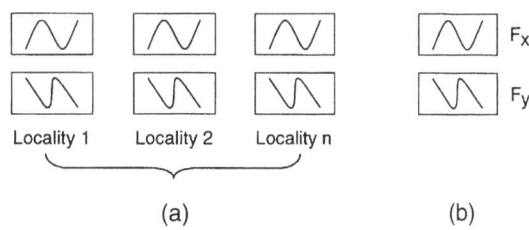

Figure 5.33   (a) Three pairs of apparently identical folds in the same order probably belonging to the same pair of sets, F_x and F_y (b).

held by some that comparability of style, order of succession and overprinting relationships are simply a matter of chance. It appears that, in their view, the relationship of structures in 'successions' is thought to be coincidental and not necessarily the result of their sequential development, the structures having formed randomly in such a way that structures of comparable style (but not necessarily of the appropriate relative ages) would be found to have comparable relationships in different successions, i.e. they would always have the appropriate 'order' in different 'successions', not as a consequence of overprinting between related structures, but as a matter of chance. Let us now further explore the possibility that coincidence can really play a significant role in this matter.

In considering the likelihood, or probability, that the identification of isolated individual folds is correct and therefore that the succession has been correctly established, three related questions need to be asked.

1. What likelihood is there that isolated single folds possessing certain characteristics can be correctly identified as belonging either (a) to the same set because they have the same characteristics, or (b) to different sets?
2. What is the likelihood of two isolated identical pairs of folds (such as those at localities 1 and 2, Figure 5.33a) **not** belonging to the same two sets? Each pair is provisionally 'identified' as belonging to two sets, say F_y and $F_{z'}$ (as in 1(a) above) because they possess the same two distinct groups of characteristics and are in the same order, but is it possible

that they did not actually develop as components of sets 'y' and 'z' but nevertheless, purely by coincidence, are always in the same order as are sets 'y' and 'z'?
3. What is the likelihood that three or more (say n) folds identified as belonging to the same three sets in terms of characteristics and relationships (as in 1(a) above) are always in the same order by chance and not because they developed in that order (Figure 5.35)?

The simple answer to these questions must be that the likelihood is not entirely certain in every case. However, consider these conditions further.

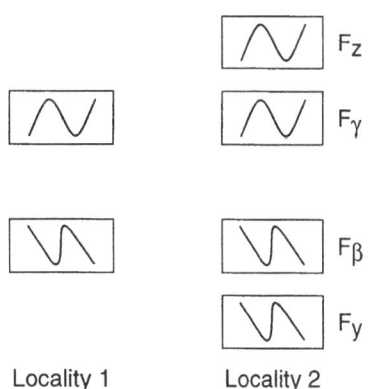

Locality 1   Locality 2

Figure 5.34   Two sets of folds (locality 1) coincidentally identical to sets **y** and **z** and **in the same order** as sets **y** and **z** but actually belonging to different sets **β** and **γ** (locality 2) which happen to have features identical to sets **y** and **z**.

Figure 5.32   Two folds with apparently identical geometry but not necessarily belonging to the same set.

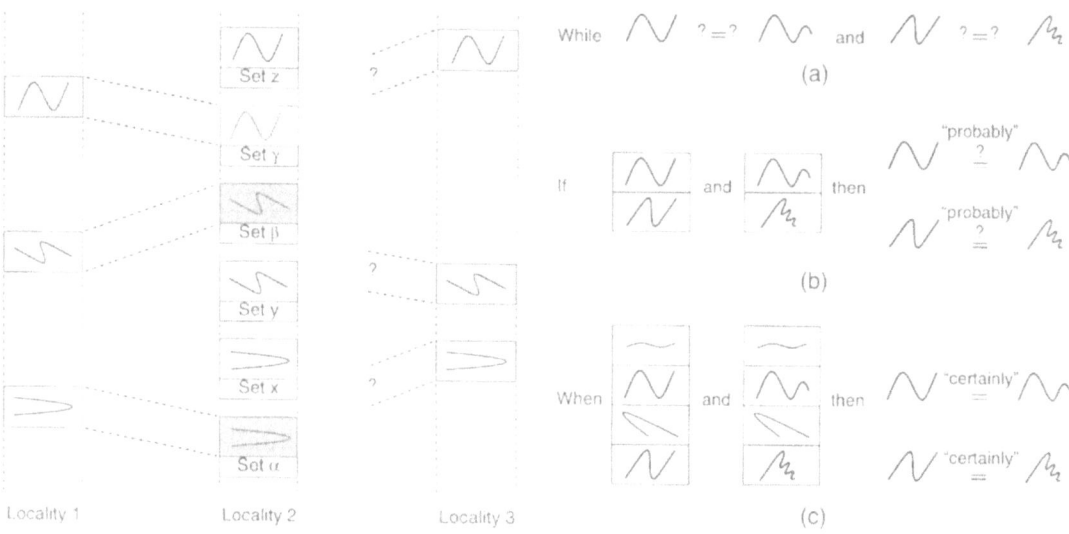

Figure 5.35 Three sets of folds (α, β and γ) at localities 1 and 3 coincidentally identical to sets x, y and z at locality 2 and in the **same** order as x, y and z, and from which they could not of course be distinguished in practice.

Figure 5.36 Possible relationships between (**a**) isolated single folds, (**b**) isolated pairs of folds and (**c**) isolated groups of three or more folds. (**a**) There is a possibility that the two upright folds might be equivalent as might the two inclined folds. (**b**) There is a probability that the folds in the two (refolded) pairs of upright and inclined folds might be equivalent. (**c**) There is reasonable certainty that the folds in the two groups of (refolded) open, upright and inclined folds are equivalent.

1. Folds such as those in Figure 5.32 with apparently more or less identical geometry, while **possibly** developed contemporaneously, might **not** belong to the same set. At least it could not be said with absolute certainty that they belonged to the same set.
2. Isolated, apparently identical **pairs** of folds, **always in the same order** (Figure 5.33a) are likely to belong to the same two sets F_y and F_z (Figure 5.33b).

However, they just **might** belong to different sets which by sheer coincidence, (i) always show the same two groups of geometrical features and (ii) are always in the same order (Figure 5.34). This coincidence, while theoretically possible, is not very likely.
3. Isolated trios or groups of three or more folds, each with apparently identical geometry and **always in the same order** must

(almost) certainly belong to the same three (or more) sets. They are certainly **not** likely to belong to some other group of three (or more) different sets which are coincidentally, (i) similar and (ii) in the same order (Figure 5.35).

These three possible relationships, involving isolated single structures, pairs of structures and groups of three (or more) structures, are summarized in Figure 5.36.

While the individual folds ($\wedge$ and $\wedge\wedge$) and ($\wedge$ and $\wedge$) in Figure 5.36a might, or might not, be equated as shown, when these folds appear in (two) pairs of folds in the order shown in Figure 5.36b, it could be said they are **likely** to be equivalent. On the other hand, when they are in equivalent positions in two groups (successions) of four folds (Figure 5.36c) they can be equated with much greater confidence, or certainty.

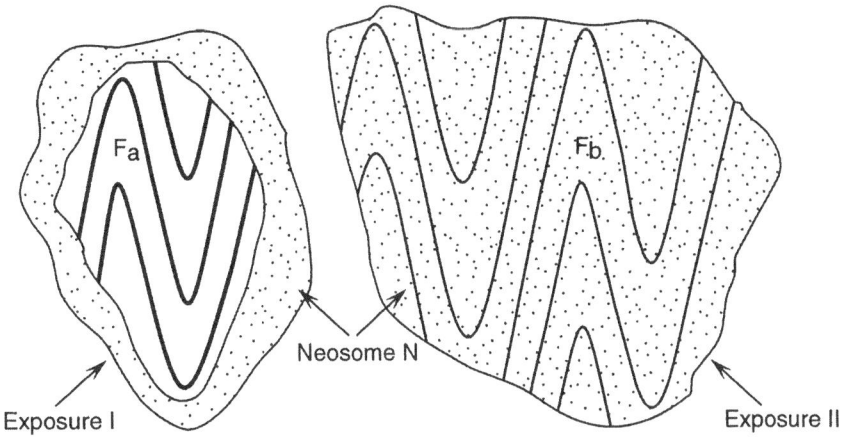

Figure 5.37   Separate exposures (I and II) of agmatite showing apparently identical folds. However the folds (F_b) of exposure II have formed **in** neosome N which can be seen from exposure I to have developed **after** folds F_a which exist as palaeosome blocks in N. Thus not only can it be demonstrated that the relative ages of the apparently identical folds (F_a and F_b) are different but it can also be shown that their origins differ. (cf. Hopgood & Bowes, 1978).

5.2.10   AGMATITES AND FOLD IDENTITY

In the same way as refolding (overprinting) relationships between one fold (structure) set and another are used, the overprinting relationships between agmatites and folds also serve to strengthen the certainty of fold identification and assignment to fold sets in migmatites (see also Figures 4.1, 4.3, 4.23, 4.24 and section 4.3.3(c) on the use of agmatites as time markers in the succession; see also sections 5.2.7 and 13.3.2).

For example, the position of a specific fold in the deformational sequence can be determined by its discordant relationship with a particular neosome Ny. In more complex situations of this kind, relationships between structures in the succession may be determined from the relationships between relict structures in the palaeosome of an agmatite and between those structures and the neosome (section 4.3.3(c) and Figure 4.38).

The presence of neosome in this context can be particularly useful for separating different fold sets that appear to be geometrically identical.

Consider the case of the two sets of similar-looking folds F_a and F_b observed in two separate exposures (I and II) of agmatite comprising palaeosome blocks in neosome N (Figure 5.37). In exposure II fold set F_b has developed in the neosome N of the agmatite and in exposure I the palaeosome block includes folds F_a apparently identical to F_b. However, set F_b are easily differentiated from set F_a because their relationship to N shows that they are clearly of different ages. Set F_b is not only different from F_a but it must also be later than F_a because F_a comprises the palaeosome of the agmatite. In other words F_a already existed before the material (neosome N) folded by F_b was generated.

5.2.11   RELATIONSHIPS BETWEEN TECTONIC STRUCTURES AND MULTIPLE AGMATITES

The preceding illustration of the role of neosome in identifying or confirming the identity

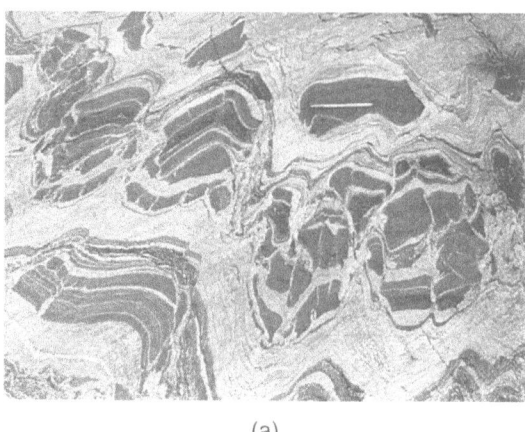

(a)

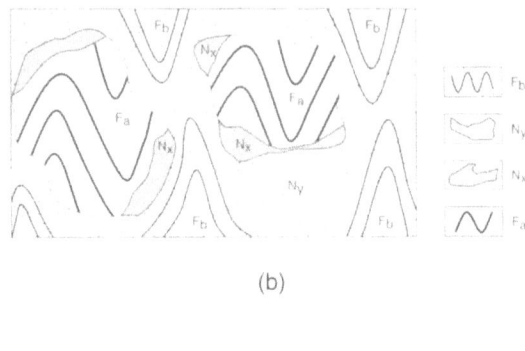

(b)

Figure 5.38 (a) Exposure of agmatite comprising palaeosome blocks of fragmented amphibolite (dark) folds (F_a) in a foliated and folded (F_b) pale grey neosome. The neosome within the blocks appears to be unfoliated and unfolded and this implies that there are two neosomes, N_x within the palaeosome blocks and N_y enclosing the blocks containing pre-N_x folds (F_a). Svecofennian migmatite, Skåldö, Jussarö region, southern Finland. (b) Sketches of a comparable setting, but here the later folds (F_b) that are developed within the later (coarse) neosome (N_y) are closely similar to the earlier amphibolite set (F_a). Nevertheless their relationship to the neosomes demonstrates that the folds (F_b) developed in the neosome (N_y) are distinguishable from F_a, even though both F_b and F_a are similar.

of fold sets is a relatively simple one. However, within the complexity which characterizes migmatite development, conditions will normally have favoured the generation of partial melt at a number of times so that several neosome bodies are likely to be present. This means that complex examples of the relationships between different sets of structures and neosomes such as those of Figure 5.37, and Figures 4.38a and 5.20a seen previously, are often encountered.

Figure 5.38a shows another example of a composite agmatite derived from folded amphibolite and gneiss. Its palaeosome is composed of separated blocks comprising folded and fractured amphibolite layers in a fine-grained grey neosome, and these are enclosed in a foliated and folded later neosome. These relationships are comparable to the association shown in Figure 5.38b and are considered in the following discussion.

Figure 5.38b shows an agmatite with folds F_b in the coarse neosome N_y and a palaeosome containing neosome N_x and folds F_a. F_a look very similar to F_b and, on cursory inspection, might be confused with them. However, fold sets F_b and F_a can be distinguished easily on the basis of their relationships to the neosomes N_x and N_y. It is clear that N_y must be later than N_x which it encloses, and that N_x must be later than F_a folds because it cuts and encloses them. Furthermore, F_b folds are developed in the later neosome N_y and so must be the latest structures in the succession, whereas F_a folds (cut by the earlier neosome N_x) are the earliest.

In this example the recognition of the fact that two sets of neosome bodies exist, and that they can be distinguished unequivocally, not only provides further information about the evolutionary history of these migmatites but also allows the distinction to be made between two similar-looking fold sets.

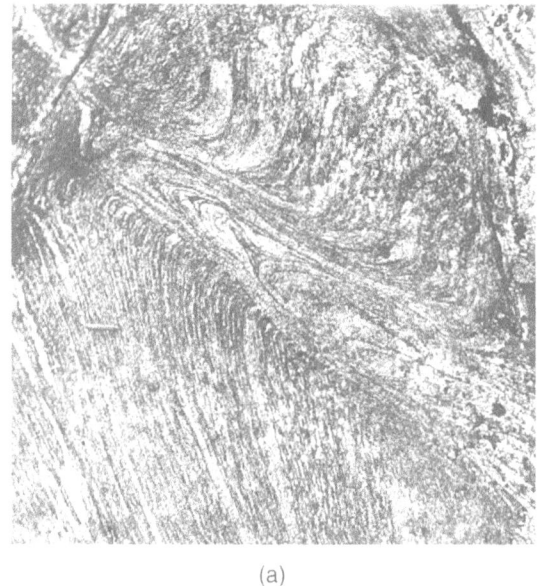

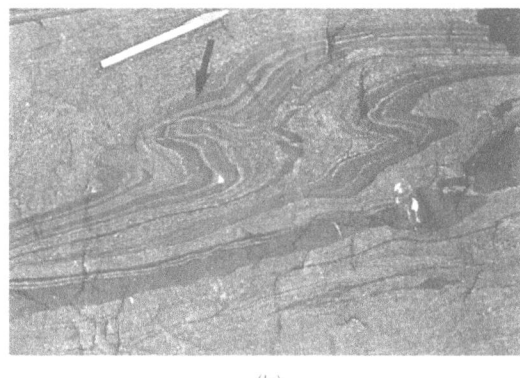

(b)

(a)

Figure 5.39   (**a**) Ductile shear with flow fold. Pre-Ketilidian migmatite, West Greenland. (**b**) Initiation of a flow fold (below the pencil) with the onset of partial melting. Svecofennian migmatites, Finland.

5.2.12   VARIATION IN EXPRESSION OF RELATED STRUCTURES IN THE SAME LIMITED CRUSTAL SEGMENT

The geometry and style of folds and other structures of the same set within a crustal segment of restricted volume are likely to vary considerably with depth because of difference in pressure, temperature and fluid content of the rocks. Migmatites often develop in an environment that is transitional between conditions favouring the formation of what might be called 'brittle' structures and structures that are 'ductile'. Structures of the same set might range between brittle shears and brittle folds, flow folds (Figure 5.39) and ductile shears (Figure 6.53). While potentially this structural variation might present a problem in correlation, such difficulties as may arise can often be obviated, or reduced considerably, by close monitoring of changes across the field. Nevertheless, this might be less easy where dismembered parts of originally closely-juxtaposed parts of the same terrane were differentially elevated and eroded after separation, so that structures exposed at the present level were formed at distinctly different crustal levels. In the absence of remnants of recognizable structural associations in adjacent areas, attempts at such correlation could be fraught with problems (Hopgood and Bowes, 1995). Normally, however, the structural succession might be expected to have a sufficient number of features in common to allow correlation to be made.

Aspects of structural variation in the same terrane are referred to in Assumption 5 of section 2.3, and variation of structural expression in general is considered in section 8.1.2.

See also the reference to changing attitudes of fold elements with time and crustal depth in section 7.2.

5.3 FOLD (STRUCTURE) SETS

As noted in section 5.2.6, 'Identification (identity) of folds (structures)', establishing a structural succession of any kind is dependent on the feasibility of recognizing groups of related structures or structure **sets** unambiguously.

A fold **set** (or structure set) is a non-genetic term and refers to a group of folds whose age relative to any other folds is the same in each case, e.g. the folds of a particular Set 'Y' are all older than the folds of Set 'Z' which are superimposed on them, and 'Y' folds are all younger than the folds of Set 'X' on which they are superimposed. 'Y' may be regarded as a family of related structures deforming the structures of earlier sets and being themselves deformed by structures of sets later in the succession 'X', 'Y', 'Z'. Comparable terms are fold 'generation' and fold 'phase'. (A 'set' may be considered as having been produced by a 'phase' of deformation.)

In much the same way as lithological units may be arranged in a (stratigraphical) succession, structural sets can also be arranged in a succession on the basis of their overprinting relationships. Also, like stratigraphical successions where there may be (some) diachronism between units in successions drawn up for different parts of a region, the folds of a single set are not necessarily of exactly the same 'absolute' (e.g. isotopic) age. Although they are likely to be very nearly the same age, they too can be diachronous. The sole criterion for their inclusion in a particular set is their age relative to that of the folds (structures) of all other sets (but see also the discussion in section 8.1.5 'Determining the structural succession: principles'). It must be stressed that the use of geometry and style alone is not a sound basis for the identification of structure sets and not an acceptable means of determining sets (see the discussion on grouping

structures on this basis in Hobbs, Means and Williams 1976, under 8.5.3 'Limitations of the Method', pp. 373–75).

At the beginning of Chapter 1 it was stated that overprinting (Sander's (1911) 'Überprägung') of structures has been recognized and referred to at least since the nineteenth century, and C.T. Clough in *The Geology of Cowal*, 1897 (pp. 23, 24), when referring to examples of early foliation crossed by strain slip (crenulation) cleavage which was itself crumpled by later movements, commented:

> Every few yards we see places where an early foliation has been crossed by early strain slips, of the type of the prominent ones in the Ardentinny Section, but these again are crumpled most variously by later 'anticline' movements.

The principle of overprinting is essential to the present approach used for defining structure sets. Implicit in this principle are the following criteria; '. . . (1) deformed structures are older than those which deform them and (2) cross-cutting features are later than those which they cut' (Hopgood and Bowes, 1972, p. 109). Broadly then, the basis for establishing the relative ages of mesoscopic fold sets is two-fold, viz. the recognition in the field of (1) the effects of refolding earlier structures by later folds and (2) the effects of folding of planar and tabular structures (such as intrusions) which have transected earlier structures prior to being folded.

Although the structural complexity resulting from polyphase deformation where folding is involved is undoubtedly much greater than that where only planar or linear structures develop, the resolution of a structural succession involving fold sets can be accomplished more easily and the results regarded with greater certainty than it can when only planar or linear structures form (see also the earlier discussion of this in section 3.1.3). This is because there exists within the composite geometry of folds a greater potential for the development of several features which, taken together, are characteristic, or diagnostic, so that the likelihood of distinguishing between different sets of structures when these are folds is considerably greater than it is where the structures are simply either planar (foliations) or linear (lineations) alone. Even where two fold sets have nearly parallel axes and axial surfaces, and even if this angular relationship is subsequently obscured by repeated deformation, the individual features of the two sets will often be sufficiently distinctive to allow discrimination between them (Figures 5.40, 5.41).

5.3.1 CHARACTERISTIC FEATURES OF FOLDS

The potential for a unique set of characteristic features in a fold is very much greater than it is in either a planar or a linear structure. This is because a fold in even its simplest form possesses three geometrically planar elements (an axial surface, S_A, and two limbs, S_{L1} and S_{L2}) and one linear element (an axis/hinge, L_H) as well as an inter-limb angle. Possibilities for unique characteristics stem from the following:

(i) some aspect of the axial surface, e.g. parallel fracture cleavage either
   a) close or spaced (Figure 5.40a) or
   b) curved or planar (Figure 5.40a) plus or minus parallel mineral growth and/or presence of neosome;
(ii) some aspect of the shape of one or other limb (S_{L1} or S_{L2}) (Figure 5.40b);
(iii) some aspect of the relative length of the limbs (Figure 5.40b);
(iv) some aspect of the fold inter-limb angle and its rotation sense (Figures 5.40b, 5.41b);

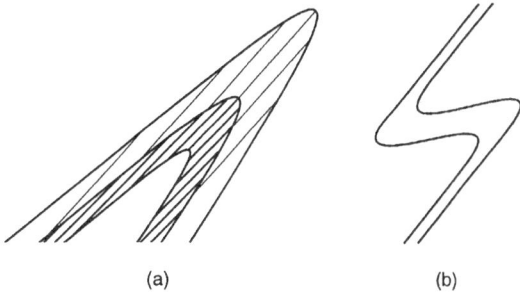

(a)   (b)

Figure 5.40 Sketches of folds with different features. (**a**) Symmetrical tight fold with differential spacing of axial planar cleavage. (**b**) Sinistral asymmetrical fold with a short thick limb and two long thin limbs.

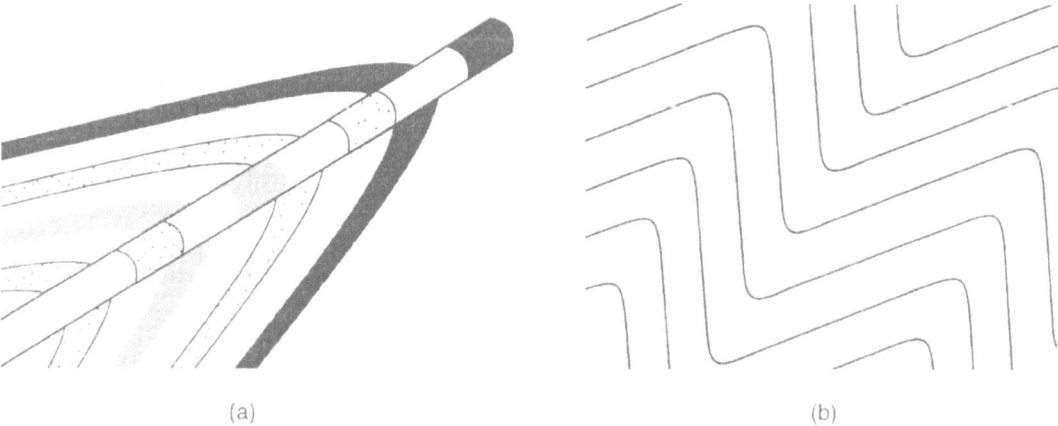

(a)   (b)

Figure 5.41 Sketches showing differences in fold profiles. (**a**) Fold exhibiting dislocation and offset of its 'core' parallel to its axial plane. (**b**) Chevron profile.

(v) some aspect of the hinge area, e.g. disharmony or dislocation of one or both limbs near the hinge (Figure 5.41a);

(vi) some aspect of the fold core (Figure 5.41a);

(vii) some aspect of the fold profile (hinge, limbs), e.g. chevron (Figures 5.40b, 5.41b).

Any one of these aspects of the fold may differ significantly from the corresponding element in folds belonging to other sets, but the potential for a combination of two or more distinctive features in folds greatly enhances the individuality of fold sets. The potential number of permutations is such that the characteristic features of the folds of any particular set commonly serve to define them clearly and allow them to be separated unambiguously from the folds of other sets. In this respect see section 5.2 on 'Identification, recognition and classification of structures'.

Where individual characteristics are not sufficiently distinctive to be definitive, it has been seen that separation of two fold sets can sometimes be effected when the earlier set has been transected by intervening features such as tabular veins (Figure 4.3 in section 4.2), or affected by neosome development and agmatization (section 5.2.10) before the later set has developed (Hopgood, 1980, pp. 60–1). On the other hand, transection by such tabular features (veins, igneous sheets etc.) is less likely to provide an easily recognizable basis for distinction between two comparable sub-parallel sets of lineations or foliations (e.g. two sets of shear surfaces, or foliations defined by coplanar mineral growth such as mica). This is because discordance with structures that are only two-dimensional (foliations) or one-dimensional (lineations) is less likely to be observed than it would be with folds which are three-dimensional (Figure 5.42). Furthermore, the emplacement of later features such as veins is more likely to be controlled by the attitudes of the pre-existing structures, i.e. parallel rather than discordant, so that distinct time relationships between the two sets of structures and the tabular veins separating them is less likely than in the case of folds whose elements have different attitudes, one

or more of which is likely to be oblique to the intervening tabular feature.

Finally, it is worth noting that confusion between fold 'style' ('structure style' in general) and fold 'phase', 'generation' or 'set' has often been the cause of considerable difficulty in the interpretation of the structure in terranes affected by multiple deformation. Although the problem is recognized, and structural geologists studying highly deformed gneisses and migmatites are keenly aware of it and take care to guard against it (Hopgood, 1971b), the tendency remains for such confusion to persist and care should be taken to discriminate between the use of these words. Terms such as 'style groups', and 'fold styles' should not necessarily be regarded as synonymous with fold sets.

Such confusion commonly leads observers to conclude that different fold sets are indistinguishable because they have formed in response to a 'single' deformation or progres-

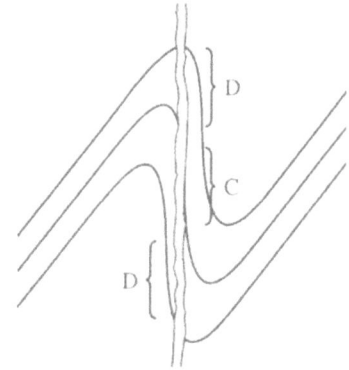

Figure 5.42 Sketch of the relationships between a fold and a later vein showing that it is possible to demonstrate discordance between them somewhere on the structure. Although the vein is concordant to the foliation at some places (at **C**), in others it is discordant (**D**). Therefore superimposition of a vein is likely to be demonstrable somewhere on a fold, whereas this would not necessarily be so in the case of later veins and planar or linear structures. See also Figures 4.8a, 4.8b.

sive deformational episode (Figures 2.6, 5.1 and 8.13). The manner in which these sets formed does not affect their identity, which is based **solely** on the criterion of **overprinting**.

If one fold (structure) overprints (i.e. has deformed) another then it must be later and, by virtue of its being recognized, it must belong to a different set.

6.1 PRINCIPLES AND BASIS OF CORRELATION

6.1.1 INTRODUCTION

The role of correlation in building up a structural succession is much the same as it is in the establishment of a stratigraphical succession in lithostratigraphy.

As stated by Whittaker *et al.*, 1991 (the emphasis has been added here), 'The fundamental procedure in stratigraphy is to **observe**, **describe** and **correlate** rock successions, correlation being the demonstration of **correspondence between geological units** in some **defined property** and in **relative stratigraphical position**. When investigating an area, the first procedure is to describe the local stratigraphical succession **objectively** in terms of mappable or traceable lithostratigraphical units present. Currently or subsequently the local succession should be correlated, as far as possible, with the international standard stratigraphical scale or with the regional stratigraphical divisions. With most Phanerozoic rocks this correlation is most commonly achieved using biostratigraphical methods.'

Two factors are fundamental to the process of correlating (fold) structures. As has already been seen (section 5.2), in order to correlate it is necessary to identify, or recognize individual folds and also to establish refolded, or overprinted relationships. Correlation itself is of two kinds and entails two distinct stages. Firstly there is correlation between **individual** structures and secondly correlation between **groups** of structures. Furthermore, there may be a wide range in the distances separating the structures to be correlated. Correlation therefore may be quite local, within a single exposure or between adjacent exposures, or it may be between widely spaced localities or areas separated by considerable distances.

The process of correlation then is one in which four stages can be recognized.

Stage 1. Establishment locally, in the first instance, of overprinted relationships, i.e. determination of the partial structural succession at a particular exposure or locality.

The process begins with the establishment on the same part of the exposure, of refolded (overprinted) relationships between fold structures (Figures 6.1, 6.2). See section 3.1.1. This process depends purely on observing the effect of one fold on another but does not depend on identifying the structures. This provides the local **partial succession** at the particular exposure or locality and the establishment of this local partial succession forms the basis for the next three stages.

Stage 2. **Recognition** (i.e. establishment of the identity) of structures belonging to particular sets produced by particular deformation 'phases'. The identity of a structure provides the basis for the next stage, viz. correlation (sections 5.2.6 and 5.3).

Stage 3. Correlation between individual folds of the specific sets so recognized. Such correlation is considerably strengthened (reinforced) where two or more structures exist consistently in the same order of succession

Figure 6.1 Overprinted relationships between two sets of folds in two dimensions. Round-hinged folds overprint (are superimposed on, or refold) isoclinally folded foliation. An intrafolial isoclinal fold hinge (at the pencil), together with its axial planar trace (F_i) and the enclosing dark layer can be seen folded around the hinge of the round-hinged fold (F_r) whose axial planar trace is parallel to the pencil. The local partial succession is F_i followed by F_r. Svecofennian migmatite, Jussarö area, southern Finland.

from one locality or exposure to another. See also the discussion in section 5.2.9, 'Certainty of identification of isolated structures – reality of the 'succession' or 'coincidence'.

Stage 4. Correlation from place to place of the ordered groups of fold sets comprising the partial successions. This enables them to be integrated to build up a 'total', or 'complete' (known) succession for the terrane under investigation (see discussions on correlation in section 12.5).

The basis for correlation of structures between separate exposures is thus comparable to that employed in stratigraphical correlation and the degree of certainty of any such correlation increases with the number of corresponding parameters to be found at each locality.

In correlating the structure from one exposure (or from one area) to another, while a greater number of structure sets strengthens the correlation, it is not so much the number of structural sets (any more than it is the number of lithological units in stratigraphical correlation) that is so important, but the arrangement and distinctive features of the sets. The critical factors are (i) the order of succession of structure sets and (ii) the pres-

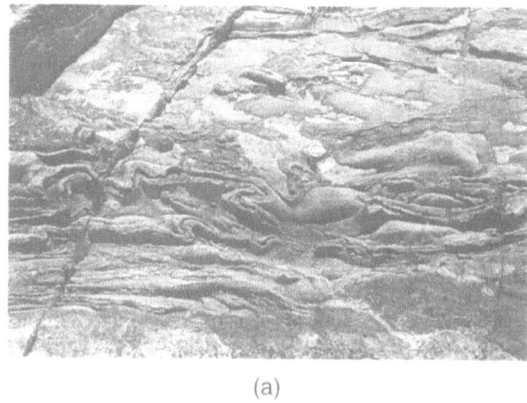

(a)

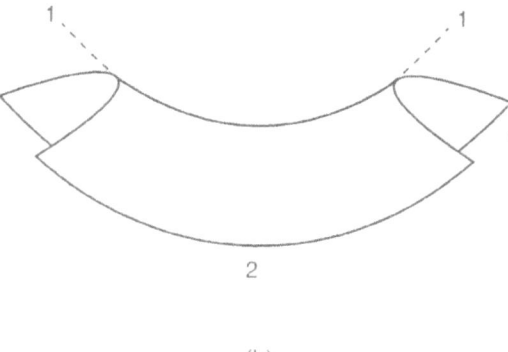

(b)

Figure 6.2 (a) Overprinted relationships between two sets of folds in three dimensions. Below the compass at the centre of the photograph the 'E–W'-trending hinges of isoclinal folds (F_i) approximately parallel to the exposure are sharply curved into a saddle form by later, 'northerly'-trending open folds (F_o) causing them to 'plunge' in and out of the exposure. (b) Sketch of a 'saddle' fold like that in (a). The local partial succession is F_I followed by F_o. Svecofennian migmatite, Jussarö area, southern Finland.

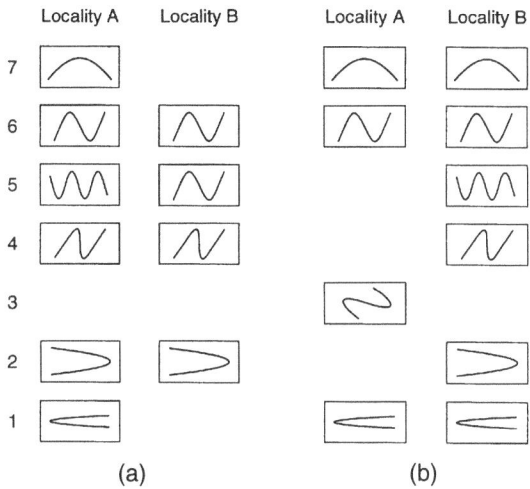

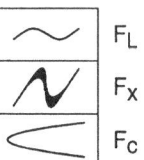

Figure 6.4 Fold profiles representing three sets of structures, each in a specific relative position in the succession and with its own stylistic features. F_x is followed by F_L and preceded by F_c.

Figure 6.3 Two examples of correlation based on comparison between sets of structures at two localities, A and B. In case (**a**) there are four sets (2, 4, 5 and 6) common to both localities and in (**b**) only three sets (1, 6 and 7) are common to both localities. In both (**a**) and (**b**) the sets are in the same order of succession.

ence within many or all sets, of identifiable characteristic features. For example, suppose that a 'total' structural succession comprises 7 sets and 5 of these sets are recognized at two localities A and B. At A the 5 sets present might be numbers 1, 2, 4, 5, 7 whereas at B, the 5 sets represented might be 2, 4, 5, 6, 7. The correlation is then based on the order of succession of those sets recognized at each locality (Figure 6.3a), as it is in lithostratigraphical correlation, and as indeed it would be if at A there were only 4 sets and at B there were say 6 sets (Figure 6.3b).

The recognition of structures so as to be able to assign them to a particular set is based essentially on style (see later section) and is confirmed in terms of their overprinting (refolding) relationships with structures belonging to earlier and later sets.

All structures of a particular set are then correlated between localities on the following bases. Firstly the structures of this set have a group of characteristics, overall constituting their particular style (section 6.2.1), which are often distinctive and, in terms of the terrane being considered, commonly unique. Secondly, this set of structures (F_x, say) lies in a specific position in the succession followed by and preceded by other sets of structures, each with its own particular style. Figure 6.4 shows this in the simple case where there are only three sets.

6.1.2 SPATIAL RELATIONSHIPS: THEIR EFFECT ON THE CERTAINTY OF CORRELATION

It is not difficult to appreciate the uncertainty which arises when correlation (particularly of single structures) is attempted between isolated exposures, perhaps remote from each other. But suppose we think about this further by considering a series of circumstances where refolded structures of different scales are compared over different distances with different amounts of exposure. One can see that correlation is not a straightforward matter of feasibility, i.e. it is not a clear-cut distinction between whether it is possible or not possible, but one of differing degrees of certainty, or confidence.

Initially, consider a single small exposure a few metres across showing relationships between the three overprinted folds, F_x, F_y and F_z on two adjacent parts of the same exposure (Figure 6.5). In this instance correlation of the three structures (F_x, F_y and F_z) between the two

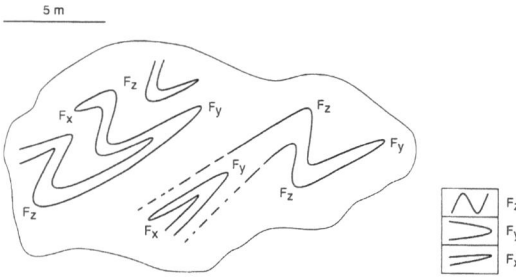

◀ **Figure 6.5** Single exposure showing refolding between three fold sets, F_x, F_y and F_z. These overprinted relationships (inset) would be accepted with little or no hesitation by most observers.

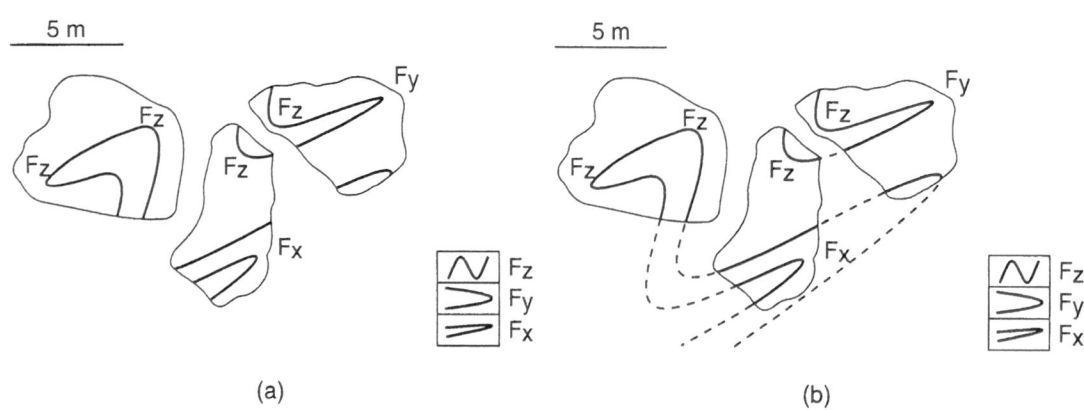

(a)   (b)

Figure 6.6   (a) Structures comparable to those of Figure 6.5 in three separate exposures a metre or two apart. (b) As (a) but showing continuity of structural trend between the exposures.

groups would be accepted with confidence by most observers.

Consider next a comparable structure exposed in three separate places spread over a distance of only a few metres, but with little or no exposure in between (Figure 6.6).

In this case, correlation between F_x, F_y and F_z is likely to be accepted by most observers but its general acceptance might be less likely, with the possibility of some reservation on the part of some observers.

Consider now a third case, where structure comparable to that of Figure 6.6 is exposed in three completely isolated places (Figure 6.7). Here the degree of certainty of correlation would be much less in the minds of some observers.

If, in the previous three cases, the same individual structures were exposed, not on outcrops a few metres across but say 100 m, or 1000 m across, on widely separated islands (Figure 6.8) then the degree of certainty would be correspondingly reduced, with fewer observers willing to accept the correlation.

The chances of accepting the correlation would be further reduced if the outcrops included only a few of the total number of structures observed at each exposure shown in Figures 6.5–6.8.

Consider now another situation. This case, like those just discussed, is also one that is influenced by the **scale** of observation and poses a closely-analogous philosophical question. This question is one that is raised time

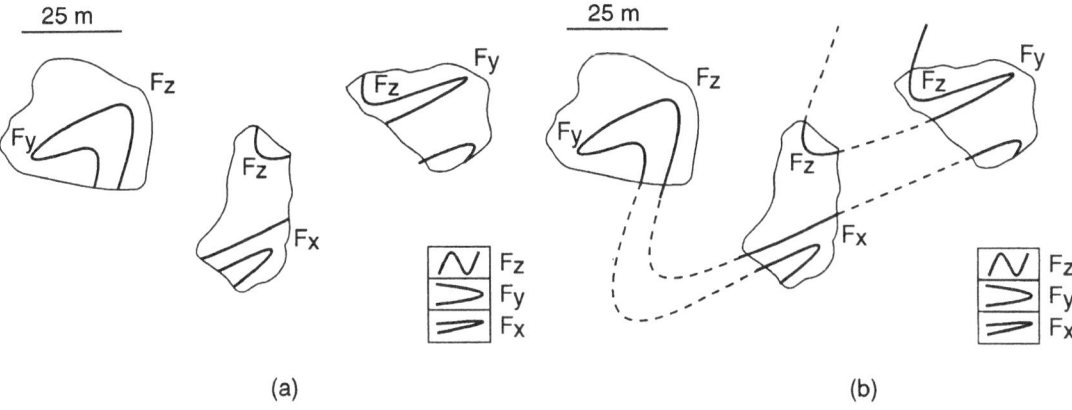

Figure 6.7 (a) Structures comparable to those of Figure 6.6 exposed in three isolated outcrops that are widely spaced. (b) As (a) but showing continuity of structural trend between the exposures.

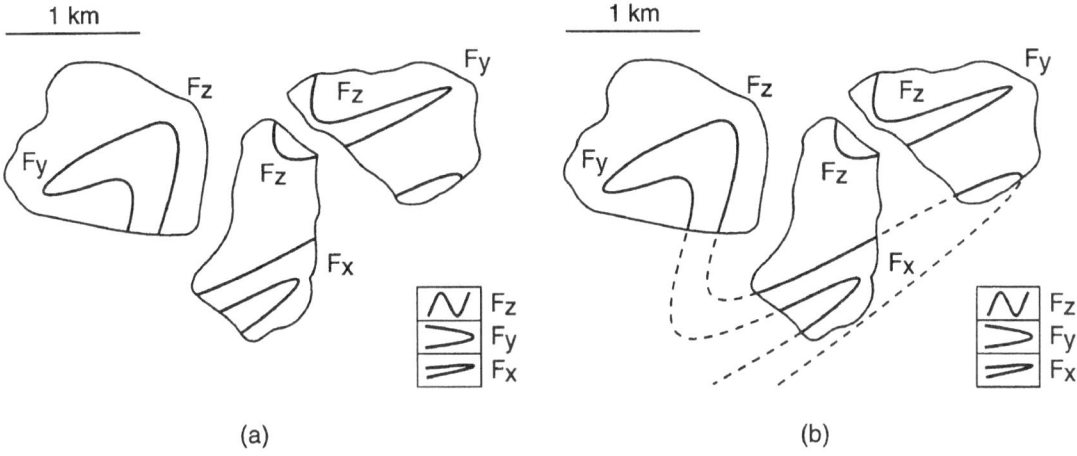

Figure 6.8 (a) Structures comparable to those in Figure 6.6 exposed on three isolated islands, widely separated from each other. (b) As (a) but showing continuity of structural trend between the islands.

and again, particularly by those who are not structural geologists.

Where a fold set (F, say), or other structure set, has been identified as part of the structural succession at a particular locality but is not seen at every exposure at that locality (in other words there is **apparently** no evidence for F at some exposures, although there could be very small-scale, and/or microscopic effects of F that are not immediately obvious), a question often asked is, 'Why is the foliation in these exposures not affected by F if indeed F exists throughout the locality?' (Figure 6.9a). In contrast, where an exposure at the same locality shows the whole of a fold F, including the hinge and limbs, and the foliation on the limbs of that fold is not (apparently) affected by F hinges, this lack of folds in the limbs is seldom (if ever) questioned (Figure 6.9b). Nevertheless the absence of hinges in the foliation of the limbs of the fold (Figure 6.9b) is exactly comparable to the absence of F hinges in the foliation of Figure 6.9a. For some reason the first condition is less readily accepted than the second, although there is no fundamental difference between the two. Why this is so is not entirely clear and perhaps can be explained in terms of the discussion relating to the previous figures (Figures 6.6–6.8). On the other hand perhaps the second case is not questioned simply because the observer, having seen many folds with apparently 'undeformed' limbs, is already familiar with the concept and unconsciously accepts the case. But with the expectation that the foliation should be affected by the F folds which are known to exist at the locality, the obvious absence of any such effect in those exposures which show only apparently unde-formed foliation immediately triggers a ques-tion in the mind of the observer (viz., 'Why is there no F fold?') because the observer is looking for just that effect. That question is answered unconsciously before it arises in those ex-posures where the presence of the hinge of the complete F fold confirms the expectation, so that the absence of hinges on the fold limbs therefore goes unquestioned. What is being observed here of course is the

effect of strain, or deformation partitioning (section 8.1.9).

The certainty of correlation in each of the above cases would be increased where the structures exhibited increasing numbers of characteristics such as associated mineral growth and unusual geometric features. The greater the number of these features then the greater would be the degree of certainty of the correlation.

The acceptance, rather than the converse, of such correlation of these structures in terms of style in the broad sense can be explained simply because it is more reasonable to do so than not to do so. If it were not accepted the alternative would be to assume an unreason-ably high degree of coincidence to account for the matching of such comparable but unre-lated structures.

Consider the following very simple example of structural correlation between two isolated localities. Among other folds, there are at each locality, folds of style 'A' with upright axial planes having the same axial trend and which have been correlated with each other.

Correlation between the two structures is soundly based, not simply on stylistic grounds (i.e. because of the similar styles of 'A' and their similar attitude and trend) but also on the

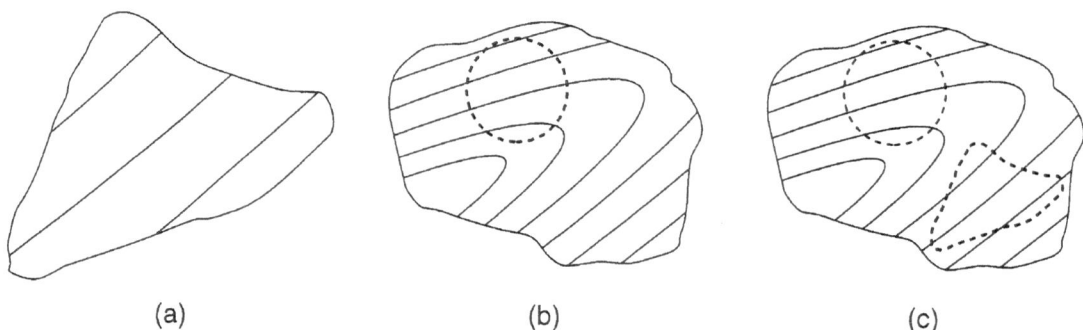

(a)   (b)   (c)

Figure 6.9 (a) Exposure showing foliation not visibly affected by hinges of the fold set **F** although this set has been identified at the locality. (b) Entirely exposed fold of set **F** showing that the foliation of the fold limbs is not visibly affected by **F** hinges. The absence of obvious **F** folds in the limbs tends to be accepted without question, although the unaffected foliation (such as that of the circled area) is exactly comparable to that of the exposure shown in (a) where the absence of **F** folds is often not so readily accepted. (c) Cases (a) and (b) compared with the area of (a) outlined.

Locality I

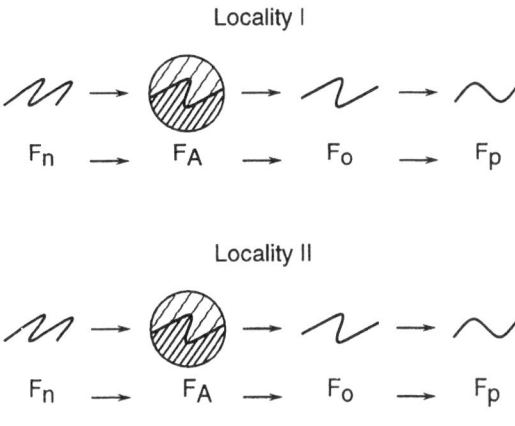

$F_n \longrightarrow F_A \longrightarrow F_o \longrightarrow F_p$

Locality II

$F_n \longrightarrow F_A \longrightarrow F_o \longrightarrow F_p$

Figure 6.10 Overprinted relationships of three folds (F_n, F_o and F_p) to the fold (structure) F_A at two localities, I and II.

basis of their relationship both to pre-'A' and to post-'A' folds (i.e. folds or structures in general) with, at both localities, pre-'A' structures having in common their own stylistic features and likewise post-'A' structures having their own particular style (Figure 6.10).

The possibility that F_A of Locality I and F_A of Locality II could be different fold sets which happen to be stylistically similar (even identical), as well as with a comparable axial planar trend, is (perhaps) not unlikely. However, the fact that at both localities these F_A structures refold other folds (F_n), in each case having characteristics different from those of F_A, renders the possibility of **coincidental** similarity between two different F_A sets less likely.

If at both localities F_A are also refolded (overprinted) in their turn by F_o folds, in each case with a similar style, then the likelihood that the two groups of F_A structures are different, but are coincidentally similar, becomes remote. Additional relationships of this kind (i.e. the even later refolding by F_p etc.) make assumptions involving coincidental similarity between the two F_A structures increasingly remote, in fact unreasonable and unwarranted.

Hence the strength of the correlation between say any three structures, F_x, F_y and F_z,

is increased (i) by the fact that there are not only common stylistic (in the widest sense) features to identify folds of a given set from one locality to another but also (ii) by the fact that these folds (a) deform other folds having features characteristic of the preceding set and (b) are deformed by yet other folds having characteristic features of a succeeding set. In other words the chance of coincidental mismatching between folds belonging to different sets is very remote, because all the characteristic features of F_x overprint all of the characteristic features of F_y which in turn overprint all of the characteristic features of F_z, so that this is not simply a case of a relationship between three folds viz. a single fold (F_y) folding and being folded by the folds of two other sets, (F_z and F_x), but one of overprinting relationships between several combinations of features viz. all the characteristic features (mineralogical association, geometry etc.) in specific, characteristic combinations of the three folds. This is important and is one of the fundamental strengths of structural correlation. The methods and principles embodied in such correlation and their application have been discussed in detail in Hopgood (1980). If necessary refer again to section 5.2.9.

6.1.3 STRUCTURAL CORRELATION OF TECTONOSTRATIGRAPHIC TERRANES

The principles involved in using complex structural successions in correlation are particularly apposite in the correlation of tectonostratigraphic terranes separated or juxtaposed along major dislocations or as a result of thrusting (Hopgood, 1971b (p. 374), Hopgood, 1973 (pp. 49, 50); Hopgood and Bowes, 1995; Hopgood *et al.*, 1995). This is because the conditions during the evolution of terranes, particularly those with a long history, favour the development of complex structural relationships.

Correlation between terranes comprising widely separated blocks of a dismembered crustal segment can be achieved in a number of ways: palaeontological, palaeomagnetic, isotopic dating, and by geochemical comparison. The use of structure as a basis may be less certain in some cases, particularly if the rocks have not been subjected to extensive deformation, but in highly deformed terranes the structure or structural pattern in the rocks can provide a powerful means of relating formerly contiguous crustal blocks. This is especially so if, as is the case in rocks such as migmatites, the structure is very complex, thereby resulting in a highly individual pattern capable of unambiguous recognition even in widely separated segments, and even when these have been rotated during and/or after separation. It also assumes greater importance where the rocks are unfossiliferous and where palaeomagnetic, isotopic and geochemical criteria are insufficiently discriminatory.

In order to gain some insight into the degree of structural complexity that could stem from a deformational history embodying structural overprinting, before, during and after fragmentation and separation of tectonostratigraphic terranes, the following hypothetical examples are used to illustrate the potential for wide variation in structural successions. They show what could be entailed in the correlation of separated terrane segments.

Consider the case of nine 'sub-terranes' derived from three terranes, I, II and III which were formerly contiguous parts of a single crustal segment Sigma (Σ) affected by both dislocation and penetrative deformation. The development of Sigma from its original undeformed state will be followed through to the final observed state where it has evolved into the nine sub-terranes. For the sake of clarity, each set of penetrative structures developed will be represented initially by a separate ornament on the figures depicting stages in the structural modification of the crustal segment and its derivative terranes. The ornament could be used to depict any tectonic structure (of deformational, igneous, meta-

morphic or ultrametamorphic origin) but in this case it is considered simply to represent deformational structures. Later in the discussion the ornament will be replaced by fold symbols. The original crustal segment is regarded as having been affected by dislocation **events** (C, D, F, H, K, N, P) causing **separation** of segments, and by penetrative deformational **episodes** (A, B, E, G, J, L, M, O, Q) resulting in **structural imprints** on the segments represented here by a different ornament in each case.

The first recorded penetrative deformations affecting Sigma (Σ) produced the structural patterns shown in Figure 6.11 in response to episodes A and B.

Subsequently, in response to event C, Σ was divided into two, Σ_1 and Σ_2 (Figure 6.12a) and later in response to event D, Σ_2 was itself divided into two, Σ_2 and Σ_3 (Figure 6.12b). This threefold subdivision of Σ gave rise to the three 'sub-terranes' (Σ_1, Σ_2 and Σ_3) which from now on are referred to as I, II and III.

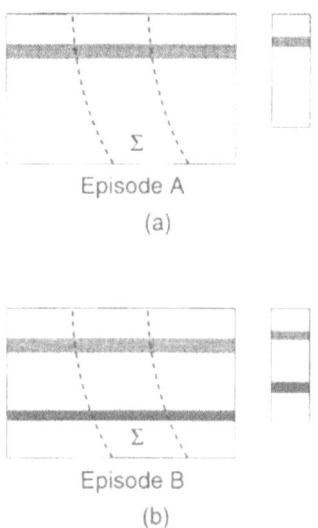

Figure 6.11 Effects of deformational episodes A and B on crustal segment Σ. (**a**) Structure imposed during episode A. (**b**) Structural imprint during episode B. The curved dashed lines represent potential dislocation surfaces.

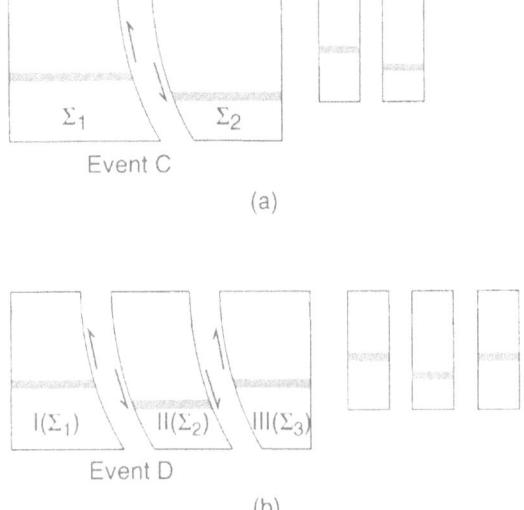

Event C

(a)

Event D

(b)

Figure 6.12 Subdivision of crustal segment Σ by events C and D. (**a**) Division of crustal segment Σ into sub-terranes Σ_1 and Σ_2 by event C. (**b**) Division of sub-terrane Σ_2 into Σ_2 and Σ_3 by event D. For simplicity Σ_1, Σ_2 and Σ_3 are denoted I, II and III respectively as shown in the figure, and henceforth are referred to by these symbols.

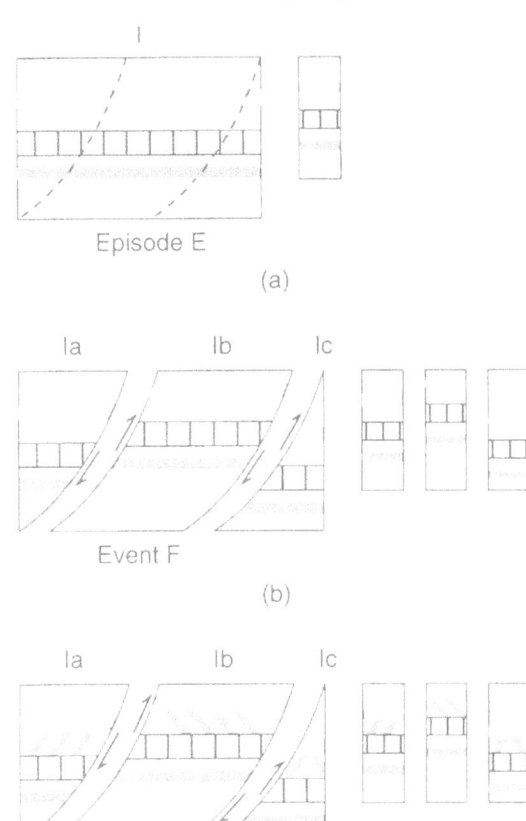

Episode E

(a)

Event F

(b)

Episodes G_a, G_b and G_c

(c)

Figure 6.13 The effects on terrane I of episode E, event F and episodes G_a, G_b and G_c. (**a**) Episode E. (**b**) Event F. (**c**) Episodes G_a, G_b and G_c.

The structural development of the three terranes, I, II and III thereafter proceeded independently.

During episode E the structural pattern of terrane I was overprinted by the structure shown in Figure 6.13a and during event F terrane I was split into three widely separated sub-terranes, Ia, Ib and Ic (Figure 6.13b). Subsequently these were affected independently by deformational episodes Ga in Ia, Gb in Ib and Gc in Ic, to produce the structures shown in Figure 6.13c which overprinted the pre-existing patterns.

In response to event H, terrane II (Figure 6.12b) also split into three sub-terranes IIa, IIb and IIc which were separated slightly by lateral faulting (Figure 6.14a) and subsequently, during deformational episode J, all three sub-terranes were overprinted by the structure shown in Figure 6.14b.

Extensive lateral translation and separation of IIa, IIb and IIc during event K (Figure 6.14c) preceded deformational episodes L_a, L_b and L_c which affected the three sub-terranes independently producing the structures shown in Figure 6.14d.

The structure of terrane III (Figure 6.12b) was overprinted during deformational episode M by the structure shown in Figure 6.15a and the terrane was split into two widely separated sub-terranes, IIIa and IIIb by event N (Figure 6.15b).

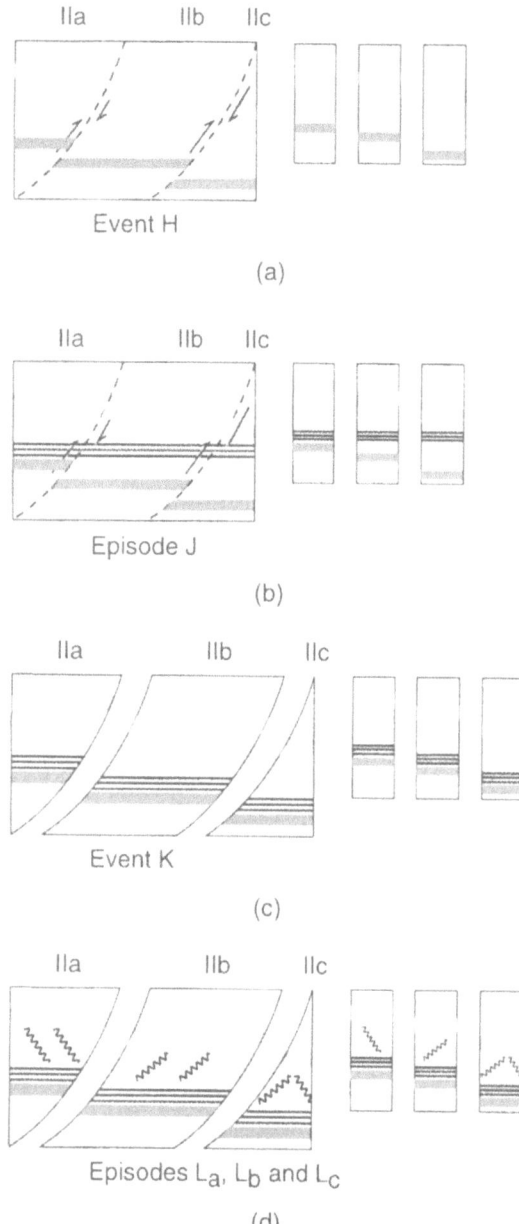

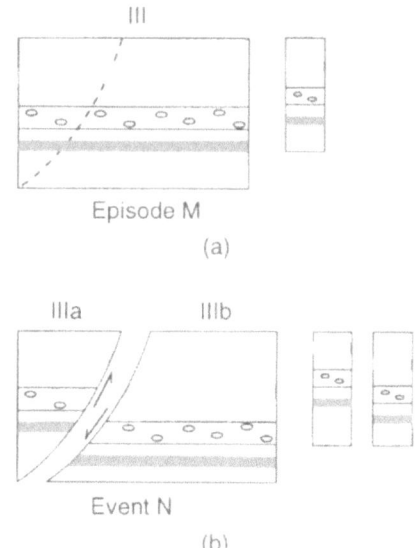

Figure 6.15 The effects on terrane III of episode M and event N. (**a**) Episode M. (**b**) Event N.

During deformational episodes O_a and O_b which affected the sub-terranes independently, the structures shown in Figure 6.16a were superimposed on sub-terranes IIIa and IIIb. As a result of a later event P, IIIb was split into two, IIIb and IIIc, by lateral faulting and considerably offset so that there were now three separate sub-terranes, IIIa, IIIb and IIIc (Figure 6.16b).

Subsequently, the structural patterns of each of the three sub-terranes was modified independently by overprinting during deformational episodes Q_a, Q_b and Q_c, to produce structures in IIIa, in IIIb, and in IIIc as shown in Figure 6.17.

In terms of 'stratigraphical' columns showing the relationships between the ornament representing the structures, the successions observed for each of the nine sub-terranes, Ia-c, IIa-c and IIIa-c, is as shown in Figure 6.18.

Structural successions for the nine sub-terranes could be something like those shown in the columns of Figure 6.19(a)–6.19(i) where the ornament of Figure 6.18 is replaced by fold

Figure 6.14 The effects on terrane II of event H, episode J, event K and episodes L_a, L_b and L_c. (**a**) Event H. (**b**) Episode J. (**c**) Event K. (**d**) Episodes L_a, L_b and L_c.

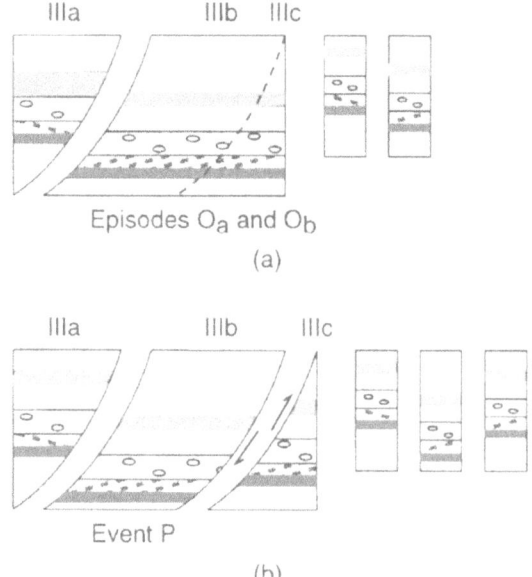

Figure 6.16 The effects on sub-terranes IIIa, IIIb and IIIc of episodes O_a and O_b and event P. (a) Episodes O_a and O_b. (b) Event P.

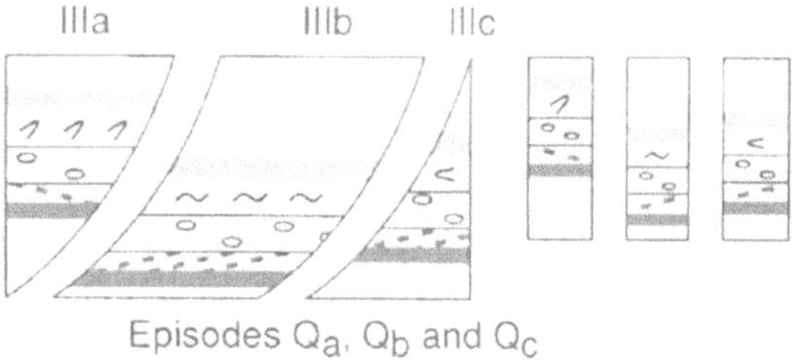

Figure 6.17 The effects on IIIa, IIIb, and IIIc of episodes Q_a, Q_b, and Q_c.

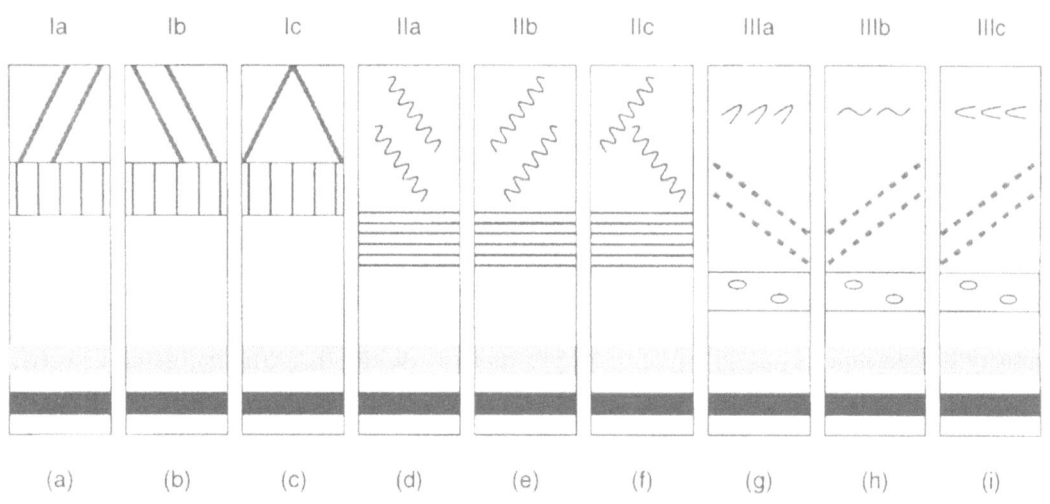

Figure 6.18a-i Structure successions Ia, Ib, Ic, IIa, IIb, IIc, IIIa, IIIb, IIIc resulting from events C, D, F, H, K, N, P and episodes A, B, E, G, J, L, M, O, Q.

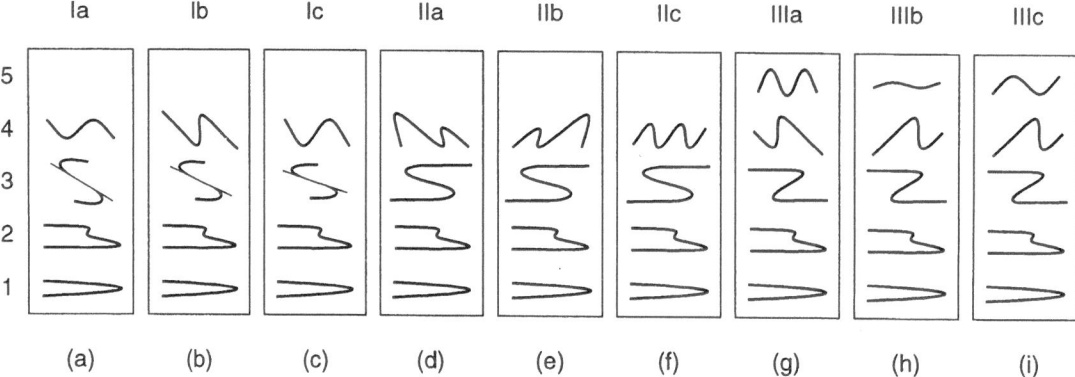

Figure 6.19a-i Structure successions in the nine sub-terranes (Ia, Ib, Ic, IIa, IIb, IIc, IIIa, IIIb, IIIc) evolved from the original crustal segment Σ.

Figure 6.20 Structure sets common to all nine sub-terranes (Ia, Ib, Ic, IIa, IIb, IIc, IIIa, IIIb, IIIc).

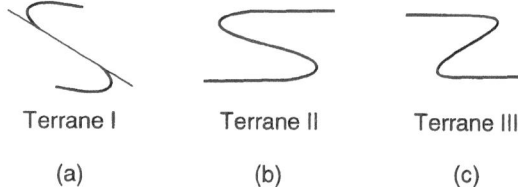

Terrane I　　　Terrane II　　　Terrane III

(a)　　　　　(b)　　　　　(c)

Figure 6.21 Structure sets confined to terranes I**a**, II**b** and III**c** and therefore diagnostic of each terrane.

symbols. The first sets of structures (Figure 6.20), shown at the bottom of the columns of Figure 6.18 and Figure 6.19, are common to all the sub-terranes, having been imposed on the original crustal segment Sigma (Σ) prior to dislocation during event C.

Each of the three sets of structures shown in Figure 6.21(a), (b), (c) is common to **one** of the three earliest-formed terranes (I, II and III), is confined to it and is unique to that terrane and characteristic of it, i.e. its presence is **diagnostic** of that terrane.

Deformation in all the terranes and sub-terranes took place either together or separately (either at the same or at different times) during the deformational history with the result that the time span over which all three terranes were deformed varied considerably. In some cases the deformation was contem-poraneous and in others it was spread over a considerable period of time. As a consequence of this the penetrative structures which have successively affected all three of the terranes (I, II and III) and corresponding sub-terranes (Ia-c, IIa-c or IIIa-c) at intervals throughout the deformational history can be grouped differently.

There are five such 'groups' in the nine successions, each group representing all the structures formed in the three terranes (I, II and III) within a particular interval between dislocation events. The five groups are shown as 'Observed successions' (numbered 1–5 on the left) at the top of the summary table of Figure 6.24.

The first and second groups each comprise one of two sets of structures that are repre-sented in all nine successions (Figure 6.19), their presence demonstrating the common origin and contiguity of the nine sub-terranes during the first two recorded penetrative deformational episodes (Figures 6.11a, 6.11b). These two sets are shown in the lower part of Figures 6.18, 6.19 and 6.24.

The third group includes three sets of structures, each confined to one terrane, I, II or III, and each one common to three corresponding sub-terranes, Ia-c, IIa-c or IIIa-c (Figure 6.21). The commonality of the sets demonstrates the contiguity of the component sub-terranes at the time when the sets were developed in terranes I, II and III (not necessarily the **same** time for all or any of the terranes). In other words, within each of the three terranes there was a common **penetrative** deformational history up to, and including, the development of the third set (during episodes E in terrane I, J in II and M in III), although their dislocation histories differed. Thereafter there were no structures common within terranes I, II and III, deformation being different in each sub-terrane. The relationships within this third group are shown (between events D and K) in Figure 6.24.

The fourth group of structures comprises eight sets formed during episodes G_{abc}, L_{abc} and O_{ab} (Figure 6.24). These are, one set in (and unique to) each of the seven sub-terranes Ia-c, IIa-c and IIIa and one set common to sub-terranes IIIb and IIIc (Figure 6.22). This indicates that while seven sub-terranes (Ia-c, IIa-c and IIIa-c) had by then separated, IIIb and IIIc were still contiguous at that time. Note once again that these eight sets are not necessarily contemporaneous.

The fifth of these groups of structures comprises three sets only, one set in each of the successions of sub-terranes IIIa, IIIb and IIIc, and these were imprinted only after the three sub-terranes came into existence as separate crustal blocks with the separation of IIIb and IIIc during Event P (Figures 6.16b and 6.24, upper right).

As well as the structure set formed during Episode O_a which is unique to sub-terrane IIIa, the latest nine sets of structures are unique to each of one of the nine sub-terranes (Figures 6.23(a)–6.23(i) and 6.24, upper). All but one of the sub-terranes therefore have a single structure set which is unique and diagnostic. Sub-terrane IIIa differs in having two diagnostic structures.

To summarize from the point of view of the geologist observing the structural evidence, the structures in the nine successions can be assigned to five groups depending on whether or not they are expressed in all or any sub-terrane. Since it is the objective of the geologist to interpret the structural patterns observed, the characteristics of each group (which depend on the distribution of the structure sets within the observed successions which in turn represent nine sub-terranes) are important

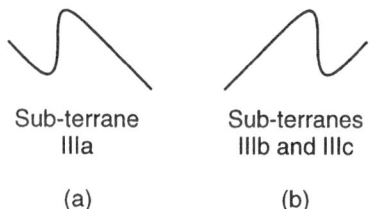

Sub-terrane   Sub-terranes
IIIa   IIIb and IIIc

(a)   (b)

Figure 6.22 (a) Structure unique to sub-terrane IIIa. (b) Structure common to sub-terranes IIIb and IIIc.

(a)   (b)   (c)   (d)   (e)   (f)   (g)   (h)   (i)

Figure 6.23 Diagnostic structures a-i. Each is unique to one of the nine sub-terranes, Ia-c, IIa-c and IIIa-c.

because they provide critical evidence relating to the deformational history of the terranes. See Figure 6.24.

Groups one and two

The presence in every one of the sub-terranes of the structure sets represented by groups one and two in the successions demonstrates that all three terranes were contiguous (or not significantly separated) up to and including the time of the second deformational episode, i.e. during episodes A and B (Figure 6.11). Up to this time fragmentation of Σ had not yet begun.

Group three

In the third group three structure sets (the results of episodes E, J and M) are represented in different successions of the sub-terranes, one set in the successions of sub-terrane I, one in the successions of II and one in the successions of III (Figures 6.13, 6.14 and 6.15). The nature of the distribution of the sets indicates that by the time of deformational episode E there existed at least two distinct sub-terranes, one (which includes Ia-c) overprinted by episode E (Figure 6.13a) and the other (which includes IIa-c and IIIa-c) where episode E structures are absent (see Figure 6.12b). Furthermore, because terrane III (like terrane I) lacks the structures of episode J (Figure 6.14b), it must have been separated from terrane II when episode J took place. For the same reason terrane II (like terrane I) lacks the structures resulting from episode M (Figure 6.15a) which overprint terrane III. The separation of terranes II and III took place during event D, as shown in Figure 6.12b.

Group four

The fourth group (Figure 6.18) comprises eight different sets, one in each of seven of the nine successions, with the remaining set common to two successions (in terrane III).

This distribution shows that terranes I and II had each split into three sub-terranes which had wholly separated from one another (Figures 6.13 and 6.14). Because terrane III was affected by only two structure sets (sub-terrane IIIa shows structure resulting from episode O_a, while both sub-terranes IIIb and IIIc show the effect of overprinting during episode O_b) it must have comprised only two sub-terranes (Figure 6.15). The structure sets of this group are different in each of the other six sub-terranes, and result from episodes G_{abc} and L_{abc} (Figures 6.18, 6.19 and 6.24).

Group five

The structures of the fifth group comprise three sets, one in each of the sub-terranes IIIa, IIIb and IIIc, stemming from deformation during episodes Q_a, Q_b and Q_c (Figure 6.17) and showing that by this time sub-terrane IIIc had split off from IIIb, during dislocation event P (Figure 6.16b). The three sub-terranes (IIIa-c) existed as distinct and separate entities, so that the subdivision of the original segment (Σ) into the observed nine sub-terranes was then complete.

Although each structural succession contains sets of structures that may be common to one or more of the other successions, the **total** structural succession, in terms of (i) the **number**, (ii) the **type** of structures, and (iii) the **order** of those structures for each terrane (or sub-terrane) is unique, distinctive and characteristic of that particular terrane or sub-terrane.

Therefore, in establishing the structural succession for a particular terrane it is theoretically possible not only to correlate that terrane in terms of its penetrative deformational history with a crustal segment from which it has become detached, but also to trace the intermediate stages of its deformational history in terms of its separation from other crustal blocks.

The translational events and penetrative deformational episodes contributing to the structural history of the crustal blocks com-

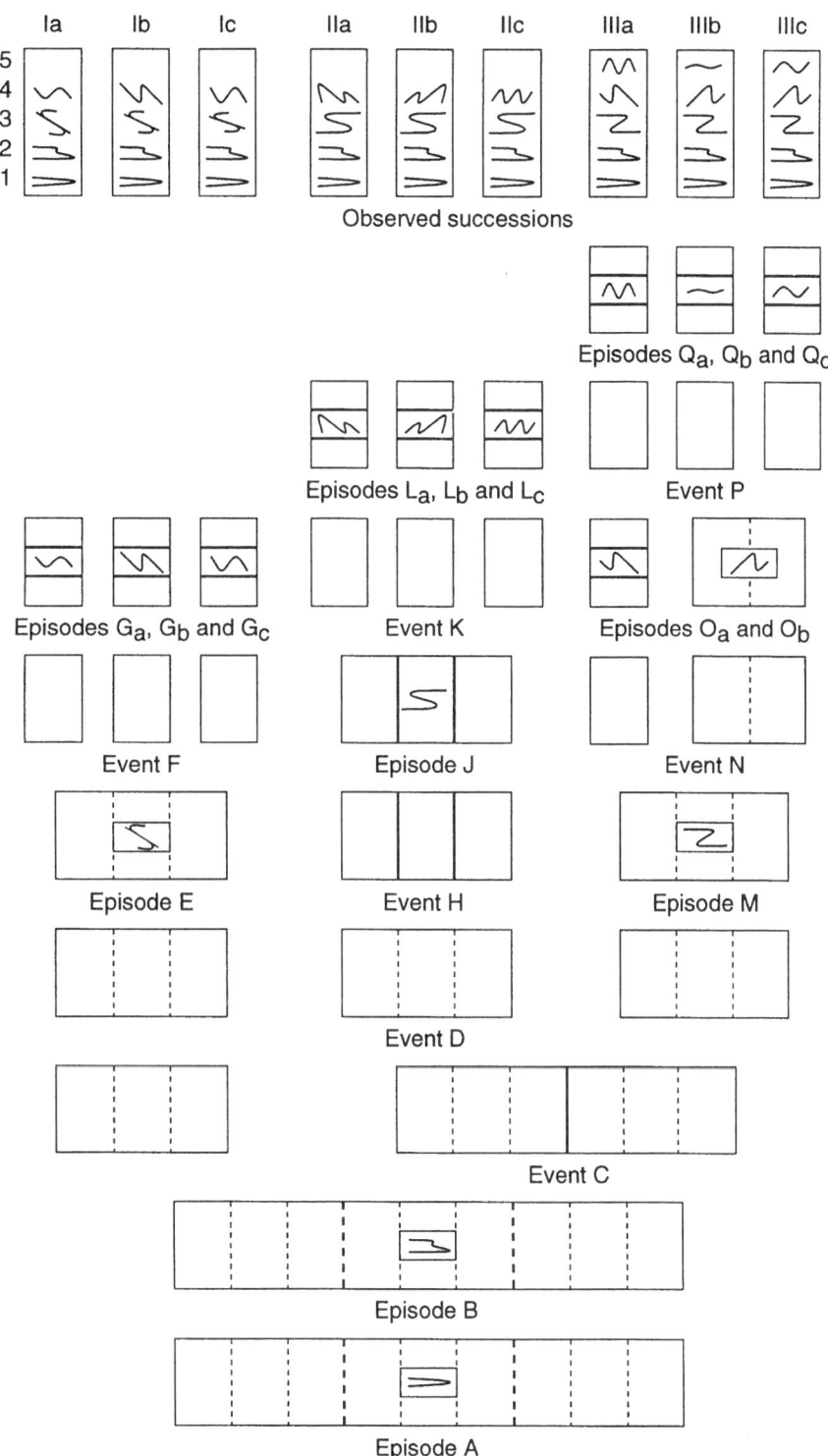

Figure 6.24 Structural development of sub-terranes Ia-IIIc from a hypothetical original crustal segment Σ.

prising the nine sub-terranes are summarized in Figure 6.24.

Examination of the structure of the nine sub-terranes and analysis of the structural successions shows that it is possible for the observer to determine the deformational history of these crustal blocks in some detail. Having first established the existence of a series of discrete crustal blocks (on the basis of the separate structural successions) and secondly, having found (on the basis of the different development of penetrative structures in different terranes) that the spatial relationships between the blocks were different at different times in their history, the observer is then able to decide what further study is necessary in order to enhance the quality of the data obtained so far. For example, if it is considered to be essential to the study, the observer is in a position to consider the selection of sites for palaeomagnetic sampling, within each of the sub-terranes identified, in order to gain some idea of the nature and magnitude of the displacement between the sub-terranes.

An indication of the structural complexity that can arise from this type of tectonic history is illustrated by the structure of the migmatites of the Cape Leeuwin–Cape Naturaliste migmatites, Southwest Australia (Hopgood and Bowes, 1995). The structural succession shows that there has been substantial translation during their developmental history and that this was both vertical and horizontal.

6.2   STYLE, ITS SIGNIFICANCE AND FACTORS INFLUENCING IT

6.2.1   THE CONCEPT OF STYLE

In the discussion on **correlation** in the preceding sections the collective importance of the characteristic features, or **style**, of struc-

tures was emphasized, as was its significance in determining their **identity** and hence their **recognition**, the essential basis for correlating structures.

There is probably more misunderstanding about the concept of style (tectonic style) than there is about any other term employed in structural geology. As stated by Turner and Weiss (pages 78, 79) in their *Structural Analysis of Metamorphic Tectonites* (1963), The term **tectonic style** introduced by Lugeon[1] (Figure 6.25) refers to the total character of a group of related mesoscopic structures which distinguish it from a group of comparable structures of another place or age, in the same way that the total character or style of a building or an art object can distinguish it from similar objects of other periods, places, or influences. Tectonic style is compounded of features which, though subtly recognizable to the experienced eye, cannot be defined precisely in geometric terms. It is properties of style rather than geometric properties that most clearly distinguish synchronous structures formed in one phase of deformation under approximately uniform physical conditions different from those of preceding or following phases of deformation.

In the present context, the word **style** when used in association with structures is taken to mean **all** features of the particular structure. It incorporates its 'total character' as referred to by Turner and Weiss (1963), i.e. it embodies the total number of attributes of the structure. It differs therefore from the more restricted definition used by Whitten (1966, p.37), and following Wegmann (1929), which appears to

[1] D.B. McIntyre ("Alpine tectonics and the study of ancient mountain chains", unpublished doctoral dissertation, University of Edinburgh, p.55, 1951) quotes an unpublished statement of Lugeon made in a letter to him (McIntyre) in 1948: "Le terme de 'style tectonique' je crois

bien qu'il est de moi. Il faudrait que je cherche dans tous mes écrits pour savoir quand je l'ai employé pour la première fois. Que cette expression me soit venue, c'est compréhensible car j'ai été le fils et le frère d'artistes soit des scupteurs." – *I think that you are probably correct in stating that it is I who introduced the term 'tectonic style'. I would have to search through all my publications to find out when I used it for the first time. That this expression should have come to my mind is understandable, for I am the son and brother of artistes and sculptors.* In fact the letter is dated 26th October, 1949 (Figure 6.25).

Lausanne le 26 Octobre 1949

MAURICE LUGEON
PROFESSEUR A L'UNIVERSITÉ

AVENUE CHARLES SECRÉTAN, 23
LAUSANNE
(SUISSE)

TÉLÉPHONE 2 94 04

Monsieur Donald B.Mc Intyre

Lect.r r of Geology
Institut of Geology
King bouildings
Weat Mains Road
Edinburgh 9

Mon cher Confrère

J'ai votre bonne lettre du i6 octobre.Des lors
je vous ai envoyé une série de tirages à part(en particulier le dernier
exemplaire de mon mémoire de I9oI) ainsi que deux cartes géologiques

Je dis toujours qu'il faut comprendre la géologie alpine avant
tout autre et je suis convaincu que vous allez ainsi trouver de belles
choses dans vos Highlands en pendant a ce que Wegmann a pu découvrir
dans le nord européen en appliquant les manières de voir et d'interprêter
les Alpins.Je ne dis pas cela par orgueil,car je sais par exemple ce qui
a été fait dans le nord de l'Ecosse et cela je le sais par ces deux
grands hommes que furent Paech et Horne

Vous me posez quelques questions: . . . . . .

. . . . . 6.- Le terme de"style tectonique",je crois bien qu'il est de moi
Il faudrait que je cherche dans tous mes écrits pour savoir quand je
l'ai employé pour la première fois.Que cette expression me soit venue,c'est
comprêhensible car j'ai été le fils et le frère d'artistes soit des
scupteurs . . . . .

Amitiés et poignée de mains.
L'arrière grand père en tectonique.

Mce Lugeon.

Figure 6.25 Facsimile of part of a letter dated 26 October 1949 from Maurice Lugeon written in reply to a letter from D. B. McIntyre in which a number of questions were asked relating to Alpine tectonics. Paragraph six is in response to a question relating to the term 'tectonic style'. Its translation is in the footnote on p.134. Reproduced by courtesy of D. B. McIntyre.

include only the profile and axial attitude of a fold.

Recognition of the fact that there is a difference between the two usages of the term 'style' is extremely important and is of considerable significance, especially where correlation of structures is concerned. The fact that the word is used in two very different ways (one limited, in folds for example, solely to profile plus axial planar attitude – assumption 5 in section 2.3) is certainly one of the contributory causes to misunderstanding relating to the interpretation of the structure of terranes affected by multiple deformation (See, for example, Hobbs, Means and Williams 1976, p.373).

The style of a fold includes the following attributes:

1. Its geometrical type (isoclinal, asymmetrical, open etc.), including the whole range of its expression (sections 5.2.2 and 5.3.1), hinge shape, relative limb lengths, thicknesses, etc.
2. Its relationship to contemporaneous planar structures (as well as their presence or absence) such as cleavage and foliation, including the whole range of their expression, depending on rock type.
3. Its relationship to syntectonic mineralogy (i.e. to metamorphic conditions), e.g. growth parallel to its axis, limbs, axial plane etc. over the whole range of their expression in different rock types.
4. Its relationship to contemporaneous linear structures, (elongate minerals, 'stretching' lineation, boudinage, mullions etc.) embracing the entire range of their expression.
5. Its attitude (orientation) with respect to earlier and later folds (especially those immediately pre- and post- the fold), particularly key structures (section 6.3).
6. Its refold or overprinting relationships with earlier and later structures (including – besides folds – cleavages, foliation, lineation, tabular intrusions, veining, agmatiza-

tion, anatexis, mineralization etc.), and particularly with **key**, or datum structures (section 6.3).

Here **all** the stylistic features of a fold (structure), when taken together, are considered to comprise its overall character, or its identity. In some cases at least, this character may be unique and thus truly diagnostic, so establishing beyond reasonable doubt the identity of the particular fold set and its position in the succession.

As was shown earlier in this chapter (Figure 6.4) and in Chapter 5, the degree of certainty attached to such identification is greatly increased if folds of a specific style are found to overprint or be overprinted consistently at various other localities by other structures each of which has its own particular style. The alternative explanation, viz. that these relationships are the result of pure chance, requires the acceptance of an unreasonably high degree of coincidence. The possibility that, by pure chance, a succession of fold sets, each with its own characteristic features, could be duplicated elsewhere throughout the region by comparable yet entirely unrelated successions of fold sets, each with the same characteristics as the first succession, and all refolding each other in the same order, is just as remote as the likelihood that unrelated sedimentary successions with similar characteristics (thicknesses, colours, grain sizes etc.) could be found at several separate localities as a result of pure coincidence. If the relationships between **all** adjacent sets of structures in the succession can be demonstrated at least in **some** places in the study region, and also if it can be shown that wherever structures of more than one set are seen together, these relationships are the same, then the only reasonable conclusion to be drawn is that the structures have been placed in the correct order by the observer and that together they comprise part, or all, of the succession for the terrane.

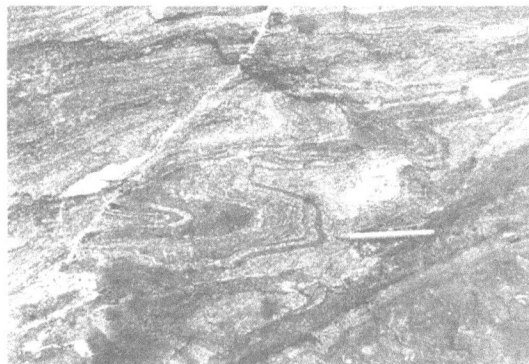

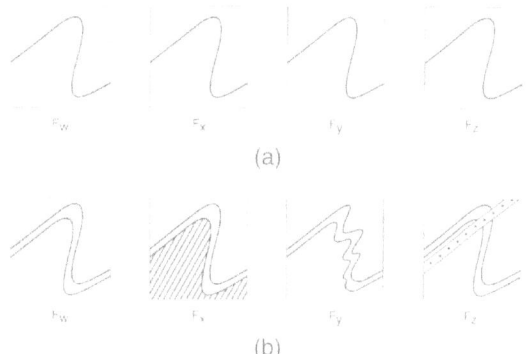

(a)

(b)

Figure 6.26 Photograph showing the profile of a singlefold can vary from open, almost square at right to isoclinal, sharp-hinged. Hence profile, particularly of only part of a fold cannot safely be regarded as diagnostic of a particular fold set. Dabie complex, Feng Huang Guan area, Hubei Province, China.

Figure 6.28 Folds (F_w, F_x, F_y, F_z) with comparable geometry (a) but belonging to different sets (b).

(a)

(b)

Figure 6.27 (a) Variation of style in different folds with comparable geometry belonging to the **same** set. See also Rast, 1963, Figure 2. (b) Changing style of folds of the same set in migmatites. Skattskär, Åland Islands, Finland.

While it might be argued that style alone, at least in the case of an isolated fold showing no overprinting relationships, may not necessarily be diagnostic of members of a particular fold set, it is nevertheless remarkable how often the total characteristics or stylistic features of a fold do in fact provide a reliable indication of the set to which it belongs. Also, although the geometry of folds of the same set may vary, even within a single fold (Figure 6.26), it is noteworthy how, in spite of the almost certain effects of differential adhesion between layers during folding of a layered medium, the geometry of folds belonging to the same set shows striking similarities and is probably largely determined by their deformational history. This is particularly so when the folds are considered collectively, as a group, even though there may be considerable departure from the 'norm' in some individual cases (Figure 6.27).

Furthermore, it is worth recalling at this stage that not only is it the case that the profiles of folds of the same set can vary but also that the converse may sometimes hold, i.e. that folds of different sets may have geometrical features in common (Figure 6.28).

So, although the expression 'tectonic fingerprint' has been used to indicate that the style of a structure is unique, (the expression is more appropriately used in the context of a **group** of mutually overprinted structures), style should always be used as a basis for classifying specific fold sets in the awareness that it can be influenced by local inhomogeneity in the rocks and may change with change in physical environment. Nevertheless it is still an extremely useful concept in structural classification if used appropriately and with a clear understanding of its limitations.

6.2.2 FACTORS INFLUENCING GEOMETRY AND STYLE OF FOLDS

(a) Effect of composite fold geometry on style

In some circumstances, folds (F_x and F_y, say) with similar geometry, and which seem to be of the same set, show differences in style. This can happen when the geometry of folds, which **apparently** formed by a single deformational 'phase', is **composite**, and requires the fortuitous coincidence of orientation of the overprinting and existing folds (Figure 6.29).

The presence or absence of associated mineral growth which could be diagnostic of one or other fold set will be governed by the lithological composition of the layers affected by the folds and if the appropriate lithology is lacking, so too will be the mineralogical means of recognizing whether the structure comprises one set or two. Where such an observed fold is not obviously compounded from the effects of two deformations, i.e. it has inherited a pre-existing fold form without showing any evidence for this overprinting (such as obvious deformation of a pre-existing fold axial plane by a later structure), it is nevertheless likely to have subtle differences in its profile compared to others of the same set which were developed in their entirety solely by the later deformation. However, without some forewarning, the observer is unlikely to recognize such subtle differences. For the most part these cases can be identified with certainty only when the emplacement of some simple structural feature (tabular vein, dyke, preferred alignment of minerals etc.) intervenes between the two deformations (Figures 4.1 *et seq.*, and Hopgood, 1980, Figure 11).

In that such cases of confusion between similar-looking folds of the same set requires coincidence, not only of the orientation but also the proximity of the actual sites of both the pre-existing and the imposed folds, these

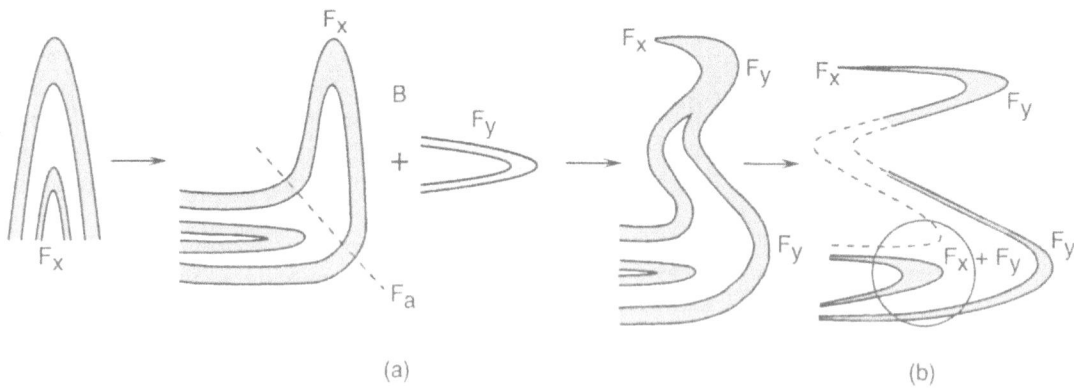

(a)   (b)

Figure 6.29   (**a**) Fold F_x reorientated by F_a and later affected by F_y. Both F_y (in **b**) and $[F_x + F_y]$, a compound structure (circled), are similar-looking structures with similar orientations and indistinguishable from one another but different in origin and therefore not folds of the same set. (cf. Figure 5.4). See Figure 12.4 and the discussion relating to it.

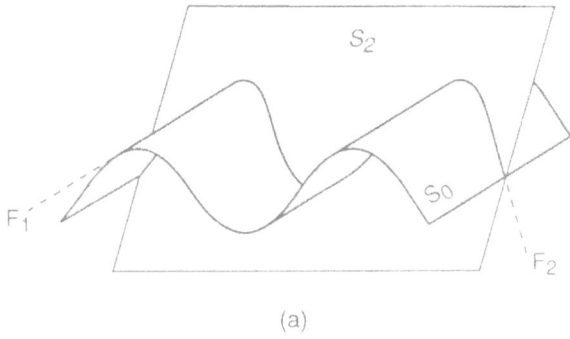
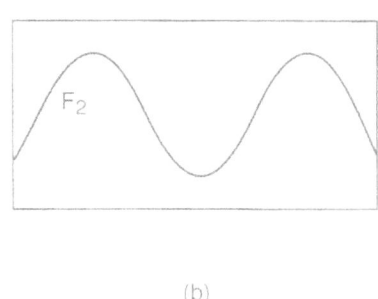

(a)   (b)

Figure 6.30   Dependence of the fold axial trend of a later fold (F_2) on the pre-F_2 attitude of the foliation. (a) Sketch of a later fold (F_2) superimposed on an earlier fold (F_1). (b) Plan showing the variation in the axial trend of F_2.

circumstances are not likely to be commonly encountered (or where they are observed, their significance is often not appreciated, i.e. they may not be recognized for what they are). The shape of examples noted (see Figures 11.13, 12.4) will be recognized as being anomalous in terms of the structures identified in the succession, and as such treated with special care. Therefore the likelihood is that their true nature will be recognized, or suspected. In any case failure to recognize them is unlikely to affect the compilation of the structural succession because F_x and F_y will have been identified at other sites.

A further point worth remembering is that it is usually possible to monitor gradual changes in style throughout a terrane (Hopgood 1971b, 1980; and Hopgood and Bowes, 1972) so that style can still be regarded as providing a useful initial indication of the identity of a particular structure, this identity always of course being subject to confirmation on the evidence of overprinting. See section 12.5.

(b)   Axial trend, axial planar orientation and order of deformation

For a discussion of the influence of pre-existing folds on the axial orientation of super-imposed folds, see Figure 3.32 in section 3.2.4.

Orientation of fold elements, fold axial plane and particularly fold hinge, of structures of comparable scales depends not only on the attitude of the stress field with respect to the foliation prior to folding but also on the geometry of the foliation, i.e. whether it is planar or has already been folded (Figure 6.30), and also of course on whether or not it has been refolded subsequently. Therefore the hinge direction is seldom a useful criterion for assigning a fold to a particular set where the structure results from multiple deformation. Nevertheless, hinge direction, i.e. its trend or azimuth, does become significant when the fold set is one of the **latest** in the succession.

Correspondingly, fold axial planar strike will be significant only if the fold belongs to a set late in the succession. The attitude of the fold axial plane, whether upright, inclined or recumbent, may be a useful indicator of the **approximate** position of the fold in the structural succession but only experience in the particular terrane can confirm this. For example, in a terrane affected by horizontal compression, upright axial planes are likely to be an indication that the folds are late stage, whereas inclined or recumbent folds are likely to have been formed early in the deformational sequence and reorientated subsequently by later folding (Figure 6.31).

(a)   (b)

Figure 6.31 (a) Refolded recumbent isoclinal fold. Lewisian complex, Rona, Inner Hebrides, Scotland. (b) Sketch showing the relationships between the folds shown in (a). The recumbent fold is overprinted by an asymmetrical fold with an inclined (and curved) axial planar trace highly discordant to the earlier isoclinal (recumbent) axial plane. This trace is at an angle averaging approximately 40° to the planar axial plane (vertical rectilinear trace) of the latest (upright) fold.

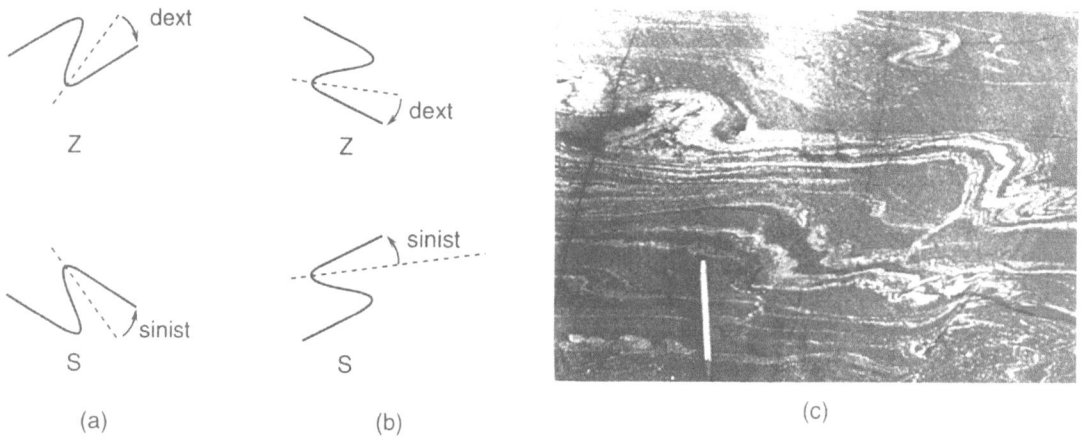

(a)   (b)   (c)

Figure 6.32 Tight ('S' and 'Z') folds are still distinguishable from one another and recognizable even though their orientation differs, between (a) and (b), as a result of later deformation. This is because their sense of asymmetry remains constant with respect to a datum such as the foliation (with an E–W trend in this case). 'S' folds always have a sinistral asymmetry sense and 'Z' folds are dextral in all cases. Note that this rule applies even when the folds are insufficiently exposed to enable the 'S' and 'Z' character of the folds to be recognized so that they appear simply as tight folds. (c) An example of 'Z' folds. The axial planar trends with respect to main foliation trend (E–W here), even of incompletely exposed examples, would still enable them to be classified as 'Z', rather than 'S' folds. Svecofennian migmatites, Jussarö region, southern Finland.

Axial planar orientation, particularly in the case of the earliest fold sets in an extensive structural succession, is unlikely to be one of the diagnostic features in fold identification. Nevertheless it should be borne in mind that in the case of any fold that is asymmetrical, the relative orientation of the axial plane with respect to some reference plane such as the dominant foliation can provide a useful indication of the identity of the structure. This angular relationship, the sense of asymmetry, can be used to distinguish between fold sets whose profiles are similar (Figure 6.32). While the size of the angle can vary because of later deformation (Figure 6.33), particularly depending on the relationships of the structure to the later stress fields, the sense of asymmetry (dextral or sinistral) or vergence sense with respect to a datum foliation can sometimes provide useful diagnostic information regarding the fold. In Figure 6.32 'S' and 'Z' folds in the dominant foliation are shown together with the trace of their axial planes. In each case, although the orientation of the folds (i.e. of their axial planes) varies, the direction sense **from** the foliation remains the same, clockwise (or dextral) in the case of the 'Z' folds and anti-clockwise (or sinistral) in the case of the 'S' folds (Figure 6.32c). Bear in mind that this applies even in those instances where deficient exposure enables only the fold closures to be observed.

The degree of modification (in folds of the same set) of the angle between the fold axial plane and the foliation by later deformation will vary, depending fortuitously on the varying attitude of particular pre-existing folds across the imposed stress field (Figure 6.33). Also, another potentially diagnostic feature which contributes to the shape of fold profiles of the same set, the inter-limb angle, will show corresponding variation.

Examples

(1) First consider the attitudes of the axial planes (S_x and S_y) of two sets of asymmetrical

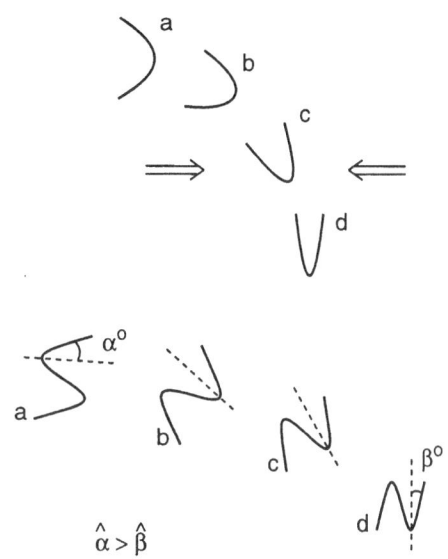

Figure 6.33 Effect on the profiles of folds (a–d) with different attitudes relative to 'horizontal' compression (above). Although angle $\hat{a}$ – $\hat{d}$ were initially (prior to compression "⟹ ⟸") the same, ultimately $\hat{\alpha}$ will be very much larger than $\hat{\beta}$ because of the difference in angular relationship between the axial plane and the direction of compression from position a to position d.

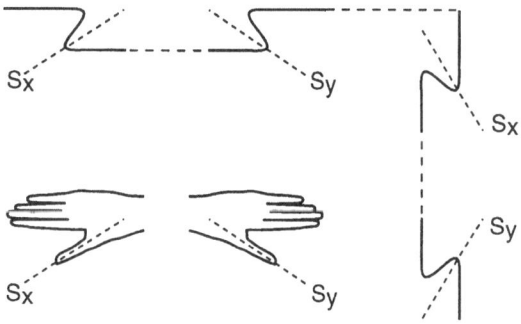

Figure 6.34 Use of the 'rule-of-thumb' mnemonic as a basis for the identification of asymmetrical folds regardless of their orientation. Where the folded foliation trends E–W, S_x has a NE trend, S_y has a SE trend, whereas in the case where the foliation trend is N–S, S_x has a SE trend and S_y has a NE trend. However, the sense of asymmetry of S_x and that of S_y (i.e. the angular relationships between thumb and fingers of the left compared with the right hand) remains constant in each case.

folds (F_x and F_y) in east-trending steeply-dipping foliation (Figure 6.34). The axial planes of both sets are vertical, and one set (S_y) trends SE and the other (S_x) NE over most of the outcrop. In places, later large-scale folds reorientate the E–W strike through as much as 90° so that the axial planar trends of the two fold sets are no longer regularly NE and SE and their axial planar trends are interchanged.

Where the two structures appear together the similarity of scale makes discrimination of their overprinted relationships difficult, but not impossible (cf. Figure 6.37). Recognition of the fact that the folds belong to two separate sets is possible however from the difference in their asymmetry. Because the sense of the obliquity of their axial planar trends to the foli-

ation strike is different, their vergence is different – sinistral in the case of the SE-trending S_y and dextral in the case of the NE-trending S_x. Even after refolding of both sets, the vergence sense for each remains unchanged, regardless of the later reorientation of the folds and foliation. Therefore the folds can still be recognized in spite of their change in orientation, in just the same way that the angular relationship between the thumbs of the right and left hand remain constant, regardless of the position in which the hands are held (cf. the 'right hand rule' of electricity and magnetism).

(2) Next consider the asymmetrical folds of Figure 6.35. Again the foliation trends broadly east. The nearly rectilinear axial traces (S_l) of the later folds have consistent (northeasterly)

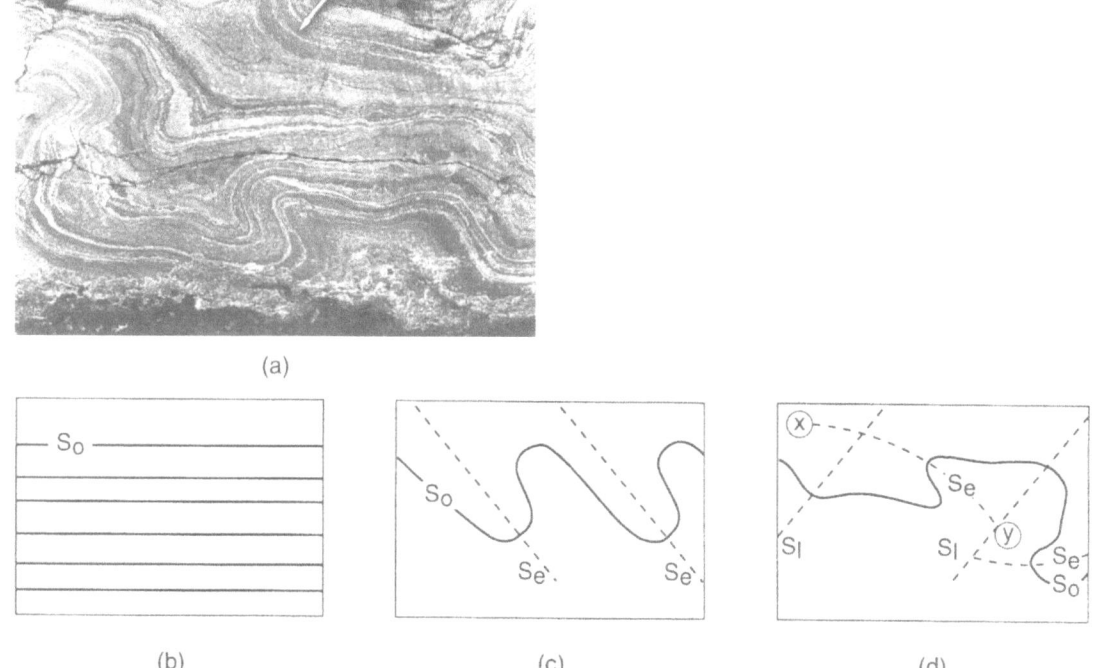

Figure 6.35   (a) Photograph of refolded folds. (b–d) Sketches showing the refolded relationships between earlier (F_e) and later folds (F_l) and the foliation and the development of the structure. (b) S_o is the trace of the original foliation. (c) S_e is the trace of the axial plane of the early fold (F_e), shown here to be rectilinear, prior to being folded by F_l (in d). (d) S_l is the axial planar trace of the later fold (F_l) and therefore is rectilinear, while S_e is curved (by F_l). Svecofennian migmatites, Djupkobbarna, Jussarö area, southern Finland.

trends and are only slightly curved, whereas those of the earlier folds (S_e) are strongly curved with trends varying between northeast and southeast. In Figure 6.35d for example, the acute angular relationships between the axial planar trace (S_e) and the mean (E–W) foliation trace ranges from approximately 10° at x, to 50° at y. However, note that the sense of asymmetry (vergence) of F_e folds is consistently anticlockwise or sinistral (Figure 6.35d), i.e. they are 'S' folds, so that in spite of the similarity of scale of both fold sets, F_e can be identified and separated from F_1 which are 'Z' folds. The sequence of events that produced the structure is shown in Figures 6.35(b), (c), (d).

(c) Effect of later fold scale on fold reorientation

While the reorientation effect on the geometry of earlier folds by later folds will be influenced to some extent by the type of later mechanism, their geometry will depend largely on the relative scales of the earlier and later fold sets and relative scales of folding as can be shown by the following simple cases (see also the discussion on 'Open folds', section 10.1.2 and Figure 10.3).

(i) Later folds the same scale as earlier

If the scale of refolding is comparable to that of the pre-existing fold (Figure 6.36) then the geometry of the earlier fold is likely to be strongly modified during refolding. This can be demonstrated by the simple case of refolding by flexure where the axis of the later fold is parallel to that of the earlier fold (see also Figure 3.5). In the general case where the axes are not parallel the fold form, compounded from the effects of the early deformation plus the later folding, is likely to be less simple than that shown in Figure 3.5.

Also (Figure 6.36), if the scales of both earlier and later sets are comparable, then the orientation of the axial plane of folds of the earlier set will vary and lack diagnostic significance. Furthermore, even the refold relationship between two such sets (the relative age of each fold) may be difficult to recognize in some cases, depending on the angular relationships of the two axial planes, especially when both sets have been affected by later deformation (Figure 6.37).

So, even if the fold sets can be provisionally identified in terms of characteristic features (style) other than orientation, this identifi-

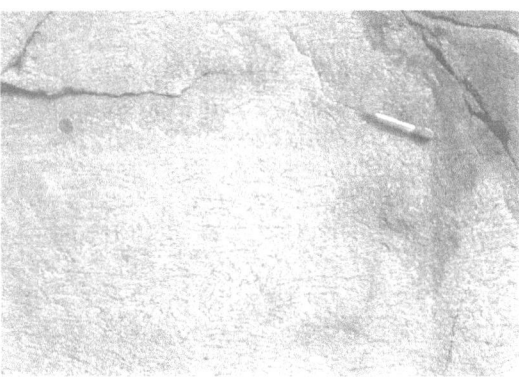

Figure 6.37 Complex pattern from refolding on similar scales. The photograph illustrates the difficulty of deciphering refold relationships in the general case where the scale of earlier and later folds is the same. Svecofennian migmatites, Skåldö, southern Finland. Compare Figures 6.32 and 6.34 and see also Figure 10.25.

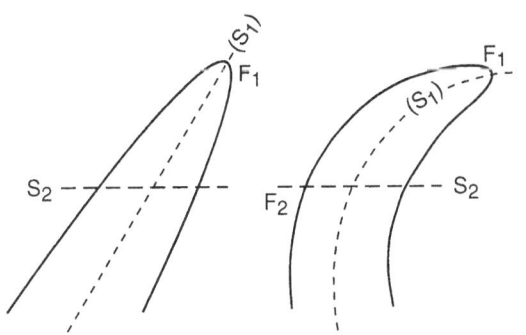

Figure 6.36 Sketch showing modification of F_1 fold geometry after F_2 flexural folding. The orientation of S_1 after F_2 folding varies and no longer has any diagnostic significance.

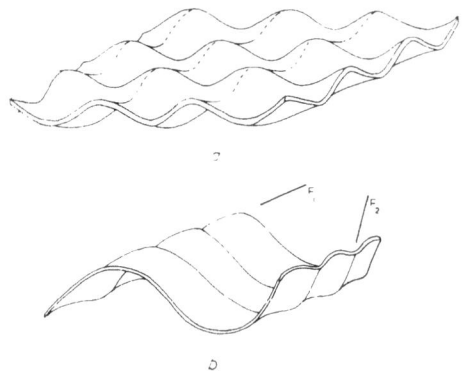

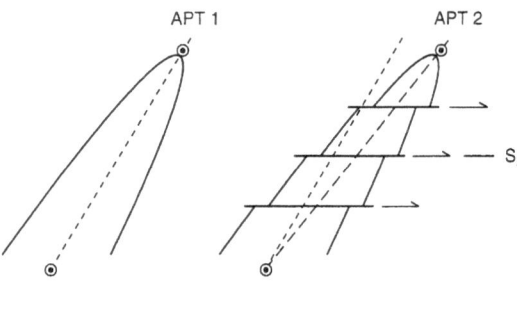

Figure 6.38 (**a**) Sketch of cross-folding on similar scales where overprinting relationships are uncertain (see also Figures 2.4, 3.16). (**b**) Refolding on different scales, the later folding on a smaller scale, where the distinction between earlier and later folds, although uncertain, is more likely to be clear.

Figure 6.39 Modification of F_1 fold geometry by relatively small-scale slip. The axial planar traces shown (APT 1 and APT 2) are the broad axial planar trends before and after deformation. Because the slip is perpendicular to the direction of the pre-existing fold axis, the fold axial directions before and after the later deformation are parallel.

cation could not be confirmed **in terms of overprinting** (refolding) relationships if the distinction cannot be made between the earlier and the later fold axial planes (Figure 6.38) (cf. Ramsay, 1967, Figures 10 and 13).

If the later fold mechanism is by slip or shear (passive folding) in **spaced** slip zones, then the effects of refolding are likely to be significant in terms of the orientation of S_1. The deformation will reorientate the axial plane. If each segment is offset consistently in the same sense (Figure 6.39), and by the same relative amount of slip, although individual segments of the F_1 axial plane will preserve their original attitude, the deformation will result overall in a net change in trend of the axial plane which will remain statistically **planar** rather than be curved as it would if the deformation had been by flexure. As in the previous example of flexural folding (Figure 6.36), because the direction of slip is perpendicular to the direction of the earlier fold axis, the axis of the modified fold remains parallel to that of the fold prior to the later deformation producing a relatively simple compound fold form.

(ii)   Later folds smaller than earlier

On the other hand, where the later folding is by slip and the sense of slip alternates (Figure 6.40a), or if the later folding is flexural and the scale of the later set of structures is significantly smaller than that of the earlier, then a series of locally-changing F_1 axial planar orientations will result. S_1 will have a 'zigzag' profile. Depending on the regularity (symmetry) of F_2, the overall (mean) orientation of both the earlier axial plane and fold axis may persist with only these local modifications (Figure 6.40b) and could still be significant in terms of identifying the earlier fold set.

(iii)   Later folds larger than earlier folds

Where folds of the later, overprinting set are much larger than the earlier folds the reorientation effects are likely to be significant and clearly recognizable. The axial planar trend of the earlier set will not be diagnostic but will vary across the later structure. As in previous examples, this relationship is illustrated by a simple example in which the axes of the earlier and later folds are parallel (Figure

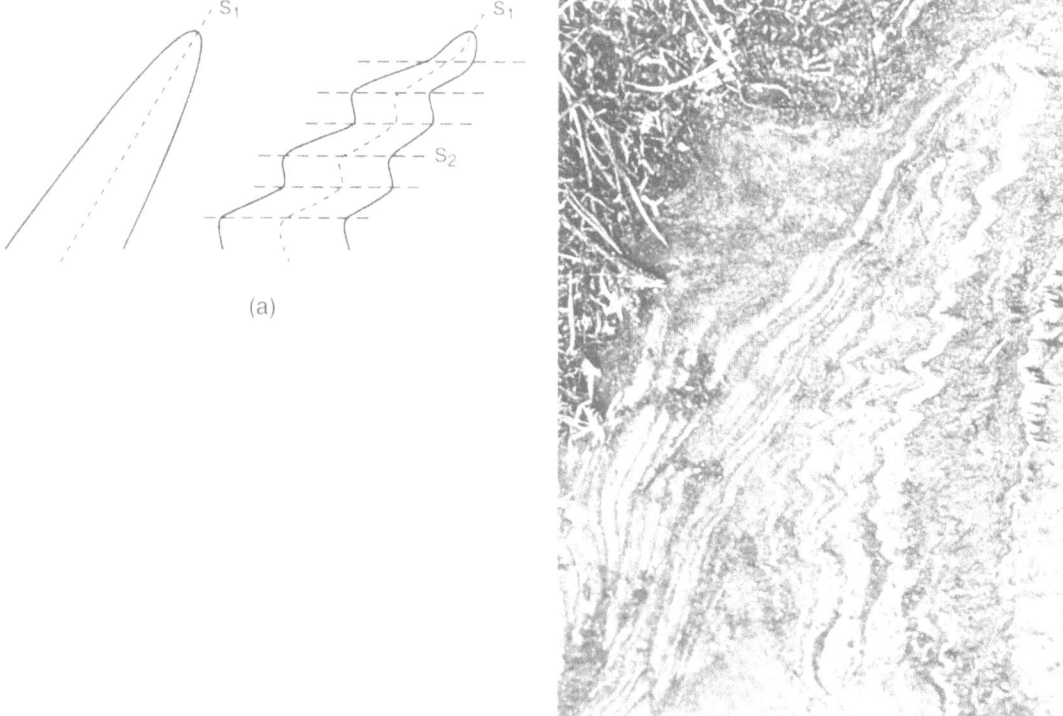

Figure 6.40 (**a**) Smaller-scale penetrative slip folding overprinting a larger-scale fold where, in this case, the axes of the earlier and the later folds are parallel. (**b**) Overprinting of fold by smaller-scale later folds. The earlier axial planar trend, and that of the fold axis, remain statistically the same. Hammaslahti, Finland.

6.41). Identification of the earlier folds will depend on characteristic features other than orientation, such as asymmetry sense.

(d) Effect of axial planar dip in fold axial trend

The attitude (with respect to that of the pre-existing foliation) of the fold axial plane of later folds is highly significant and has a very important control on the degree of regularity of its trend. Even if the fold set is very high in the structural succession (i.e. one formed in response to deformation late in the sequence) a decrease in axial planar dip from 90° will likewise decrease the consistence (regularity) of its trend (Figure 6.42). This of course has an

important bearing on the diagnostic significance or relevance of orientation as a basis for the identification of folds.

6.3 KEY STRUCTURES

6.3.1 INTRODUCTION

During reconnaissance, or as work progresses during the initial stages of the study, it may become evident that the structures of one set in particular are distinctive and readily distinguishable from others. This may be for a variety of reasons, for example, because the nature of deformation and metamorphism was such that certain aspects of the geometry of the structure, and/or the associated mineral growth

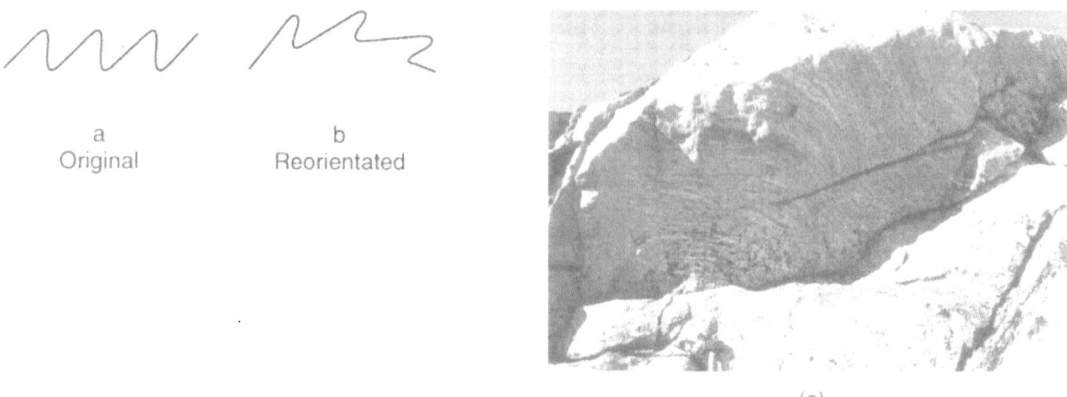

Figure 6.41 (a) Refolding of small-scale folds by larger-scale. Although the axial planes of the earlier folds are strongly reorientated, in this example the earlier and later axes of folding are parallel. (b) Dispersion of the attitudes of axial planar traces of small angular folds by later folding around a larger open fold. (c) Example of case (b). The axial planar attitudes of the smaller folds range from inclined above the hammer to vertical on the right. Pre-Ketilidian migmatites, West Greenland. See also Figures 9.12b, 10.5a and 10.10b.

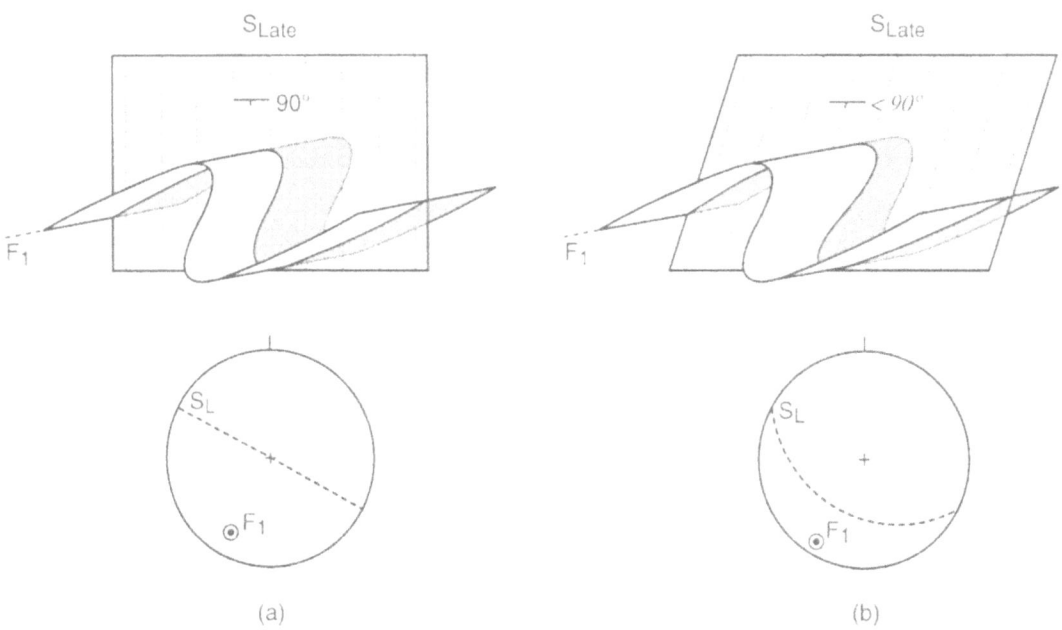

Figure 6.42 Sketches showing the influence of axial planar dip of a superimposed fold on the plunge and trend of the later fold (F_L). (a) A vertical later fold axial plane results in constant trend for the later fold (F_L) whose plunges lie along the diameter of the stereo plot (S_L). (b) An inclined later fold axial plane results in a varying later fold axial trend (F_L plunges lie along the curved (inclined) great circle (S_L) on the stereo plot).

and neosome development etc., were more strongly imposed on the rocks than in other cases. Such datum structures or **key structures** provide useful markers in the structural succession (Hopgood, 1980, p.63). Comparable to key or marker horizons in lithostratigraphy, they represent specific stages in the tectonic or deformational sequence.

The process of synthesizing the total succession of fold sets in complex terranes becomes easier if during preliminary reconnaissance it is possible to identify at least one distinctive set of folds which are well developed more or less everywhere and whose style is sufficiently characteristic to enable them to be recognized easily throughout in the area being investigated. These key structures effectively divide the total succession into two shorter, more manageable successions; one comprising folds prior to the development of the key structures and hence deformed by it, and the other consisting of folds formed after the key structure and which therefore deform the key structures (Figure 6.43). This is especially important where the succession is a long one resulting from a complex history of deformation. Recognition of the relationship between a particular fold and the key structure will quickly establish its approximate position in the succession, whereas establishing its relationships to other fold sets whose positions in the succession might not yet have been determined, will be less informative. Recognition of the generally pre-, or post-key structure relationships of a fold constitutes a useful preliminary step towards its ultimate placement in the exact position in the structural succession.

In some cases there may be more than one set of structures that will act as key structures, in which case the succession is divisible into three or more groups.

Clearly it is important that a key structure possesses distinguishing features that are sufficient in number to allow it to be recognized unequivocally. Therefore, the criterion for a key structure is that besides being widespread and fairly easily recognizable, it is a multi-

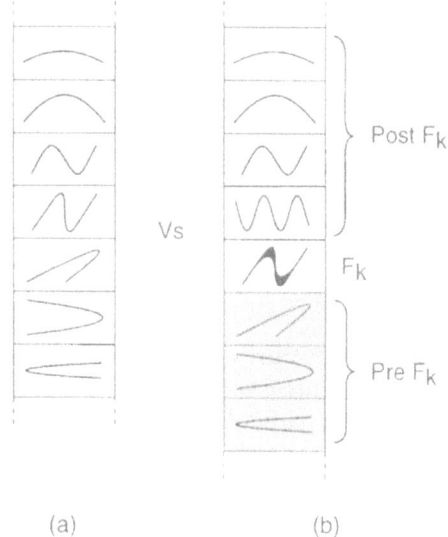

(a)   (b)

Figure 6.43   (a) Extensive succession of structures. (b) The same succession in which a key structure (F_k) has been identified. Post-key structures (which have refolded F_k) are readily identifiable as being later in the succession and pre-key structures (which have been folded by F_k) are readily identifiable as being earlier in the succession.

Figure 6.44   Well-developed crenulation cleavage axial planar to, and therefore associated with, a fold in amphibolite. Lewisian complex, Rona, Inner Hebrides, Scotland.

parameter structure. It must be distinctive in the study area in terms of such parameters as geometry, attitude of its elements (axial plane, axis, limbs), syntectonic mineral association, syntectonic veining and association of veining with respect to the fold geometry, and/or have clear-cut relationships to particular igneous or metamorphic events.

6.3.2   EXAMPLES OF KEY STRUCTURES

The kinds of features that characterize a good key structure are both distinct and distinctive. Examples are particular types of (axial planar) cleavage such as the crenulation cleavage strongly developed in amphibolite, and parallel to the axial plane of the fold in Figure 6.44. The cleavage forms a distinctive structure and its axial planarity allows it to be related to the particular fold set shown.

Preferred mineral orientation associated with the fold geometry is another potential key structure. In the fold shown in Figure 6.45a, ellipsoidal garnets in the amphibolite layers (Figure 6.45b) can be seen to have their long axes parallel to the fold axial plane (particularly below and to the right of the pen) thus demonstrating their syntectonic growth and fixing their position in the structural succession with that of the fold.

Axial planar cleavage without obvious preferred orientation of mineral growth is another potential key structure. The strong axial planar crenulation cleavage in folded tuffaceous rocks shown in Figure 6.46 could be sufficiently distinctive to warrant considering it as a key structure.

Sometimes the style of deformation is such that cleavage is developed asymmetrically with respect to the fold axial plane as in the example shown in Figure 6.47. Here the cleavage is parallel to the fold axis but not to the fold axial plane, being convergent towards the fold closure, and is more strongly developed on the right of the structure. Where this relationship is consistent, the fold could be used as a key structure.

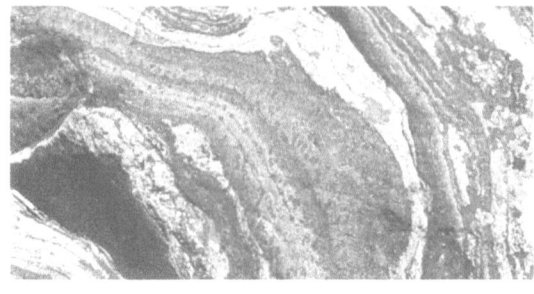

(a)

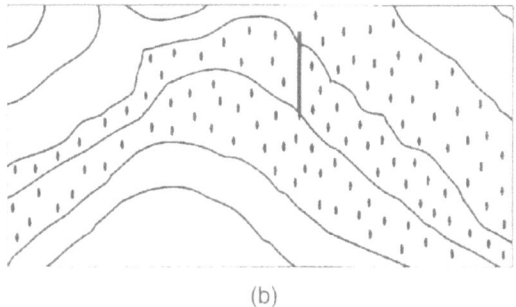

(b)

Figure 6.45 Mineral growth associated with folded amphibolite. (**a**) Ellipsoidal garnets (small dark grey ellipses) with long axes parallel to the fold axial plane. (**b**) Sketch of the structure in (**a**), with a pen added to show the scale. Svecofennian migmatites, Pohja, Finland.

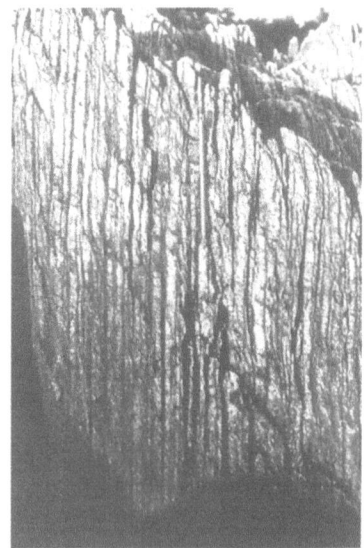

Figure 6.46 Strongly developed axial planar crenulation cleavage. Indian Bar, Maine, USA.

Figure 6.47 Fold showing asymmetrical expression of cleavage development. In addition to pervasive axial planar cleavage there has been preferentially stronger development of a planar fracture inclined to the axial plane, and on the right of it. Hammaslahti, Finland.

Figure 6.48 Development of an irregular leucocratic vein within the axial planar zone of a fold. The curvature of the vein demonstrates the effect of later open folding. Svecofennian migmatites, Pohja, Finland.

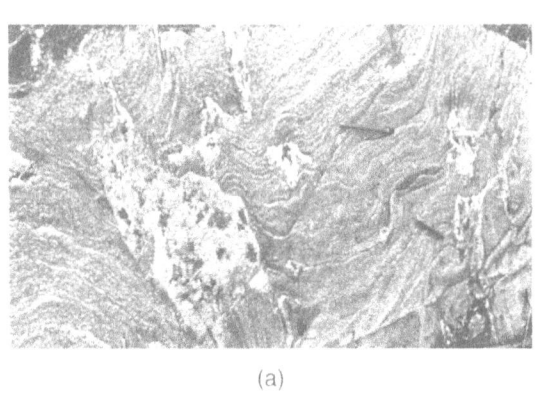

(a)

(b)

Figure 6.49 (**a**) Discontinuous veins and 'tabular' pockets of leucocratic neosome parallel to the limbs of open folds. Svecofennian migmatites, southern Finland. (**b**) Discordant, near-tabular leucocratic vein replacing the limb of a chevron fold. Lewisian complex, Isle of Lewis, Outer Hebrides, Scotland.

Axial planar leucocratic (or other) veining, 'smudging' or 'blotching' caused by partial melting often develops in a particular manner that is sufficiently characteristic to enable the associated folds to be used as key structures (Figure 6.48). In the fold shown, an irregular leucocratic vein is developed off the crest of the fold and (prior to later deformation) parallel to the fold axial plane, forming a distinctive structural relationship likely to be suitable as a key structure.

Veining, or potential veining, is often associated with folds in migmatites and this can form parallel to one or other (or both) limbs of a fold and can vary considerably in its regularity, from discrete irregular lensoid bodies (Figure 6.49a) to more or less tabular veins (Figure 6.49b), depending on the style of deformation and physical conditions obtaining at the time of deformation. Again, if the relationships are characteristic and consistently developed the associated structure will form a valuable datum in the succession.

At times the development of leucocratic neosome is more diffuse, sometimes within only one limb of a fold (Figure 6.50a), but commonly parallel to both limbs (Figure 6.50b), causing 'smudging' or blurring of the structure of the limbs. It varies considerably in expression and can form discrete pockets within the fold limbs (Figure 6.50c).

Other possible distinctive stylistic features might be some particular kind of consistent asymmetry of the fold, round or angular fold hinges, or some peculiar form of crumpling in hinge zones, all with or without associated leucocratic material. An example of this is shown in Figure 6.51 which shows extreme

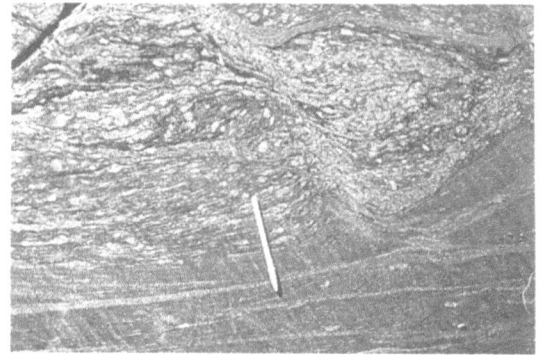

(a)

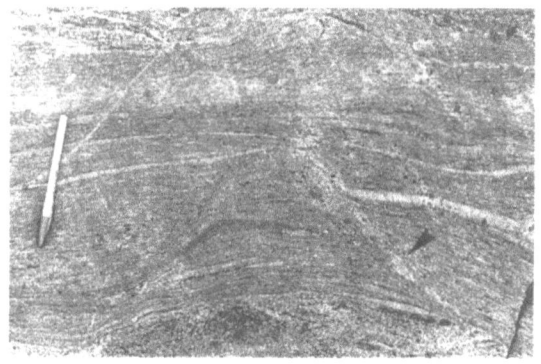

(b)

Figure 6.50 (a) Development of diffuse leucocratic neosome causing 'smudging' parallel to one limb only of an open fold. (b) Diffuse leucocratic 'smudging' parallel to both limbs of an open fold. Svecofennian migmatites, Jussarö area, Finland. (c) Contrast the discrete leucocratic 'veining' more or less parallel to both fold limbs shown here with (a) and (b). Belemoride migmatites, USSR.

(c)

Figure 6.51 Crumpled fold hinge zone. Inter-layered carbonate and quartzofeldspathic gneiss in Svecofennian migmatites. Lilla Kvarnskär, Åland Islands.

Figure 6.53 Isoclinal flow fold in migmatite. In some cases, folds such as this can serve as a key structure locally. Vícenice, Czech Republic. See also Figures 5.39 and 8.17e.

Figure 6.52 Folds defined by offset, on surfaces associated with leucocratic veins, of pre-existing open fold hinges, whose profile can be seen in the segments between the veins. In contrast to the very open form of the earlier (pre-offset) folds, the new folds have a distinctive, easily recognizable, 'rectangular' cross-section. These are folds of set F_{bb} (Hopgood, 1984). Svecofennian migmatites, Jussarö area, Finland.

crumpling in the hinge zone of the fold, another type of potential key structure.

Another example is that shown in Figure 6.52 where the style of deformation has been such that associated neosome development followed offset of the fold limbs resulting in a very distinctive structure. Here leucocratic neosome development and slip were parallel to the fold axial plane imparting a 'rectangular' profile to an otherwise (formerly) open, round-hinged fold. Structures of this type constitute ideal key structures where they are extensively developed.

However, it must be stressed again that care is needed in using fold profile alone as the basis for selecting a key structure, especially where only part of a single fold is seen, as in the isoclinal fold profile of Figures 2.13 and 6.26.

Extending the use of key structures to wider areas within the crustal segment under examination must always be treated with some caution. This is because enlarging the field in which the key structure is used increases the likelihood that the rocks encountered will have been subjected to a wider range of the physical conditions that influence the style of the structures. This would increase the chance that the style of the folds will vary across the region. In such cases subtle changes in the style of the key structure would need to be monitored continuously and carefully across the study area in order to ensure its recognition everywhere. For example, folds formed during extreme cases of compression during conditions favouring anatexis sometimes develop by flow to produce a diapiric aspect which is locally distinctive. One such example is the round-hinged isoclinal fold in gneiss shown in Figure 6.53. This might serve as a key structure, but only over a limited 'area', because the fold form might be consistent only locally, whereas folds of the same set within the crustal segment under examination, which developed at some distance away under different physical conditions, are likely to have a very different style. In this respect see again the discussion on 'Variation of expression of related structures in the same limited crustal segment', section 5.2.12, and Figure 8.2 in 'Spatial variation in structural expression', section 8.1.2.

CHARACTERISTIC FEATURES OF STRUCTURE SUCCESSIONS

7

7.1 GENERAL FEATURES OF FOLD SUCCESSIONS

Experience shows that the early folds in extensive structural successions are invariably isoclinal, the earliest of these tending to be intrafolial. They are often preserved only as isolated rootless hinges detached from the limbs because of extreme flattening and shear associated with deformation subsequent to their initiation. In many cases, particularly where there has been a long deformational history, the present shape of these structures probably owes little to the processes that formed them initially and is more likely to represent the effects of modification stemming from later deformation.

Because repeated deformation is usually accompanied by compression, the form of early folds which originated as open structures becomes modified causing them to become progressively tighter in profile with diminishing inter-limb angles. Many presumably formed as concentric folds in response to flexure and continued to develop by shear and flattening during later folding. Nevertheless recognizable evidence of overprinting of one isoclinal fold by another is still likely to be preserved regardless of any such modification.

7.1.1 IDENTIFYING FOLD MECHANISMS BY USING STEREOGRAPHIC ANALYSIS

Where a fold set affects a clearly defined linear structure (whatever its nature – fold hinge, intersection lineation or mineral growth) that has survived as a relict feature, it might be possible in some cases to use the geometrical analysis of its behaviour during the folding to identify the fold mechanism (see Figure 7.1; Figure 2.3 in section 2.2.2; the discussion in section 2.2.4, and Figure 2.10). The degree of success of any such analysis would depend of course on the degree of complexity of the structure.

Bearing in mind the factors that can exert a controlling influence on fold development there are certain features (i–v below) that are likely to be common to most fold successions.

(i) The earliest folds typical of an extensive structural succession include detached intrafolial isoclinal hinges which often impart a coarse linear structure to the rocks (Figure 7.2). Also they are often associated with a strong axial planar fabric.

(ii) Later isoclinal folds are more likely to be entire, although this is not always the case (Figure 7.3), and it depends to some extent on the relative competence of the folded layers as well as the 'degree of intensity' of the deformation. Isolated examples may be distinguished provisionally (prior to confirmation in terms of overprinted relationships) from earlier isoclinal folds by their more open, intact profiles, even when they are overprinted by later folds and/or by subsequently deformed veins. They may, or may not, differ from the earliest intrafolial folds by lacking lineation and axial planar foliation associated with, and transecting, their hinges (Figure 7.4a). The association of axial and axial planar structure is not of course restricted solely to tight folds

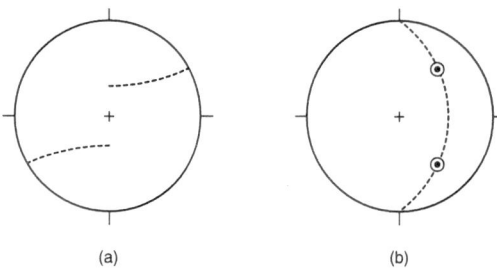

(a)   (b)

Figure 7.1 Distribution on stereoplots of pre-fold lineation after (**a**) flexural folding – dispersed on small circles, and (**b**) after slip folding – dispersed along a great circle, the fold axial plane. Distinction between pre-fold lineations and post-fold lineations parallel to the axial plane is possible because the later lineations would form point maxima as shown (two sets of lineations shown in this case) in contrast to the spread shown by pre-fold lineations (cf. Figures 5.9b and 5.10a).

Figure 7.2 Pronounced coarse linear structure associated with (now detached) isoclinal fold hinges. Lewisian complex, Inishrahull, Eire. See also Figure 3.30.

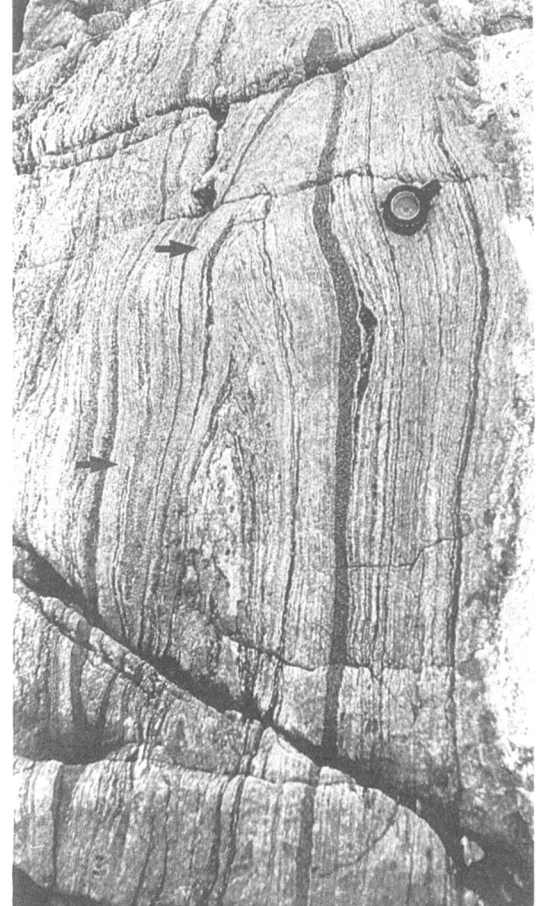

Figure 7.3 (**a**) Earlier intrafolial isoclinal folds folded by isoclinal folds, and surviving as isolated hinges (arrowheads) on the 'left' limb of the large fold between the large transverse crack in the foreground and another smaller crack at the compass. Lewisian migmatites, Rona, Inner Hebrides, Scotland.

early in the succession, as is shown by the mineralization associated with the open fold of Figure 7.4b.

(iii) There is a tendency for folds somewhat later to intermediate in the succession to range from round-hinged isoclinal or tight folds to recumbent or inclined asymmetrical folds (Figure 7.5).

(iv) Later still in the succession the folds tend to become more open, and less 'ductile' in appearance, with sharper hinges and more steeply-dipping axial planes (Figure 7.6) with the potential to form upright folds where the foliation is close to horizontal (see also the discussion on potential fold forms late in the succession and total structural complexity,

Figure 7.3 (**b**) Isoclinal folds still surviving as 'entire' structures even though they have been modified by flow and intruded by coarse quartzofeldspathic veins and overprinted by asymmetrical folds. Svecofennian migmatites, Jussarö area, southern Finland. ▶

section 7.4 and Figures 7.22–7.29). Open, spaced cleavage becomes more prominent.

(v) The latest folds are often 'brittle' in character, tending to be structures (such as the conjugate folds of Figure 7.7a) which lie close to the boundary between folds and fractures. Other folds are very open warps on upright axial planes (Figures 7.7(b), (c)) which, in highly deformed terranes, are often not only difficult to define quantitatively, but are indeed often difficult to detect amongst the general structural complexity. Where stereographic plotting of foliation attitudes is not a practical means of recognizing warps, it might be necessary to resort to measuring and plotting the dispersion of the attitudes of other structures deformed by the warps in order to identify such open folds with any degree of certainty.

In some instances, where the relationships between more than two sets of folds, and also perhaps their associated cleavage, are determinable on a single three-dimensional exposure, part of this progression from early isoclinal (intrafolial) folds through tight then recumbent folds to upright structures can be seen at a single locality. Examples of such relationships are shown in the sketches comprising Figures 7.8, 7.9 and 7.10, drawn from photographs (cf. Figures 11.13 and 12.4).

Figure 7.8 shows the relationships between a tight fold (F_t) which has been refolded by a more or less recumbent asymmetrical fold (F_a), with a horizontal axial plane (dashed trace) which was then deformed by an open upright fold (F_u). Figure 7.8 represents the structure shown in the photograph of folds in Lewisian gneiss in Figure 11.13a.

In Figure 7.9 an isoclinal fold (F_1, with axial planar trace S_1) has been folded by a tight fold F_2 (axial planar trace S_2) then refolded by a recumbent structure (F_3, axial planar trace S_3). All are refolded by an open upright fold (axial planar trace S_4) which has caused curvature of the recumbent axial plane. The folds (F_1–F_4) in this figure demonstrate the commonly-seen evolution from isoclinal to more open, and become progressively more upright. This figure represents the structure of the photograph of Figure 8.12 in which the recumbent structure can be seen to be associated with a strongly-developed, sub-horizontal, coarse axial planar cleavage.

A more or less complete range of fold types in the progression from early isoclinal to latest upright open warps is represented diagramatically in Figure 7.10 which summarizes the fold succession in the Lewisian complex of the island of Inishtrahull, Donegal, in Eire (Figure 1b, Bowes and Hopgood, 1975, p.373).

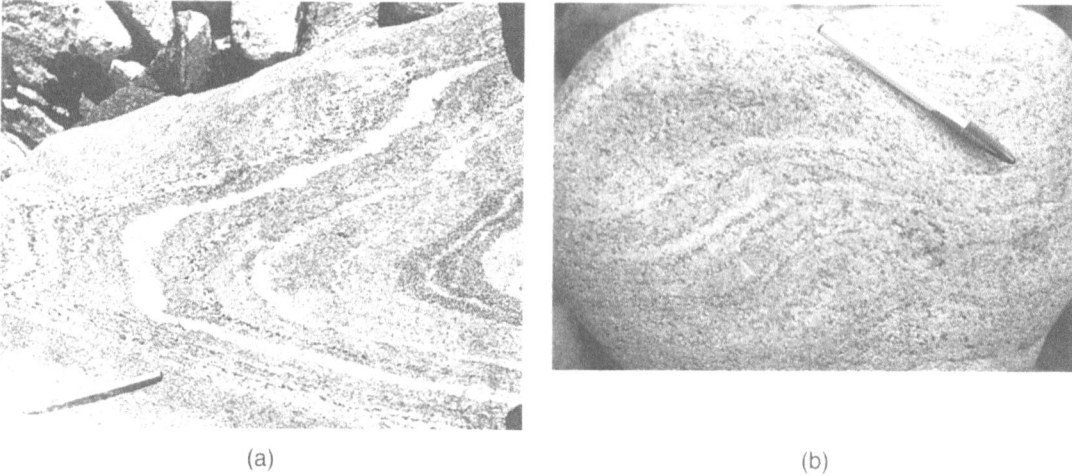

(a)   (b)

Figure 7.4   (a) Tight fold with an associated axial planar mineral alignment. Pre-Ketilidian complex, West Greenland. (b) Open fold associated with axial planar mineralization. Leeuwin Block gneiss, Cape Leeuwin, Western Australia.

(a)   (b)

Figure 7.5   (a) Tight, round-hinge fold. Belemoride migmatites, Soviet Karelia. (b) Recumbent asymmetrical folds, Lewisian complex, Rona, Inner Hebrides, Scotland.

In the light of what has just been discussed it is useful when studying a complex terrane to keep in mind some simple yet reasonable assumptions (1–4 below) relating to the structures in the succession. Some of these have already been noted in section 2.3. Remember though, because they are assumptions, they should be confirmed whenever possible.

1. Earlier folds formed in a long or complex sequence of folding are normally tighter, often isoclinal in form and sometimes reduced to rootless intrafolial hinges. This is because with time, repeated deformation, involving as it does both compression and intense shear, is likely to result in the gradual tightening of the folds (by flexure) and attenuation (with

◀ **Figure 7.6** Folds on an upright axial plane with round to angular hinges and almost planar limbs. These folds are characteristic of structures formed late in the succession. Karelia, Finland.

(a)

(b)

(c)

Figure 7.7 Examples of folds typically later in the succession. (**a**) Conjugate fold. Dalradian, Scotland. (**b**) Upright broad warp. Note that in poorly exposed terrain, the existence of such a gentle fold might not be recognized at all when for example only part of one or other of the limbs was visible. Lewisian complex, Lewis, Outer Hebrides, Scotland. (**c**) Very open, upright fold. In this case where the structure is smaller than that in (**b**) the chance is greater that enough of the fold would be exposed to enable it to be recognized. Leeuwin Block gneiss near Cape Leeuwin, Western Australia.

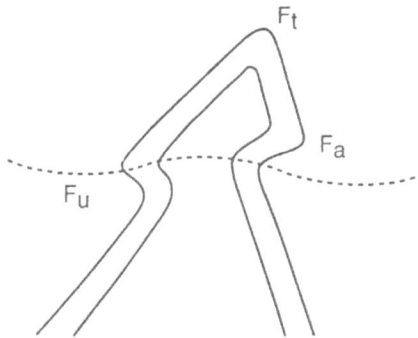

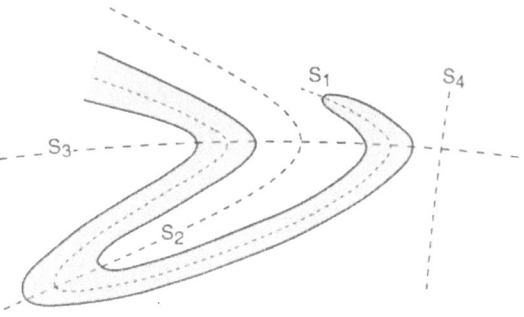

Figure 7.8 Sketch of the refold relationships between three sets of folds, tight (F_t), recumbent asymmetrical (F_a) and open upright (F_u). F_t was followed by F_a (see Figure 3.6 in section 3.1.2) and this was followed by F_u because it deforms the F_a axial plane (S_a). Drawn from folds in the Lewisian complex, Rona, Inner Hebrides, Scotland. See Figure 11.13a.

Figure 7.9 Sketch of four sets of folds. F_1, F_2, F_3 and F_4. Drawn from folds in the Lewisian complex, Rona, Inner Hebrides, Scotland.

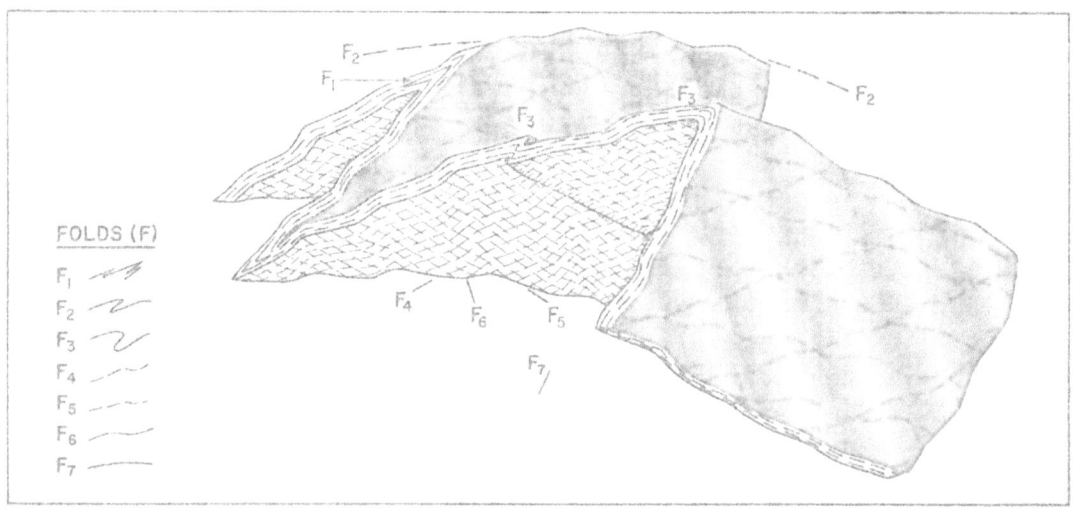

Figure 7.10 Composite ('summary') structure comprising the seven fold sets in the succession recorded in the Lewisian complex, Inishtrahull, Donegal, Eire. (Figure 1b, Bowes and Hopgood, 1975, p. 373. Reproduced with the permission of The Royal Irish Academy.)

flattening and simple shear) resulting in detachment of hinges from limbs.

2. Earlier structures often have a more 'ductile' appearance with attenuated limbs and sometimes thick, lobate hinges (Figure 7.14), whereas folds formed late in the sequence often have a more 'brittle' aspect associated with fractures and shears. This is controlled not only by strain rate but can also be the result of modification of early structures by subsequent deformation or because the first structures were formed in partially lithified sediments (see section 3.2.5).

Furthermore, as the total structural complexity is compounded, progressing from unfolded layers to the observed complex three-dimensional pattern, then for a given layered sequence the resistance of the rock body to deformation will tend to increase because of its increased strength and rigidity (that is, provided there has been no fundamental alteration to the environment of deformation; change in depth, temperature, pressure etc.). Also the nature of its response to a given stress field will vary because, with increased rigidity, it is likely to respond in an increasingly brittle fashion (assuming no significant change in strain rate), this response being of course modified by the changes that might have taken place with each successive deformational 'pulse'. The effect is analogous to the 'work hardening' of metals and comparable to the increase in mechanical strength produced when a sheet of material (cardboard or metal say) is corrugated. In the case of 'angle iron' (a sheet of iron (steel) either bent at right angles or bonded at right angles to another sheet of the same thickness), the strength (i.e. resistance to bending) of either sheet will largely depend, not so much on he thickness of the sheet but on the width of the cross-section formed by the 'angle piece' (Figure 7.11). Further modification will be controlled by any changes in the overall physical conditions (pressure, temperature, pore fluids etc.) of the environment of deformation.

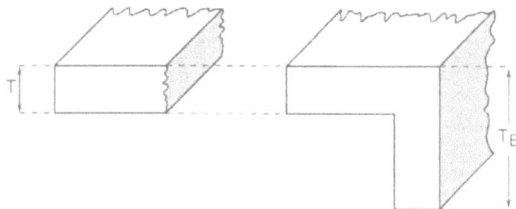

Figure 7.11 Sketch to show how the effective thickness (T_E), will increase the strength of an angled layer of thickness T (i.e. its resistance to bending from stress parallel to T_E).

3. Later structures, as well as being open, tend also to have more upright axial planes (cf. Figure 6.42). The axial plunge (0–90°) on the other hand, depends of course on the attitude at the onset of folding of the surface that is folded, so that folds imposed on, say, the steep limb of an existing fold with a horizontal axis could have steep axial plunges whereas those formed on its crest or trough could be horizontal, depending on the orientation and attitude of the superimposed fold axial plane (Figure 7.12).

4. Commonly too, anatexis with quartzo-feldspathic partial melting and 'granitization' tends to be earlier in the deformational sequence and gives way later to more discrete, 'igneous-type' discordant emplacement and veining by pegmatites etc. Consequently there is a characteristic tendency for discrete 'igneous' bodies to be associated with folds that are later in the sequence while earlier folds will be associated with more diffuse veining.

This pattern of development is consistent with the concept of an orogenic episode which develops to reach a climax and then gradually wanes, with later structures becoming progressively weaker and more 'brittle'. In effect this is comparable to the development of structures (folds) by the superimposition of a series of deformational 'pulses' with differently orientated stress fields, probably in conditions of gradually varying pressure and temperature. However, the changing style and attitude of folds throughout a structural

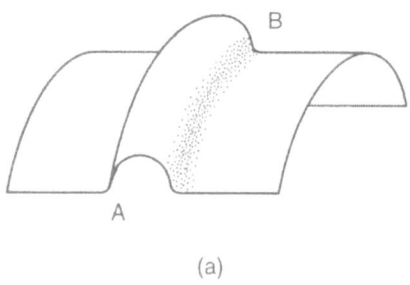

(a)

(b)

Figure 7.12   (a) Sketch to show variation in plunge (steep at A and gentle at B) of a superimposed fold with an upright axial plane. (b) An example of a steeply plunging late fold (above the hammer) superimposed on the limb of an upright fold equivalent to site A in (a). Lewisian complex, Uist, Outer Hebrides. Although the axial plunge is steep, the fold axial plane is more or less vertical as is often the case with late structures. Compare Figure 9.5.

succession indicates that the movement associated with their development also changed throughout the deformational sequence. This could be because of changes in the directions of local stress components such as those stemming from collision of irregular, rather than parallel, plate margins (cf. Figure 3.29). Although the overall external stress field remained more or less constant in orientation, the internal response to this would vary directionally (see Figure 8.9 and compare the overprinting effects shown in Figure 2.6 in section 2.2.4, and Figure 8.13 in section 8.1.6(c)). Even if the same, or nearly the same, external stress field operated throughout, there could have been different internal directional responses to this stress because of the changing internal structure of the rock body caused by folding for example. See also Ramsay, 1967, Chapter 10, p.518, where causes of overprinting listed include crossing orogenic belts separated by a long time interval, and successive deformations in one orogenic cycle (very common) etc.

Therefore the structural succession represents the response to a given set of stress directions (more or less constant in orienta-

tion) of surfaces whose orientation changes during deformation as they deform (fold). This change in orientation stems from internal variations in friction and viscosity within the deforming rock body because of local differences in temperature, water content etc. Conditions such as these could account for the development of the curved hinges of the Jura Mountains (Figure 2.7) and the superimposed folds of Figures 2.6, 5.1 and 8.13 where the hinge of the evolving fold became curved such that parts of it became orientated nearly parallel to the direction of maximum compressive stress. In the case of the structural successions shown in Figure 7.35 in section 7.6, inhomogeneous strain led to the development of several of the structures under a more or less consistently orientated stress system.

Following this line of reasoning, the tightening and overturn of an upright fold initiated by flexure, for example, would represent the early stages of structural evolution which included flexural slip, followed by passive folding involving slip oblique to the initial axial direction, followed by flow. An example would be the changes (from folds named F_{br} to those named F_{bb}) in the Jussarö succession in the Svecofennian migmatites of southern Finland (Hopgood, 1984). Following F_{br} folding, continued stress resulting in slip on discrete surfaces associated with leucocratic neosome could have contributed to the distinctive form of F_{bb} (Figure 6.52). However, close examination of the developmental stages of F_{bb} suggests that a more complex deformational history was involved and the likelihood is that significant changes in movement directions and local stress systems were also involved in shaping the overall deformational sequence in the migmatites, from the earliest folds (named F_{aa}) to the latest (F_{late}) (Hopgood, 1984).

The significance of fold axial planar attitudes in metamorphic rocks has frequently been the subject of debate, as has the variation with time and crustal depth of fold attitudes. See the discussion relating to variation in structural expression in successions (section 8.1.2) and that on variation in the expression of related structures in the same limited crustal segment (section 5.2.12). See also Figures 7.35–7.38.

7.2 ILLUSTRATIONS OF TEMPORAL CHANGES OF STYLE IN FOLD SUCCESSIONS

The photographs comprising Figures 7.13 to 7.20, of structures from the succession in the migmatites of the Jussarö area of the Finnish Archipelago, illustrate the geometrical varia-tion with time of structures within an evolving orogen, from tight, 'ductile' folds to open, upright and 'brittle' structures. These changes in expression between one set and another are **temporal** variations and are distinct from those referred to in the preceding paragraph between related structures (i.e. between structures of the same set) in the same crustal segment which are **spatial** variations (section 5.2.12). The illustrations begin with the earliest structures in the succession, intrafolial isoclinal folds (Figure 7.13).

7.3 SUMMARY OF PRINCIPAL FEATURES OF FOLD SUCCESSIONS

In the broadest general terms, then, an extensive fold succession will tend to show a progression from early isolated, attenuated intrafolial fold hinges, through isoclinal folds with hinges with decreasing amounts of flattening, to asymmetrical folds which have increasingly upright axial planes and which become more open and symmetrical until they are little more than broad warps. At the same time the structures exhibit an overall change in aspect from 'ductile' to 'brittle', with spaced, open cleavage tending to become more common in the later folds (Figure 7.18).

While the earliest folds are usually isoclinal and often intrafolial it must be borne in mind that geometry alone, particularly as regards folds later in the succession, cannot necessarily

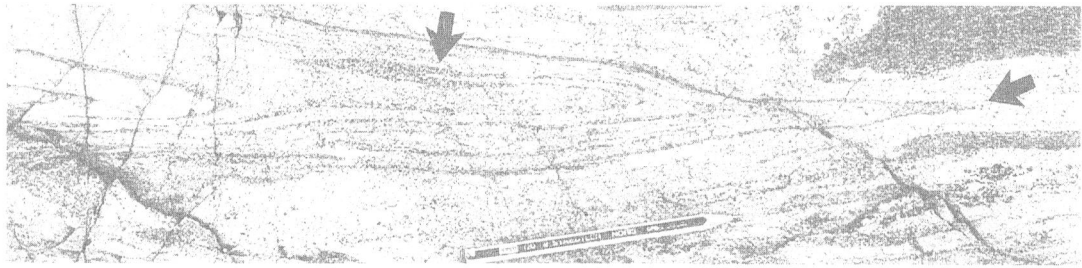

Figure 7.13 Intrafolial isoclinal folds. The hinge at the top centre (arrowhead) has been folded by the fold whose hinge shows at the right (arrow) and has an axial planar trace trending across the centre of the photograph. Structures typical of those in the earliest part of a succession. Svecofennian migmatites, Finnish Archipelago.

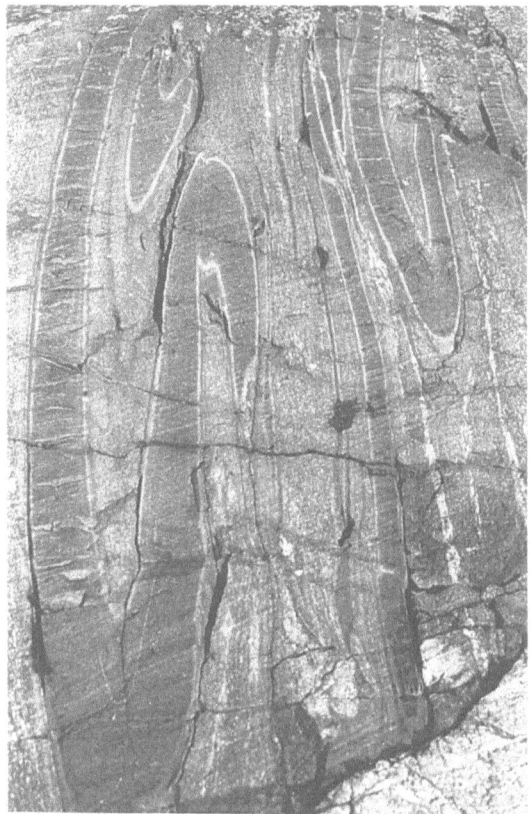

Figure 7.14 Isoclinal fold hinges showing thick, round hinges and relatively thin limbs. These are structures typical of the earlier part of a succession. Svecofennian amphibolite, Skåldö, Finland.

Figure 7.15 Asymmetrical round, to angular-hinged, fold. A structure common in mid-succession. Svecofennian migmatites, Finnish Archipelago.

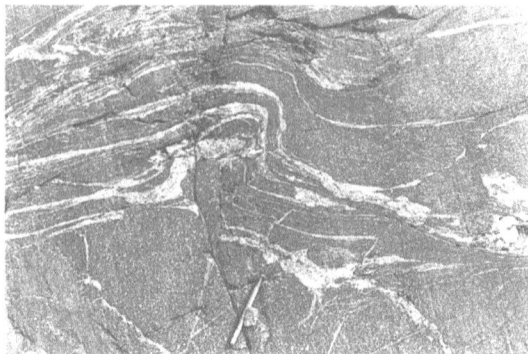

Figure 7.16 Asymmetrical, angular-hinged fold with 'steep' axial plane. The form of this structure suggests a position in the latter part of a succession. Svecofennian migmatites, Finnish Archipelago.

Figure 7.17 Folds with round to angular hinges and upright axial planes. These folds are typical of structures later in the succession. Lewisian complex, Rona, Inner Hebrides, Scotland.

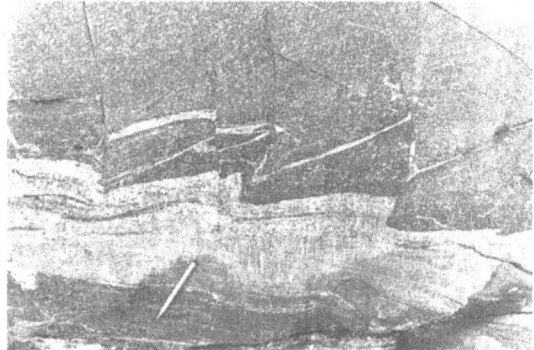

Figure 7.18 'Brittle' structure. Folds merging into fractures. Structures indicating a position late in the succession. Svecofennian migmatites, Finnish Archipelago.

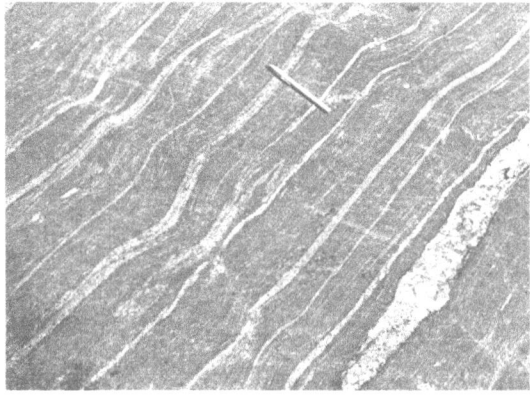

Figure 7.19 Very open late folds. Svecofennian migmatites, Finnish Archipelago.

Figure 7.21 Stylized sketches showing the changing profiles of folds through the succession, ranging from early isoclinal to late upright folds.

be considered a reliable indication of relation to any particular fold set (Figure 6.26 in section 6.2). The folds of more than one set may have similar geometry, and conversely folds of the same set may possess different geometries (see Figures 6.27, 6.28 in section 6.2).

On the other hand the totality of all aspects of folds, shape, inter-limb angle, associated mineral growth, neosome development etc. (i.e. the overall style) is often a useful pointer to the fold set to which they belong. It is rarely that in the succession resulting from multiple folding there are not a number of sets which possess distinctive styles (see also section 6.3).

7.4 VARIATION WITH TIME OF TOTAL STRUCTURAL COMPLEXITY

Confusion sometimes arises over what is meant by the statement that 'late' (or the 'latest') structures in a succession are relatively simple, whereas the latest observed structure is in fact **complex**. What are being referred to here are the latest **imposed** structures and, where they can be isolated from other structures, it can be seen that **individually**, the latest-formed structures are in fact simple because they have not been modified by subsequent deformation (see section 4.1). Of course what is in fact normally observed, however, is the **total** structure, or structural pattern which is complex by virtue of its being compound, the **sum of all the structures imposed on the rock**.

It is worth discussing further this difference between the potential form of the latest structure and the observed structural pattern.

Figure 7.20 Broad warp. This is a structure that may be difficult to detect. It is often observed to reorientate the structures shown in the preceding figures and is therefore amongst the latest in the succession. Svecofennian migmatites, Finnish Archipelago.

A. Consider the case where, from time to time, simple open folds are formed from initially planar layers in a sequence of deformational events stemming from comparable stress fields (see again section 4.1). Supposing that prior to **each** such event there happened to be within the rock at least one planar layer which could fold, and furthermore, during each event this layer always had the **same orientation with respect to the stress field responsible for the deformation** (Figure 7.22).

In these special (and unlikely) circumstances folds (say F_a–F_n) with the same geometry would develop during each deformational event (Figure 7.23).

Thus the succession (F_a–F_n), comprising each structure **in its original form** (but those structures **only**), would be one of a series of **identical** structures (Figure 7.24).

B. A more likely situation is one where, as the second structure (F_b) in the succession (F_a–F_n) is formed, the preceding structure (F_a) will be affected also and its shape will be modified accordingly. This will apply throughout the deformational sequence and the cumulative effect will be that the shape of all but the last structure will be modified by succeeding deformational events, with the earliest (F_a) being most affected and the penultimate structure (F_{n-1}) least modified. F_n will be the only fold to have the simple form shown in Figures 7.23 and 7.24. In other words successive folds would be modified to a decreasing extent by succeeding deformational events. In the case of F_a the effect would be to produce something like that shown in Figure 7.25.

This will result in the kind of succession commonly observed in complexly deformed terranes where, because of continued (or repeated) deformation, the form of the earliest-formed structures became increasingly complex compared to the less-modified, latest-formed structures with all or most structures being compound (Figure 7.26).

In reality, of course, conditions are likely to be significantly different. The stress field operating is likely to vary and change in orientation with time. Even if the stress field were to remain constant in orientation, its angular relationship to the deforming surfaces in the developing structures will certainly change so that instead of the simple arrangement depicted in Figure 7.26 a more realistic developmental sequence would be that shown

Figure 7.22 Sketches showing the effect of compressive stress parallel to an initially horizontal planar layer.

Figure 7.23 Sketches of the development of folds F_a–F_n with the same geometry, derived in each case from a planar layer having the same angular relationship to the imposed stress field, during deformational events D_A–D_N.

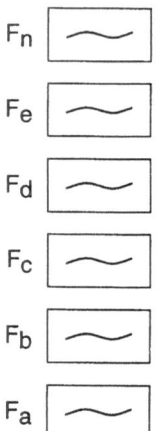

Figure 7.24 Sketch of the structure succession formed by events D_A–D_N.

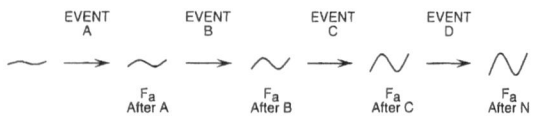

Figure 7.25 Sketches showing the modification of fold F_a (formed during D_A) by subsequent deformational events D_B–D_N.

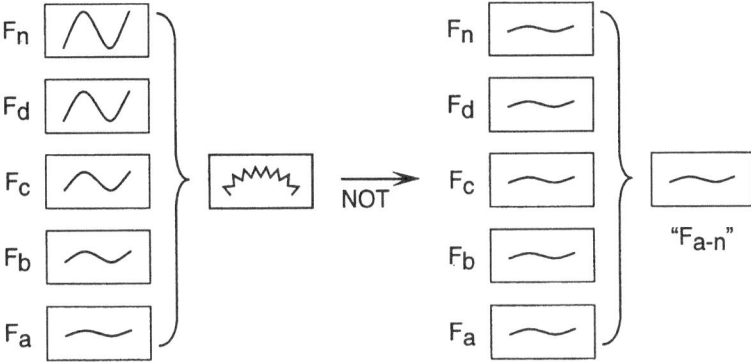

Figure 7.26   Sketches showing the structure succession formed by D_A–D_N compared with the special case of Figure 7.24.

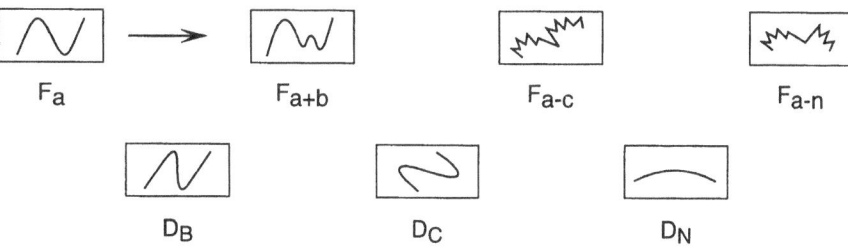

Figure 7.27   Sketches showing the modification of an initial fold (F_a) as a result of its changing orientation during subsequent deformational events D_B–D_N.

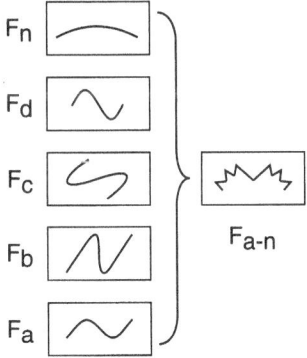

Figure 7.28   Sketch showing the possible geometry of structures in the succession resulting from deformational events D_A–D_N.

in Figure 7.27, with a final structural succession something like that shown in Figure 7.28.

Therefore, while individual structures of the late sets are simple, the latest (i.e. final), integrated (i.e. compound) structure is, of course, obviously complex, such as in the case of the compound structure shown in Figure 7.29. Here the resultant complex structure is compared with the structural succession from which it is derived and with a column showing the stages of increasing structural complexity with the successive superimposition of structures stemming from successive deformational events, beginning with an intrafolial isoclinal fold, followed by an inclined tight fold, then an inclined asymmetrical fold, then another inclined asymmetrical fold with the opposite vergence sense leading finally to the latest observed structure with the overprinting by a simple open upright fold. The attitudes and shapes of these structures are of

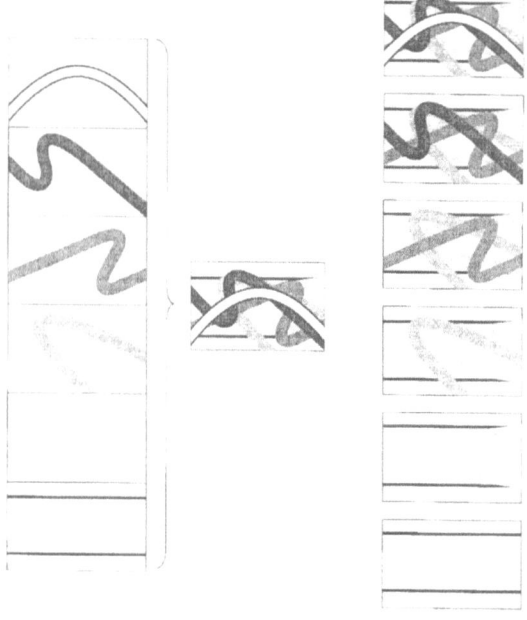

Figure 7.29 Complex structure (centre) resulting from the integration of all six structures of the succession in the left column. The right column shows the progressive increase in structural complexity upwards with time as the structure existing at each stage is overprinted by successive deformational episodes. It corresponds to the increasing structural complexity shown in Figure 7.27. Compare this succession with those of the figures in section 6.1.3.

course not only the response to the stress field at the time of their formation, but also the result of modification by subsequent deformational 'events'. See also section 6.1.3.

7.5   PARADOX OF APPARENT SIMILARITY OF FOLD SUCCESSIONS

One of the so-called problems of fold successions is their superficial similarity and this has been referred to in the preceding pages. It has led some observers with limited experience of analysing the structure of complex terranes to conclude that all (polyphase) successions

are more or less 'the same' (Hopgood 1973, 'Discussion', p.48). In fact careful observation shows that successions from polyphase deformed terranes are no more 'the same' than are folds in general all 'the same'. The similarities between, or features common to, unrelated fold successions are caused by the fact that there must of course be broad limits to the number of ways in which layered rocks can deform successively in response to applied stress. Comparable limitation must apply to any layered material. Furthermore, repeated deformation is almost certain to lead to the gradual tightening up of early folds even if these did not begin as tight structures. However, structural successions are by no means exactly the same, especially when the effects of other complicating factors such as igneous activity, metamorphism, migmatization, agmatite formation etc. are taken into account. They are no more 'the same' than different groups of people are 'the same', by virtue of the fact that all members of a group possess particular physical features, styles of dress etc. which identify them as individuals. These features contribute to the attributes of a group and together comprise its collective character, making it distinctive and unique. In the same way every structural succession has its own individual characteristics or style, and is unique.

7.6   EXAMPLES ILLUSTRATING THE SUPERFICIAL SIMILARITY OF FOLD SUCCESSIONS

The following sketches of folds in migmatite successions recorded from migmatite complexes in Western Australia, southern Finland, West Greenland, Northwest Scotland and southwestern Uganda (Figures 7.30–7.35) demonstrate the progressive change in the form of structures discussed, from early to late in the fold succession. Although the forms have been simplified and stylized (orientations are omitted) to show only essential characteristics, they also illustrate the broad but superficial similarity of different successions, a reflection of the changing response of layered successions to stress systems in evolving orogens.

Although cursory inspection of the partial successions just figured shows that they all bear some similarities to one another, there are differences between them and examination of the following example, that of a more complete structural succession in Svecofennian migmatites of the Jussarö area of southern Finland (Figure 7.35 and Figure 7.36), confirms the fact that the similarity is indeed only superficial. The comparison of the *partial* succession from Finland shown in Figure 7.30 with that in Figures 7.35 and 7.36, demonstrates that in detail, the structural attributes of structural successions collectively confer on them a character that is distinctive and unique.

The unique character of the Jussarö succession can be even better appreciated by studying the profiles in photographs of the structures from each set in the succession and comparing them (Figure 7.36).

7.7   SUMMARY OF FOLD VARIATION IN SUCCESSIONS

In the discussion earlier it was noted that because repeated deformation is (almost) certain to be accompanied by compression at some stage, the form of early folds becomes modified during successive deformational episodes. As a consequence, while the earliest folds observed in a structural succession are usually isoclinal and often intrafolial, those same folds may have originated as open structures which have responded to later deformation by developing progressively tighter profiles as their inter-limb angles diminished.

However, continue to bear in mind that the geometry alone of a structure, particularly as regards folds later in the succession, is not an indication of its relationship to any particular fold set (see again Figures 2.13 and 6.26). The folds of more than one set may have similar geometry while conversely folds of the same set may possess different geometries (see

again Figures 6.27, 6.28 in section 6.2). It is worth reiterating at this stage also that, on the other hand, **all** aspects of a fold, its shape, inter-limb angle, associated mineral growth, neosome development etc. taken together, i.e. the overall **style**, often provides a useful pointer to the set to which it belongs.

As an extension to this it is important to remember that, while there are certain basic

Figure 7.30 Stylized fold profiles from part of the Svecofennian structural succession near Skåldö, in southern Finland. Adapted from Figure 20, Hopgood, 1980. Reproduced with the permission of the Royal Society of Edinburgh.

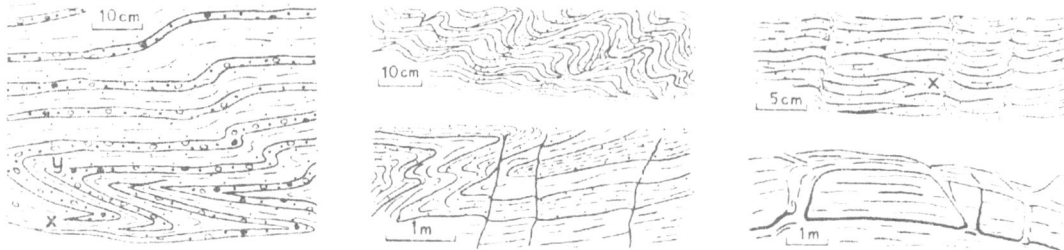

Figure 7.31 Stylized, simplified fold profiles from the structural succession in migmatites of the Leeuwin Block, southern Western Australia. Earliest structures (intrafolial folds folded by isoclinal folds and modified by later, open folds) are on the left, followed by asymmetrical folds (centre) with the latest (upright warps) on the right. (After Hopgood, 1973, Figure 3 and reproduced with the permission of the Geological Society of South Africa.)

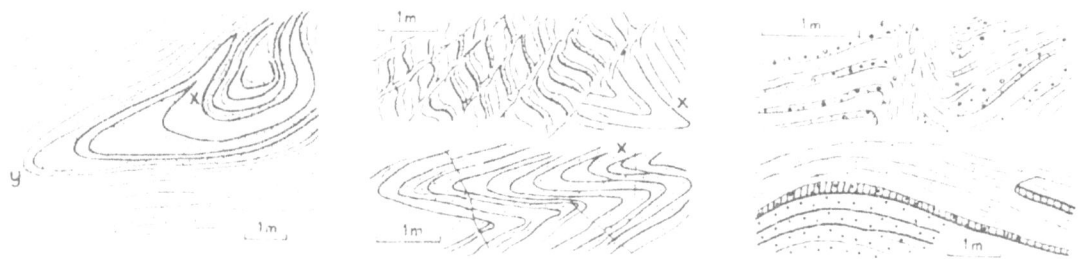

Figure 7.32 Simplified fold profiles from the Pre-Ketilidian migmatite structural succession in southern West Greenland. Intrafolial isoclinal folds folded by isoclinal folds with axial traces curved by later open folds (left and centre) are the earliest structures. Later structures include slip surfaces associated with leucocratic neosome veins (centre) with open upright folds associated with diffuse veining and upright warps (right) the latest structures. (After Hopgood, 1973, Figure 2 and reproduced with the permission of the Geological Society of South Africa.)

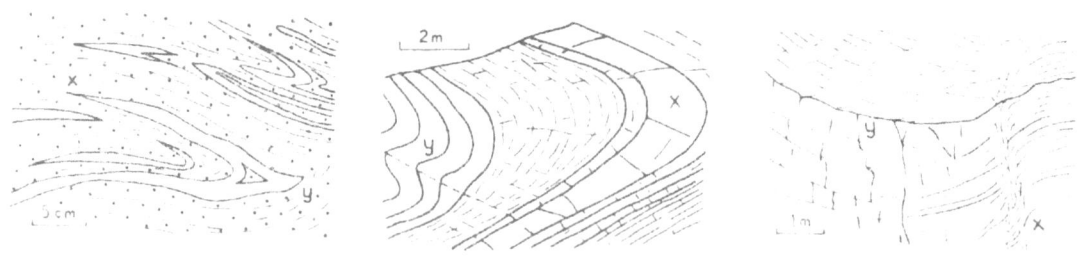

Figure 7.33 Fold profiles from the succession in migmatites of the Lewisian complex, Outer Hebrides, NW Scotland. The earliest structures shown (on the left) are intrafolial isoclinal folds refolded by isoclinal folds. These are followed by recumbent tight folds affected by inclined folds (centre) and finally by upright, very open folds (right). (After Hopgood, 1973, Figure 1 and reproduced with the permission of the Geological Society of South Africa.)

Figure 7.34 Simplified fold profiles from the structural succession in migmatites adjacent to the margin of the Western Rift Valley, Ankole, southwestern Uganda. The earliest folds shown (again on the left) are intrafolial hinges folded by 'fragmented' isoclinal folds. Later structures (centre) are inclined asymmetrical folds and the latest (right) are upright open folds. (After Hopgood, 1973, Figure 4 and reproduced with the permission of the Geological Society of South Africa.) ▲

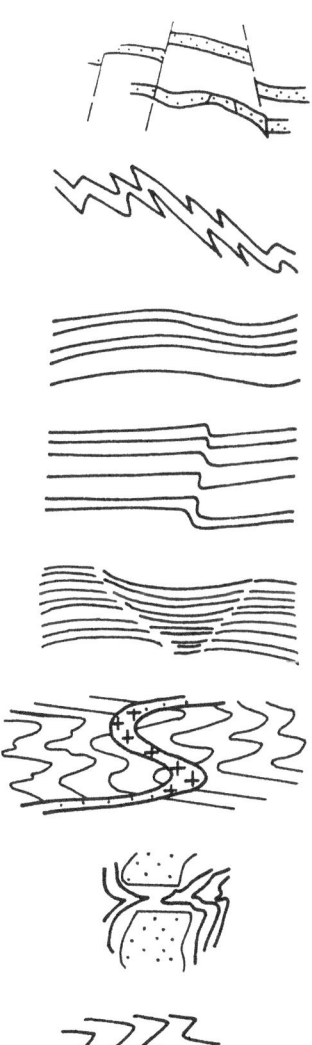

Figure 7.35 Succession of structures (profiles stylized) from the succession in Svecofennian migmatites, Jussarö region, southern Finland. Adapted from Table 1, Hopgood, 1984. Reproduced with the permission of the Royal Society of Edinburgh. ▶

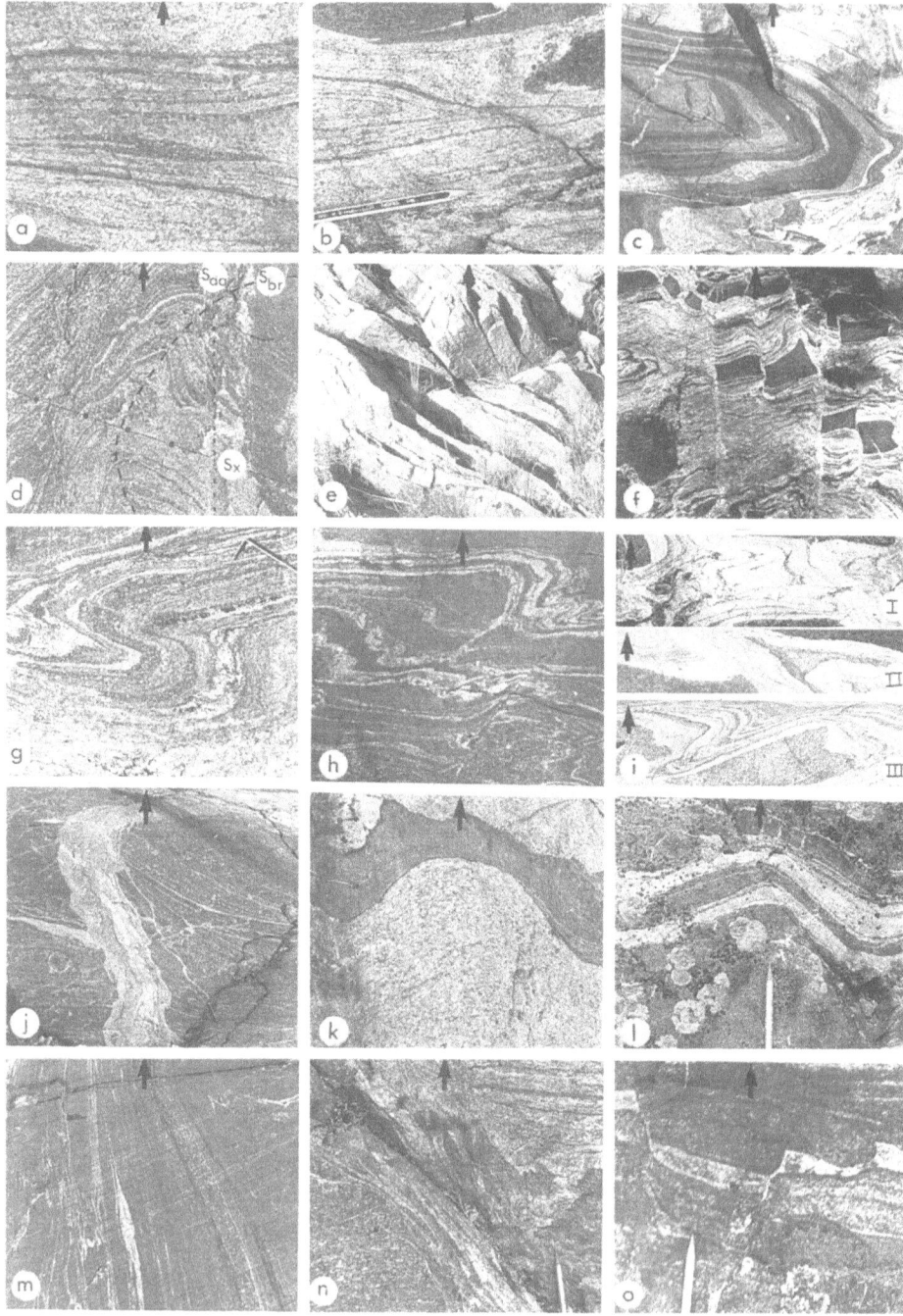

Figure 7.36   Photographs (**a–o**) of the structures in the succession from Svecofennian migmatites grouped to allow comparison of the changing style from earliest at the top left to the latest at bottom right. Jussarö, Finland. From Figure 5, Hopgood, 1984. Reproduced with the permission of the Royal Society of Edinburgh.

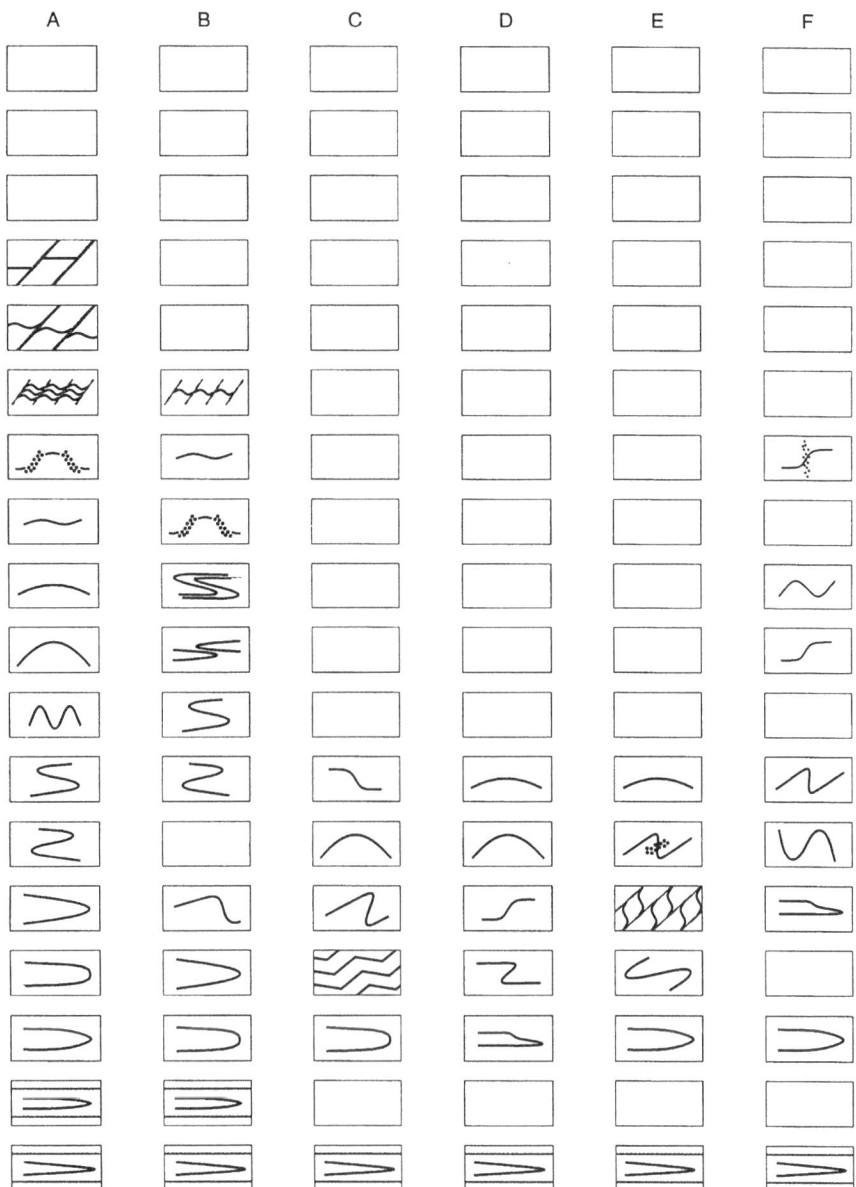

Figure 7.37   Comparison between successions from different terranes. Each succession, although simplified, can be seen to be distinctly different from the others, particularly when factors such as syntectonic metamorphic mineral growth and composition of associated partial melt (neosome) and igneous veining are taken into account (these are omitted here). However, each has in common a broad temporal change in structural style, with fold geometry ranging from tight isoclinal intrafolial through recumbent isoclinal to asymmetrical inclined, tight upright to open upright followed by upright warps, as well as increasing discordance to any associated emplaced igneous melts. (**A**) A stylized summary 'succession' ('open ended' at the top) of all structure types shown. (**B**) Succession in Svecofennian migmatites, Jussarö region, southern Finland. (**C**) Succession from southwestern Uganda. (**D**) Succession from the Lewisian complex, Outer Hebrides, Scotland. (**E**) Succession from the Pre-Ketilidian of Southwest Greenland. (**F**) Succession from Sharyzhalgay migmatites, Lake Baikal, eastern Siberia.

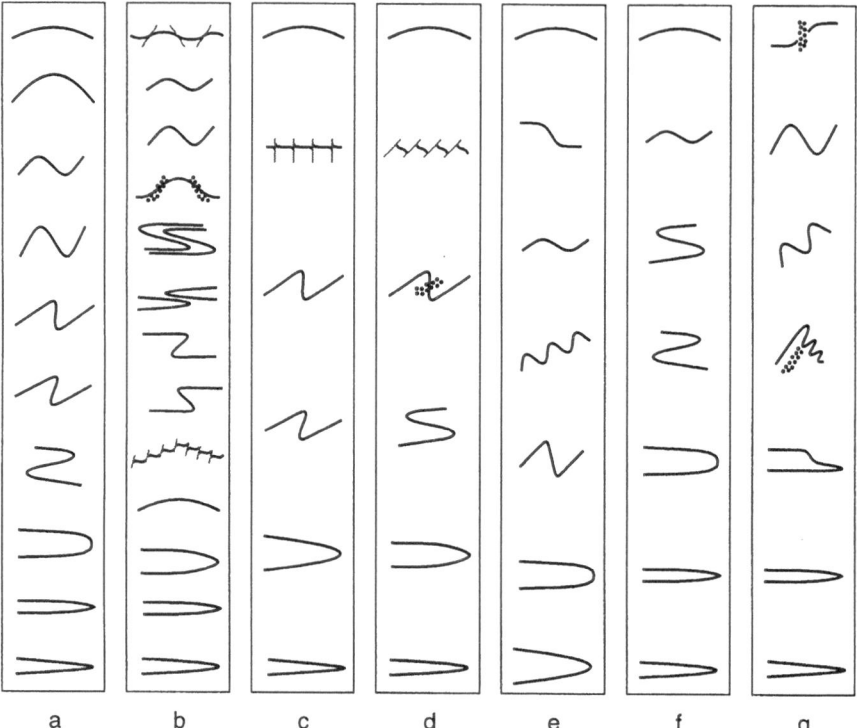

Figure 7.38 Successions of structures recorded from different terranes, shown in stylized form to enable their similarities to be contrasted with their differences more easily. Note that the similarities between the successions stem from the fact that they all begin with isoclinal folds which are generally intrafolial (i.e. they lie within the dominant foliation) or very tight folds, and progress to more open and upright folds and brittle structures. On the other hand the differences between the successions arise from the fact that their composite structure (from all the structures comprising the successions) is such that each succession is quite distinct from the others. As well as being stylized, the structure profiles in the successions shown are modified in some cases from the originals, but the order of succession is unchanged. (a) Theoretical succession of fold structures from a 'typical' orogen. (b) Fold succession in the Svecofennides in the Jussarö region, southern Finland. (c) Fold succession in the Leeuwin complex, southwestern Australia. (d) Fold succession in the Pre-Ketilidian of Southwest Greenland. (e) Fold succession in the Lewisian complex, Outer Hebrides, Scotland. (f) Fold succession in gneisses from southwest Uganda. (g) Fold succession in migmatites, Lake Baikal, Eastern Siberia.

similarities in structural successions in terms of the way their structures evolve with time, when all their characteristic features are taken into consideration each succession will itself be seen to be **distinct** and **distinctive** (Figures 7.37 and 7.38).

PRINCIPLES OF SUCCESSION DETERMINATION

8.1 DETERMINING THE STRUCTURAL SUCCESSION: THEORY

8.1.1 CONCEPTS

When setting out to establish the structural succession in complexly deformed rocks such as migmatites, it is advisable to restrict the approach to one which, as far as possible, is observational rather than deductive. This is important because when working with such structural complexity, any deductions made from a premise based on insufficient evidence are almost certain to lead to incorrect conclusions.

One problem associated with a deductive approach stems from the likelihood of unconsciously observing only those features that appear to fit the initial premise (which anyway might not be entirely correct) while other significant relationships could be missed. This provides the potential for further error because, when it comes to the interpretation of those observations, this will be done in terms of this premise, so leading to incorrect conclusions because of the limitations of these observations.

Consequently the inaccuracy of conclusions arrived at in this way from limited observations would be compounded, not only by any misinterpretation of the data, but even more so if the initial premise proved to be entirely incorrect.

For this reason it is advisable not to adopt a 'hypothesis-led' approach such as one which, for some reason, makes an assumption regarding the number of structural sets in a succession, or the number of deformational phases involved in the structural development of this succession. Failure to confirm such an assumption could lead to (fallaciously) attempting to 'explain' this failure by 'showing' the 'inadequacy' of some particular argument or technique used to demonstrate the initial premise, a premise which might in fact have been incorrect. Such an instance might be the 'demonstration' of the futility of analysing fold relationships based on the false assumption that polyphase-deformed rocks have been affected by only a single deformational episode, or by only a small (specified) number of episodes.

The resolution of the structural complexity of a terrane depends essentially on the ability of the investigator to use the evidence available to establish overprinting relationships between the geological structures observed. In the widest sense this will entail relating structures that are not only tectonic but also igneous and metamorphic in origin, because in most cases a complex tectonic history is typically associated with high-grade metamorphic, or ultrametamorphic, events. While this is not always the case (e.g. in Britain the metamorphic grade of the strongly deformed rocks of the Scottish Southern Uplands is low and likewise the strongly deformed Carboniferous rocks exposed along the Fife coast in Scotland are essentially unmetamorphosed), metamorphism forms an integral part of the production of migmatites and gneisses whose structural study forms the basis of this work.

8.1.2 SPATIAL VARIATION IN STRUCTURAL EXPRESSION

The events that contribute to the structural succession are likely to have differing expressions over a wide region, depending on the physical conditions prevailing during deformation, and will vary according to the depth at which they were formed as well as laterally (see also the discussion on variation of **related** structures in section 5.2.12). Structures of the same set will change in style and a particular structural set might be represented by folds in one place and by, say, ductile shear zones in another, whereas in 'transitional' environments both may coexist (Figure 8.1 and Figure 5.39b). Such change on a large scale, from deep-level folds to superficial thrusts, has long been recognized in regions such as the European Alps and has been the inspiration for much research, including, more recently, that of Epard and Escher (1996) involving geometric modelling to explain the change in style of basement and cover deformation. In this connection see the discussion in section 8.1.9; the comments in section 7.2 regarding temporal as well as spatial variation in structural expression, and also the references in section 5.2.12, 'Variation in expression of related structures in the same limited crustal segment'.

Each structural set can be regarded as the response to an event which is temporally related to other events, earlier or later, in the history of the terrane. It is the aim of the investigator to determine the succession of 'structures' which represents the response of the rocks to the sequence of these 'events' contributing to the geological history of that terrane.

As stated in the introduction to Chapter 1, the principle embodying the use of overprinting as a criterion for distinguishing temporal relationships is by no means new, having been employed long before the work of Sander (1930) became generally known. It has been recognized for at least two hundred years,

Figure 8.1 Fold associated with ductile shears in migmatites. An example of a structure formed in a 'transitional' environment where folds and ductile shears coexist. Dabie complex, Lutian, southeastern China.

since Hutton's 1795 recognition of it by his reference to a transecting granite vein (section 1.1) and later in the 19th century when Clough commented on refolded cleavage in *The Geology of Cowal* (1897). (See Chapter 1 and especially section 5.3.)

Structural overprinting includes refolding, folding of planar structures such as cleavages, and overprinting of cleavage and folds by later cleavages etc. (see the illustration based on work published by Rutland and Etheridge (1975) in section 10.1.8). While for the most part straightforward, such relationships are sometimes complicated where the cleavage 'transects' the folds. However, in areas of reasonably extensive, good exposure, inconsistencies stemming from such 'transection' should soon become evident (section 3.2.2). Refold relationships are usually less prone to misinterpretation, and furthermore folding tends to be more common than cleavage in the complex rocks of high-grade terranes (see also section 3.1.3).

On the basis of the simple principle of overprinting, one that is fundamental to geological thinking, structural relationships can be established directly by field observation. This is because deformed structures are earlier than those that deform them, and transecting (cross-cutting) features are later than those they cut (see 'Principles of structural analysis', section 2.2.2, and 'Principle of overprinting', section 3.1.1). Therefore structures, metamorphic mineral growths and igneous intrusions that are observed to post-date other structures etc. can be traced (i.e. correlated by using characteristic features already noted as a means of recognizing them) to other outcrops where they in turn may be observed to have been deformed by later structures or transected by later intrusions etc. (section 5.2.5). In this way the succession of structural, igneous and metamorphic relationships within the rocks of the crustal segment under study can be built up.

Folds belonging to a particular position in a structural succession are grouped as a set and structures of a particular set are related to those of other sets by the same overprinting relationships. In some cases it may be necessary (particularly in the early stages of a study) to resort to using style in order to make a provisional classification of a particular structure, especially when exposures are isolated from one another (see section 5.2.3, 'Importance of early provisional classification of structures'). But classification on this basis is always subject to confirmation in terms of overprinting relationships.

Simple general rules can also be laid down as an aid to the initial classification of struc-

tures. For example, overall consistency of trend of a structural feature implies that it is one formed late in the deformational sequence and will therefore be high in the structural succession. Similarly, late structures may be more clearly defined because they have been little modified by later events (see section 7.4). These conditions are referred to in section 2.3, 'Assumptions inherent in the approach' and in the assumptions discussed in sections 7.1 and 7.1.1.

The approach used is dependent on a number of factors, especially inhomogeneity of deformation. If all structures were equally well developed in terms of scale and intensity throughout the rock body then it is likely that it would be very difficult, if not impossible, to do the work effectively. However, equality of structural development is seldom, if ever realized because local inhomogeneity of the layered succession causes inhomogeneous transmission of stress throughout the body. This in turn results in different responses to the same external stress in different places and further structural inhomogeneity is imposed because of this strain (deformation) partitioning (Figure 8.2). Different parts of the rock body are deformed (folded) to different degrees, such as in the hinge zones of existing

Figure 8.2 High strain zones (ductile shears and fold) separated by zones of low strain (unfolded) with more or less planar foliation. Lutian, southeastern China.

large-scale folds where the pre-existing foliation is perpendicular to the compressive stress, while some parts are not folded at all.

The hinge of a pre-existing fold provides a range in foliation attitudes (up to 90°) with respect to a stress direction acting perpendicular to the fold axis (Figure 8.3). Furthermore, crumpling in the hinge zone of a large fold provides many more sites where the foliation attitudes can present widely differing angular relationships to a particular stress direction during a particular deformational event. This means that the chances that each and every deformational episode will be recorded in the hinge of a fold are much higher than they are on the limbs, particularly in those cases where the folds are tight and the limbs nearly parallel (Figure 8.4). On the other hand, resolution of refold relationships of small-scale folds is more likely to be possible in the limbs, rather than in the crumpled hinge, of the large fold. This is because there are likely to be fewer fold sets recorded here (Figure 8.5). Therefore the structural complexity will be correspondingly less than in the hinge so that the relationships between the sets is clearer. The overprinting relationships between only two or three folds (Figure 8.6a), rather than between three or more (Figure 8.6b), is normally a comparatively simple matter to resolve on the outcrop by direct observation.

It is also worth noting that complex structural relationships are often easier to resolve

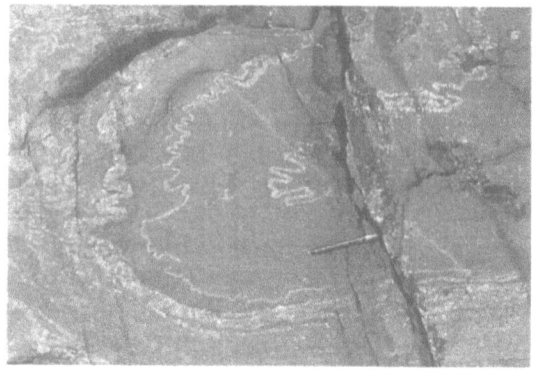

Figure 8.4 Round-hinge fold showing varying attitudes of folded thin leucocratic layer in the hinge in contrast to the parallelism of this layer with the fold limbs. Svecofennian migmatites, Jussarö area, southern Finnish Archipelago.

on two-dimensional exposures than on those that are three-dimensional. This is because in the case of two-dimensional exposures where the structural relationships on each reference plane are consistent, the observed relationships are generally simpler, whereas in three dimensions where the structural relationships are being viewed on different, often varying, surfaces, the basis for comparison between different structures is not a consistent one.

Besides the fact that variation in scale and geometrical expression helps in the distinction between fold sets where structural relationships may otherwise be complex and difficult to decipher, the association of, or absence, of

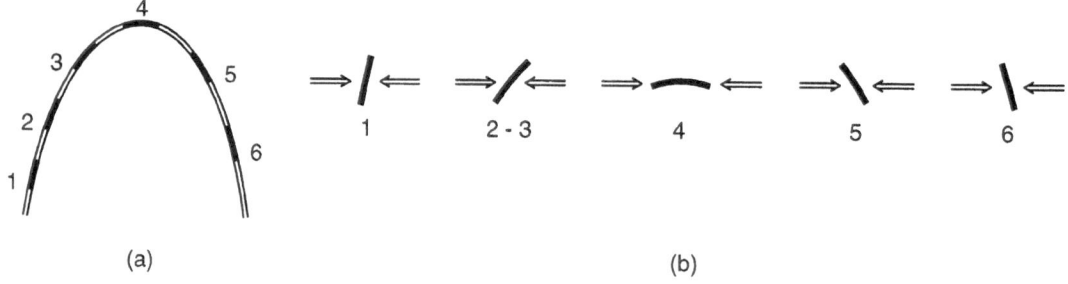

(a)   (b)

Figure 8.3 (a) Sketches to show the varying angular relationships between the direction of (horizontal) maximum compression and a curved (folded) surface. (b) Sectors 1–6 in a cylindrical fold hinge each present a different aspect to an applied (horizontal) compressive stress direction, ranging from perpendicular in the case of sector 1 to parallel in the case of 4. Compare Figures 8.4 and 8.5.

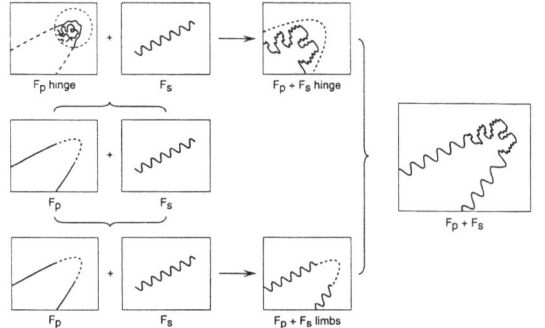

Figure 8.5 Sketches explaining why the recognition of small-scale folds (F$_s$) superimposed on the limbs of pre-existing folds (F$_p$) is relatively easy, in contrast to recognizing folds superimposed on the hinges. The complexity of the structure in the hinge zone makes identification of individual superimposed structures considerably more difficult. (cf. Figures 8.4 and 10.1–10.4).

mineral development, cleavage etc. and the nature of any cleavage or mineralization can also serve to separate fold sets (see the discussion in section 6.2, 'Style, its significance and factors influencing it').

The 'total', or total **observed** structural succession is synthesized by piecing together 'segments' of the structure comprising the various partial successions determined from different structural settings in different parts of the terrane. These partial successions might well differ from one another because of differences in structural settings, and because of their differences each will contribute to the building up of the 'total' succession. As has been seen the differences in these settings are controlled by factors such as:

(i)   the local attitude of the foliation with respect to the stress acting at the onset of a particular deformational episode (e.g. the foliation in the limb rather than in the hinge segment of a pre-existing, steeply plunging fold with an N–S axial trace is more likely to respond to N–S directed horizontal compressive stress to produce folds with E–W axial traces);

(ii) the presence of relatively large-scale pre-existing folds so that the relationship of the smaller folds superimposed on different segments of the earlier folds can readily be determined, and

(iii) the presence of relatively small-scale folds initially which will be dispersed around the hinges of later large folds.

8.1.3   TEMPLATE, OR MASTER KEY CONCEPT

From the outset, the structural study at each locality should be treated individually. There are two reasons for this. Firstly the structure might differ significantly from that at the other localities for reasons given earlier in the discussion on 'Structural correlation of tectonostratigraphic terranes', section 6.1.3. Secondly, there is the very real possibility that one or more sets of structures could predominate over others in the succession at any one site and so influence the structural expression locally. The effect of this could be to make it appear initially that the 'total' structure (i.e. the structural succession) at this particular site is fundamentally different from that at other localities, an impression that is apparent rather than real.

The succession having been determined for a particular locality, it is later compared with successions from other localities. For example, the structure at any one place may be characterized by, or dominated geometrically by, say F$_x$ or F$_d$ or F$_q$ etc. whereas at other localities the expression of those particular structures may be minimal or insignificant. Even when some parts of the succession are not recognized at all in a particular area, and this condition tends to the rule rather than the exception, those parts of the succession that have been identified will always be in the same order and will always fit in the correct position in the same succession elsewhere (Figure 8.7).

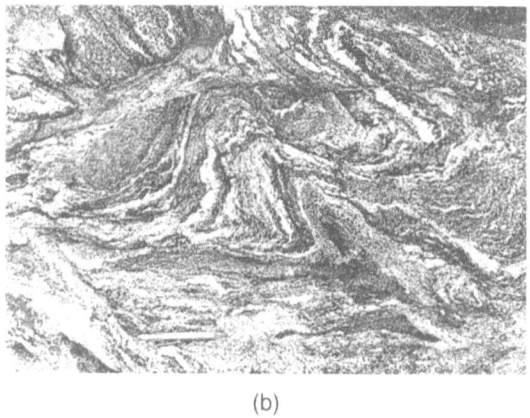

(a)

(b)

Figure 8.6   (a) Comparatively simple interference pattern from which the refolded relationships can be determined easily. Migmatites, Lewisian complex, Harris, Outer Hebrides, Scotland. (b) In contrast, this complex interference pattern resulting from refolding such as might take place in the hinge of a pre-existing fold, is much more difficult to resolve. Migmatites, Dabie complex, Lutian, southeastern China.

This is in some way comparable to having a master key that will fit any lock in the system, or strictly it is more closely analogous to the converse situation, viz. that of a series of locks which are adaptable to the same master key. Each local partial succession is like a piece of a larger template or three-dimensional jigsaw puzzle into which it will fit (Figure 8.8). See also section 6.1.3, section 8.1.9, section 12.5 and section 13.4.

The structural successions at localities 1–3 and 1a–3a of Figure 8.8b will have much in common but are quite likely to display local differences in response to variations in stress fields. These variations could arise from in-homogeneity affecting the local orientations of the principal stress directions and possibly also in response to variations from place to place in the degree of obliquity of edges of impinging lithospheric plate margins during collision (Figure 8.9a(iii)).

Successions are likely therefore to have some structural features that are consistently found at each locality as well as local variants which are specific to each locality, i.e. there will be those structures that are common to all

localities and those that are restricted to only one locality, or to a few localities (Figure 8.9b).

8.1.4 TESTING THE VALIDITY OF OBSERVATIONS

The test of the validity of the structural association that has been established lies in the fact that comparable associations are found repeatedly throughout the study area. At each locality or exposure the structural relationships, no matter how extensive or limited in their expression, should consistently match the established succession.

Besides the consistent **association within** specific fold sets of characteristic features such as mineral growth, cleavage, fold geometry, veining and other stylistic features (i.e. the very features which at the outset draw the attention of the observer in the field to the probable existence of these fold sets), confirmation of the existence of the sets is shown also by the consistent **relationship between** the sets, such that at localities throughout the terrane, (i) structures of the same set possess the same stylistic features and (ii) structures of particular sets always have the same relationships to one another. This means that a group of structures, folds say, of Set A for example, always bear the same relationships to folds of Set B and to those of Set C, or to those of Set N.

In just the same way as any scientific premise is 'confirmed'[1] by the consistency of some particular relationship, the recognition of comparable structural successions consistently in exposures throughout a terrane

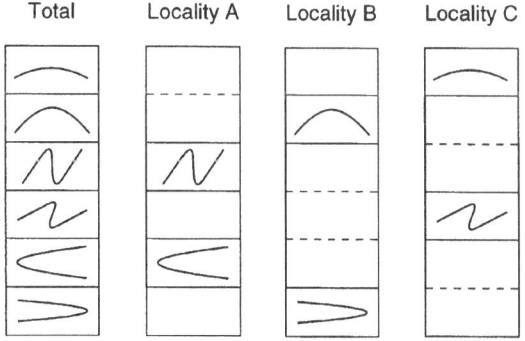

Total	Locality A	Locality B	Locality C

Figure 8.7 The 'total' succession (left) from the combined local partial successions at localities A, B and C. Although the partial successions at localities A, B and C are far from complete and also different from one another, the relative order of the structures in the successions is the same, and the same as that in the 'total' succession.

[1] Or strictly, is **accepted** because of this consistence, at least until such time as evidence is found to the contrary. This is the case in any scientific study. As stated by Stephen Hawking, 1988, on p. 11 of his *Brief History of Time*, "Any physical theory is always provisional, in the sense that it is only a hypothesis: you can never prove it. No matter how many times the results of experiments agree with some theory, you can never be sure that the next time the result will not contradict the theory." In the present case of the study of migmatite structure, 'observation' would replace the word 'experiment'.

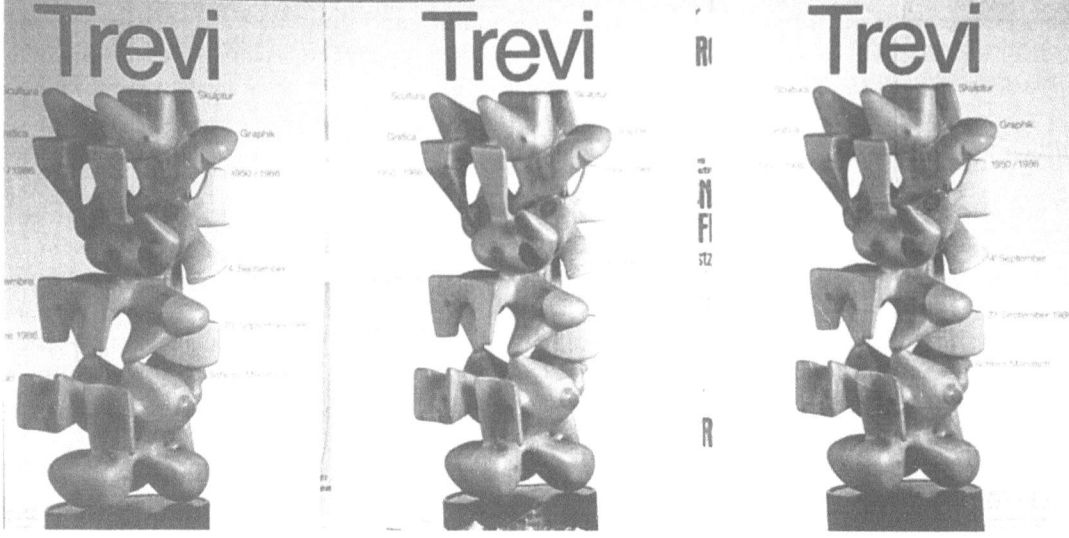

(a)

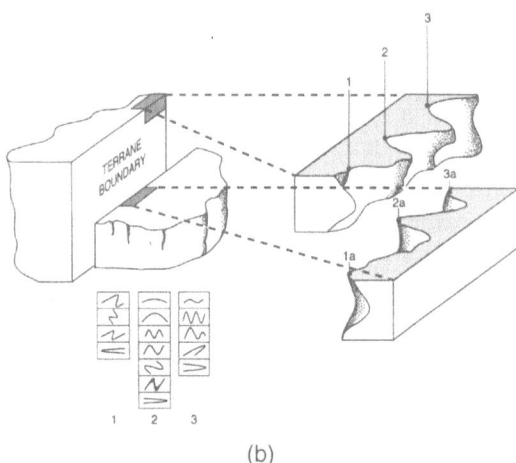

(b)

Figure 8.8 (**a**) Poster showing photographs of sculptures, Merano, northern Italy. These 'figures' are analogous to three-dimensional jigsaw puzzles where matching of contiguous parts is rigorously controlled by the intricacy of the pattern. Precise matching of the component parts of the sculpture is possible and only those parts that match correctly can be fitted together. (**b**) Structural matching along a tectonostratigraphic terrane boundary in terms of the comparability of local structural patterns, i.e. of the local structural successions. Structural successions can be matched between 1 and 1a, 2 and 2a, 3 and 3a etc., so providing an unambiguous match of the boundary segments at each of these localities. (Compare Figure 8.7).

provides confirmation of the fact that for the whole of that terrane there exists a particular, single order of structural succession. This succession is the product of the deformational sequence which affected the terrane, with local variants in the succession stemming from differential local responses to deformation, for whatever reasons, e.g. such as those discussed above (Figure 8.9).

8.1.5   DETERMINING THE STRUCTURAL SUCCESSION: PRINCIPLES

The approach to the structural investigation of gneisses and migmatites will vary somewhat depending on the nature of the deformation and the structures produced – the relative scale of the folds of different sets, their relative orientation, the presence or absence of distinctive features, and especially the number of fold sets.

In structural analysis one, or both, of two different approaches can be used and the

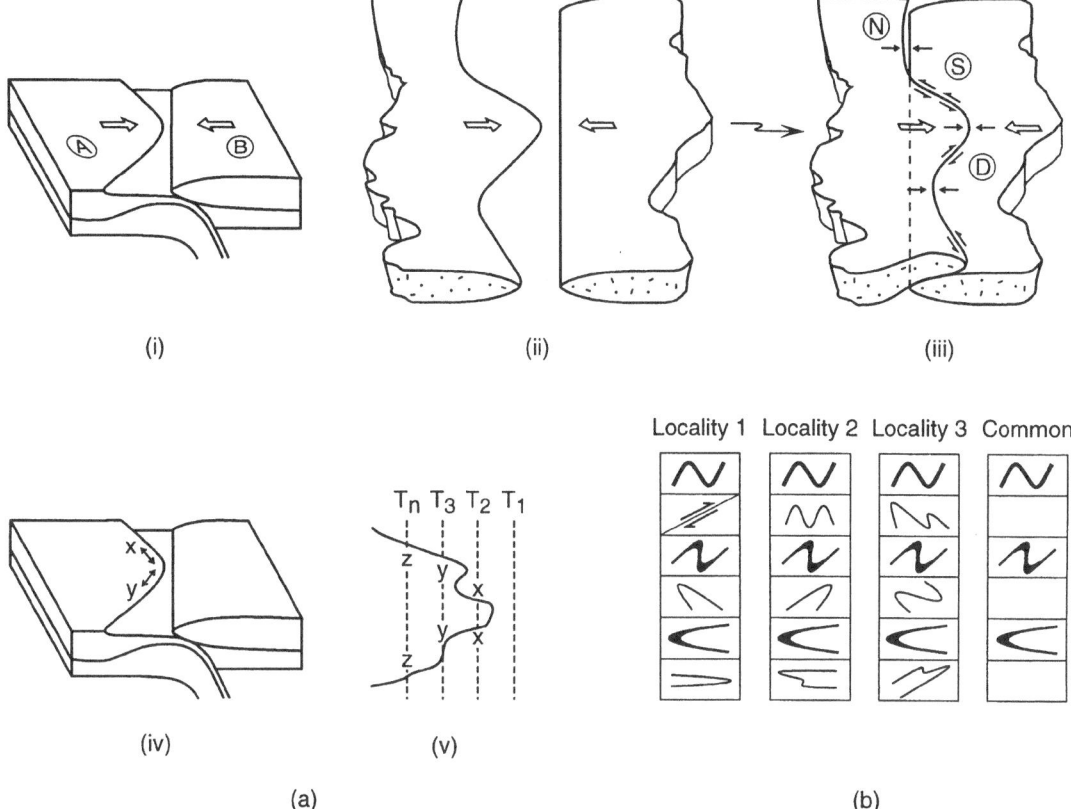

Figure 8.9 (**ai** and **aii**) Convergence of lithospheric plates A and B showing the potential for differing deformational effects and timing of events arising from obliquity of impinging irregular (non-parallel) plate margins. (**iii**) The range in stress orientation during collision varies between normal (at N) to dextral oblique (at D) and sinistral oblique (at S). (**iv**) The fold axial trend ranges between SE at X and NE at Y. (**v**) Diachronism of tectonothermal events. (**b**) Structural successions from three localities (1, 2, and 3) along the plate boundary of (**a**), showing features common to all localities (bold and in column 4, 'common'), as well as features specific to each locality.

choice depends primarily on the number of fold sets affecting the rocks.

One approach relies on the use of stereographic plotting of structural data to analyse the temporal relationships between folds in terms of the way in which the data is distributed on the stereo plots. As was said earlier, this is normally confined to situations where deformation has been such that the structures are limited to no more than say two or three fold sets. This might be the case where the relationships between only the latest two or three sets has to be determined, such as when these sets can be distinguished from the preceding (earlier) sets 'by inspection'.

On the other hand, a different approach to the investigation of the fold relationships is needed where (i) it is necessary to be able to determine the relationships between fold sets **prior to** the latest two or three (or four) sets, and (ii) in those instances (the general case) where the latest sets cannot be readily distinguished from the preceding sets of folds. In other words a different approach must be adopted where the succession comprises several sets of structures, as is the case with migmatites. This entails studying the refold relationships on the exposure directly.

Experience has shown that with few exceptions it is unusual to find exposures where the relationships between as many as three or four sets of folds can be determined by direct observation. In most cases there might not be any apparent overprinting at all, or at best there might be evidence for refolding relationships between only two sets. Clearly then it will seldom, if ever, be possible to establish a full structural succession for the terrane being studied solely on the basis of direct observation. Determining the succession must necessitate some form of integration of 'fragments' of observational information, some of it overlapping, collected from several localities. Each of these fragments embodies the relationships between structures in a segment of the succession and if the relationships in all the segments are found then it becomes possible to construct the full structural succession for the terrane.

Before discussing the process of assembling the data from different localities to produce the 'total' structural succession it is necessary to consider carefully the significance of what is seen on the outcrop and examine in detail the meaning of the relationships observed.

Supposing overprinting relationships have been observed between three sets of folds. Strictly all that has been observed is simply the **relative** refold relationship between the three sets, and not the **absolute** relationship. What is known then is the **order** of overprinting but what is **not** known is:

(i) whether the three folds represent **all** the fold sets affecting the rocks;

(ii) whether other fold sets **precede** or **succeed** the three sets observed, and

(iii) whether there are other fold sets **between** the three sets observed.

Consider an isoclinal fold with a curved axial trace crossed by a set of angular folds (Figure 8.10).

Fold set F_z is earlier than set F_x and both are earlier than F_y. But it is not possible to say whether F_z, F_x and F_y represent F_1, F_2 and F_3 (i.e. F_{1-3}), F_2, F_3 and F_4 (i.e. F_{2-4}) or F_1, F_3 and F_6 for instance, or some other combination of fold sets. So, while the order of overprinting is known, the true position of each fold set has yet to be determined.

It must now be clear from what has been said that in order to build up the succession from several exposures it will be necessary at some stage to depart from the purely observational aspect of the study, namely one where relationships are determined between a group of structures seen together on the outcrop. Inevitably this departure from pure observation entails two distinct steps,

(i) the recognition of folds of a particular set, to enable

(ii) their correlation between one isolated exposure and another.

This stage of the procedure requires particular care in its execution and is one that has given rise to much of the confusion about the method as well as to many of the doubts concerning the feasibility of establishing structural successions in complexly deformed rocks, the bases of which have already been discussed in section 5.2.2 and section 5.2.5.

Consider next the simple case, shown in Figure 8.11, where there are three folds exposed together, an isoclinal fold, F_n (? = F_1), an asymmetrical fold, F_x and a very open fold or warp, F_y.

The three folds are F_n, initially and **provisionally**, considered to represent the first set (F_1), F_x and F_y. Exposures of the overprinted relationships between the three sets show only the relationship of F_x to F_n (F_1) or of F_y to F_n (F_1), i.e. the paired relationships of F_n with F_x and of F_n with F_y. Suppose, however, that overprinting between the three folds, F_n, F_x and F_y (i.e. 'F_1', F_x and F_y) always shows both F_x and F_y to be **later** than F_n ('F_1') so that the following relationship holds: F_n is followed by (F_x & F_y), or to put it another way, either:

1. $(F_n + F_y)$ are followed by F_x or,
2. $(F_n + F_x)$ are followed by F_y.

If $n = 1$ then (1) above could be any of the following, depending on whether one or more sets exist between either or both F_n and F_x, and F_x and F_y:

1a. If there are no other sets between them, F_n is followed by F_y then by F_x or,
1b. If there is a set ($f_?$, say) between F_n and F_y, then F_n is followed by $F_?$, by F_y and then by F_x or,
1c. If set $f_?$ lies between F_y and F_x then F_n is followed by F_y, followed by $f_?$ and then by F_x or,
1d. If there is a set $F_?$ say between F_n and F_y and a set $f_?$ between F_y and F_x, then F_n is followed by $F_?$, which is followed by F_y, which is followed by $f_?$ and then by F_x etc.

Similarly, such variants could apply to (2) above. Although on the basis of observational experience (see section 7.1), since the open form of F_y suggests it is likely to be a later fold than the tighter asymmetrical F_x, the succession is more likely to be:

2a. F_n then F_x then F_y or
2b. F_n, $f_?$, F_x then F_y etc. The present example is one where the relative ages of F_x and F_y could be confirmed by examining the distribution in stereographic projection of F_x folds compared to that of F_y folds. If, as is suggested, F_x precedes F_y, sterographic plotting of its axial direction should show a relationship consistent with dispersion about the axis of F_y.

Hence it is necessary first to determine which of F_x and F_y is the earlier set and second to discover whether or not F_x and F_y are the two sets **immediately** succeeding F_n ($? = F_1$), viz. F_2 and F_3, or whether they happen to be F_3 and F_5, or F_4 and F_7 etc. In other words it is necessary to know whether the three sets comprise the whole succession ($F_{(n,n+1,n+2)}$, or simply only part of a succession such as $F_{(n,n+2,n+4)}$ or $F_{(n,n+3,n+6)}$ etc., with the relative

order of F_x and F_y to be determined for whichever of the cases is correct.

So to summarize, if F_x precedes F_y then (2) above might become:

2a. F_n, F_x, F_y
2b. F_n, $f_?$, F_x, F_y
2c. F_n, F_x, $f_?$, F_y
2d. F_n, $F_?$, F_x, $f_?$, F_y

And, if $n = 1$ and F_x succeeds F_y immediately, so that $x = 2$ and $y = 3$, simply: F_1, F_2, F_3.

Therefore, there are two aspects to be considered when establishing the relationship between fold sets in order to construct a fold succession. Firstly, their order of overprinting must be known. In the case discussed above of

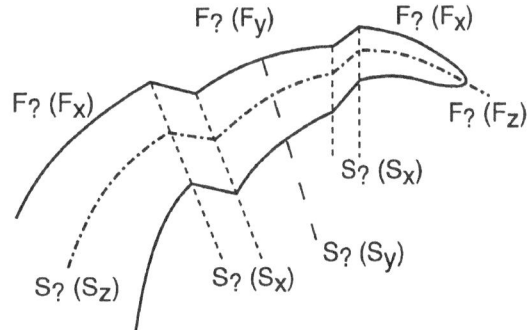

Figure 8.10 Sketch showing the relationships between three fold sets (F_x, F_y and F_z) where F_z precedes F_x and F_x precedes F_y but the precise positions in the successions of F_x, F_y and F_z are not known.

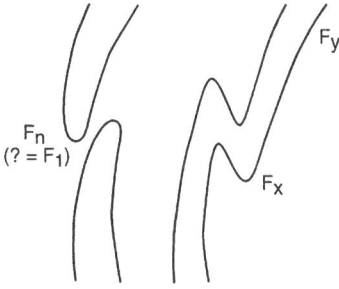

Figure 8.11 Sketch of three fold sets, F_n ($? = F_1$), F_x and F_y.

three fold sets F_n, F_x, and F_y (Figure 8.11), if the two sets F_x and F_y are distinctive it is likely that they will be recognized in at least one other place where it **is** possible to observe their relationships (say F_x before F_y). With the order of overprinting of the three sets being known, viz. F_n, F_x, F_y, one aspect of the problem is resolved. Secondly, it is necessary to discover whether the three sets F_n, F_x, F_y comprise the whole succession or only part of it, and if sets are missing, to determine how many of these sets there are, and where they belong.

Where the overprinting relationships between all three fold sets can be observed directly on a single exposure (Figure 8.12), determining the structural succession is easier than in the case just discussed because the order of folding (F_p, F_q, F_r, here) is known. Nevertheless, several uncertainties still remain to be resolved. The situation is summarized below using this example in which F_p is equivalent to F_n of the earlier case, F_q is equivalent to F_x and F_r is equivalent to F_y.

In this example, NEITHER is (a) the **position** of (F_p, F_q, F_r) known in the overall succession, in which case the following possibilities exist:

$$F_p + F_q + F_r + F_4 + F_5 + \ldots . F_n$$

i.e. they are the EARLIEST fold sets in the succession,

$$F_1 + F_2 + F_3 + \ldots . + F_{(n-3)} + F_p + F_q + F_r.$$

They are the LATEST in the succession.

$$F_1 + F_2 + F_3 + F_p + F_q + F_r + F_7 + \ldots . F_n.$$

They are INTERMEDIATE in the succession.

NOR is (b) the **number of fold sets between** each or any of F_p, F_q and F_r known, in which case the following possibilities exist:

$$\ldots + F_p + F_? + f_? + \ldots + F_q + F_r + \ldots$$

$$\ldots + F_p + F_q + \ldots + F_? + f_? + \ldots + F_r + \ldots$$

$$\ldots + F_p + F_? + f_? + \ldots + F_q + F_?' + f_? +$$

$$\ldots + F_r + \ldots$$

With hindsight, while the conclusions of the preceding discussion might seem obvious, experience has shown that this can be far from the case while the exercise is being undertaken in the field. This will be recalled by many observers who have been through the experience of having to unravel the relationships between two or three fold sets from convoluted interference patterns, especially when these relationships are far from clear, whether because of paucity of outcrop or because of complexity of structural pattern. The sense of achievement, at times even elation, at having successfully resolved the relationship between structures F_x, F_q, F_p in such circumstances often tends to override questions of whether there are other, intermediate structures involved in the succession. Furthermore, uppermost in the mind of the observer immediately following the determination of the relationships is a concern to seek confirmation (or otherwise) of the relationships so determined. With the satisfaction engendered on discovering such confirmation the need to establish whether or not other structures exist in that part of the succession may be overlooked or relegated, and temporarily at least, forgotten.

8.1.6 NOMENCLATURE: SYMBOLS FOR STRUCTURES AND EVENTS: PRINCIPLES

Before considering the practice of choosing names for structures sets, first consider the theoretical aspects of this.

As the preceding discussion implies, because the procedure requires continual review of the overprinting relationships determined at each exposure it is advisable at the outset to avoid the use of numerical terminology, such as F_1 to F_n, or indeed any nomenclature having a sequential connotation. This is because at some stage or stages later in the investigation a fold F_x might well turn out to be $F_{(x \pm y)}$ following the discovery of new information. It may be possible and convenient to name the fold sets on the basis of localities, shape etc. using suitable abbreviations as suffixes. It is certainly advisable, if not essential, to avoid any such system as 1, 2, 3. . . .n or a, b, c, . . . z. This is not so much because of the inconvenience of having to change the names of sets as the succession is revised, nor because it can give the wrong impression to outside observers at intermediate stages in the study, but more seriously because it avoids the almost inevitable situation where the observer unconsciously and uncritically accepts a position in the succession for a specific fold set which subsequent information may show to be wrong, and it also avoids the consequences of this uncritical acceptance.

(a) A theoretical system of nomenclature

Strictly, even when the order is unknown at the outset, a theoretical system something like the following might be considered where such a rigorous approach is desired that the nomenclature in no way influences the observer. Ideally what is needed is a combination of symbols that do not imply a sequential connotation e.g. D x II n i a # d § d + β & ç etc., or any symbols having names which can be referred to in discussion. Those listed here could be used for a succession of as many as 14 sets such as that described from the Jussarö area of southern Finland (Hopgood, 1984). For shorter successions, a A a i I 1, mathematical symbols, or the upper and lower case letters of different alphabets would be adequate.

Instead of using the upper and lower case of a single alphabet, a mixture of one, or two non-consecutive Roman and Greek letters might be used, together with apostrophe marks, inverted commas, asterisks etc. such as the following: X, X', X", X#, x, x', x", x# + P, P', P", P#, p, p', p", p# + s, s', s", s#, S, S', S", S# + d, d', d", d#, D, D', D", D#. This combination for example would combine to provide as many as 32 symbols, sufficient to cope with a fairly extensive succession such as that described from the Dabie complex, China (Hopgood *et al.*, 1989). However, in cases of

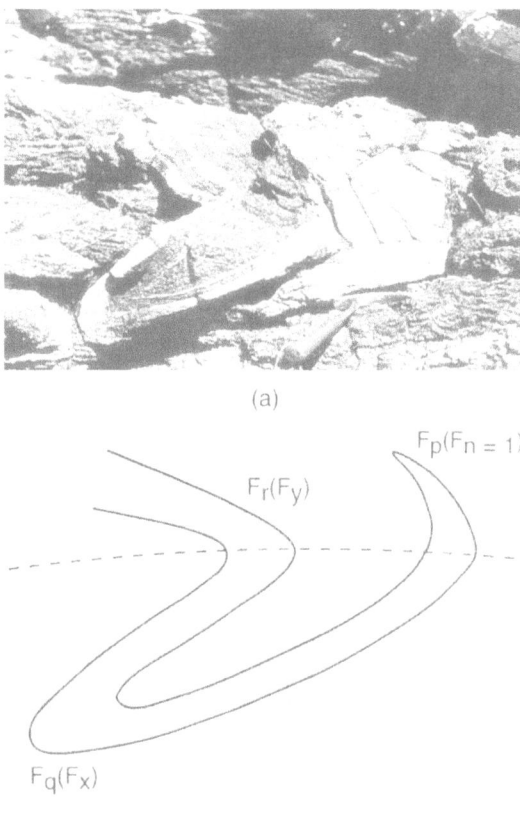

(a)

(b)

Figure 8.12 (a) Refold relationships between three fold sets, F_p, F_q and F_r in inter-banded gneiss and amphibolite. Lewisian complex, Rona, Inner Hebrides, Scotland. (b) Explanation of (a).

extensive successions similar to that of the Dabie complex the practicality of correlating all or even most of the structure sets over a wide area using symbols such as these might be severely restricted simply because of the time it would take.

(b) Nomenclature allowing expansion of the succession

Where the order of succession has been established, as in Figures 8.11 or 8.12b, it would be theoretically advantageous to adopt a form of nomenclature for the fold set using subscripts of a kind such that, while there is an implicit **sense of order**, this does not necessarily imply that the structures are **consecutive** (i.e. that the succession is a complete one). For example the use of subscripts as in, A, A′, A″, a, a′, A″, and B, B′, B″, b, b′, b″, and C, C′, C″, c, c′, c″, etc., so that Fold 'n' (F_n), or Fold 'p' (F_p) of the examples becomes $F_{a''}$, or F_A, and F_x (or F_q) becomes F_b, or $F_{b'}$, or $F_{b''}$, and F_y (or F_r) becomes $F_{c'}$, or $F_{C''}$ etc. This not only allows the terminology to be extended to accommodate new fold sets as they are identified, but also preserves the sense of order.

In this way a succession initially comprising F_a, $F_{b'}$, F_C, could be extended to F_a, F_B, $F_{b''}$, $F_{b'}$, F_C, if say two sets newly discovered between F_a and $F_{b'}$ were labelled F_B and $F_{b''}$. The succession has thus been expanded with the sense of order retained.

This approach to fold set nomenclature (allowing expansion of the succession) differs from that proposed earlier where, by using unrelated symbols (a # d § d + β & ç etc.) a sequential sense is avoided deliberately. It is useful because it provides flexibility in cases where, although the order between some fold sets may be known locally, the order between sets from different localities may not. The use of the nomenclature proposed in the present instance is valid because it relates to fold sets whose position, as they are discovered, is known. Its application is illustrated in Table 8.1 which shows an example where, as a result of further work, newly discovered sets, F_B, $F_{b''}$ are added to an initially observed structural succession, F_a, $F_{b'}$, F_C.

(c)   Nomenclature relating deformational, igneous and metamorphic structures

Because of the importance of nomenclature in recording successions of structures and sequences of events, it is useful at this stage to consider in more detail the principles used in the consistent labelling of metamorphic, anatectic and igneous events, as well as deformation structures. Firstly it is important to

Table 8.1   The structural succession of **(i)** modified by the addition of new observations **(ii)** which show the relationships to **(i)** of $F_{b''}$ and F_B to give the revised succession shown in **(iii)**.

[i]	[ii]	[iii]
F_C $F_{b'}$ F_a	$+$ $\quad$ $F_{b''}$ F_B	$=$ $\quad$ F_c $F_{b'}$ $F_{b''}$ F_B F_a

note that in the interests of systematics, while comparable symbols may be used for structures formed (folds, linear structures and planar structures – F_n L_n S_n) and associated metamorphic events (M_n), during a particular deformational episode (D_n) such numerical equivalence of symbols may not always be valid. In such cases not every structure so developed (L_e, say) is necessarily associated with an equivalent metamorphic event (M_e, say) so that perhaps only one of the, say, three folds (or other structures) so formed would be labelled F_e. In this connection see Bowes *et al.*, 1992, p. 38 and Hopgood *et al.*, 1995.

Secondly, during progressive deformation it is possible for more than one fold set to form with overprinting relationships during a single deformational episode such as in Figure 8.13 where fold sets F_o and F_c, with F_c overprinted on F_o, could form during a deformational event D_n, say.

Table 8.2 shows an example of the effects of a hypothetical deformational sequence (D_{1-n}) a metamorphic sequence (M_{1-3}) and an anatectic (neosome-forming) event (N) with the corresponding succession of structures and products of metamorphism and neosome-formation. Here some structures are syn-metamorphic while others are not so that the succession does not show numerical 'equivalence' of symbols.

This type of potential relationship has been one of the sources of controversy over what constitutes a 'fold phase' and has caused

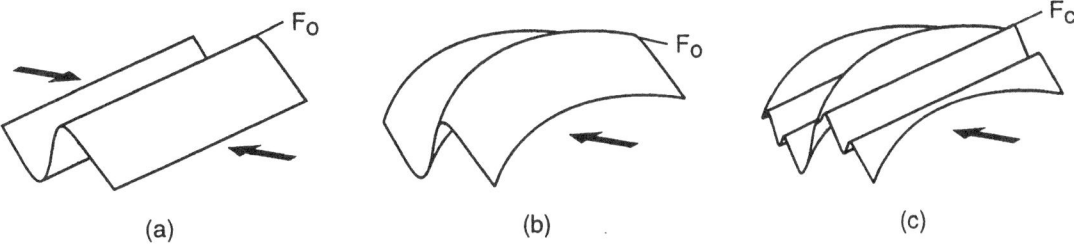

Figure 8.13 Two sets of folds (F_o and F_c) formed during a single progressive deformation event (D_n). Both F_o and F_c began with the same orientation but differential movement accompanying the development of F_o caused say, by frictional differences, resulted in curvature of F_o hinges prior to the superimposition of F_c. The resultant angular discordance between the two hinge sets provides clear evidence of overprinting of F_o by F_c, during D_n.

Table 8.2 Relationships between the nomenclature of metamorphic products and synmetamorphic structures and that of structures **not** associated with metamorphism. F = fold, S = foliation, L = lineation.

Deformational event	Metamorphic event	Anatectic event	Structural succession
D_1			$F_1 \, S_1 \, L_1$
D_2			$F_2 \, S_2 \, L_2$
D_3			$F_3 \, S_3 \, L_3$
D_4	M_1		$F_4 \, S_4 \, L_4 \, M_1$
			(**Not** $F_1 \, S_1 \, L_1 \, M_1$
			or $F_4 \, S_4 \, L_4 \, M_4$)
D_5			$F_5 \, S_5 \, L_5$
D_6	M_2		$F_6 \, S_6 \, L_6 \, M_2$
			(**Not** $F_2 \, S_2 \, L_2 \, M_2$
			or $F_6 \, S_6 \, L_6 \, M_6$)
D_n	M_3	N	$F_n + N \, S_n + N \, L_n + N \, M_3$
			(**Not** $N_3 \, F_3 \, S_3 \, L_3 \, M_3$
			or $N_n \, F_n \, S_n \, L_n \, M_n$]

considerable misunderstanding about this approach to structural analysis. The fact that its basis, namely analysis in terms of overprinted relationships, is not affected by genetic factors such as these means that it does not matter whether F_2, F_3 and F_4 formed during a single deformational 'phase' (cf. Figure 8.13) or as the results of three 'phases': **their overprinting relationships are the same in either case, as is their order of succession.**

With the order of overprinting known between fold sets that have been identified early in the study, and the realization that possibilities exist for successions of the kinds just discussed above, the need for the systems of

fold set nomenclature described above should be clear. They are rigorous, and avoid unwarranted implications regarding (a) the length of the succession, (b) whether or not any of the fold sets immediately succeeds another and (c) the position in the succession of specific fold sets, yet they have the flexibility that allows the succession to be extended without destroying the sense of order of fold sets already incorporated.

As stated earlier, the initial succession $[F_p + F_q + F_r]$, which is sequential but not necessarily immediately consecutive or complete, can be expanded if necessary as more fold sets are identified, to something like

$F_p + F_{p'} + F_{p''} + F_p + F_{p'} + F_{p''} + \ldots + F_Q + F_{Q''} + F_q + F_{Q'} + \ldots . + F_{r'} + F_r + F_R + \ldots$ etc. by suitable choice of subscripts. The order of overprinting has been preserved, the succession extended with potential for further extension but the precise position (i.e. set number) in the succession has not been implied as it would be if a numerical system (i.e. $F_1, F_2, \ldots F_n$) had been used.

The discussion so far has been concerned essentially with the purely theoretical aspects of nomenclature and while much of what has been said can be applied in many cases, in practical terms it is desirable to find a system that is simple, and unlikely to be prone to error, particularly during discussion or when being used in correlation (see 'Labelling structure sets in practice' below). Therefore, because different letters are more readily identified visually than the superscript 'ticks' such as a′, a″ etc. above, these will be used as fold set subscripts (F_a etc.) to make the discussions easier to follow. Indications that a succession based say on observations at a single location is likely to comprise more than just three fold sets such as the $F_p + F_q + F_r$ discussed earlier, are usually to be found from refold relationships at other localities. This is especially so where the newly observed structures are markedly different from the three sets observed at the outset, i.e. different to the extent that they can be distinguished from the others with a high degree of certainty. If these localities are nearby or adjacent exposures, they are likely to be part of the same terrane (i.e. not separated by major terrane boundaries), so that the overprinting relationships can be taken as good evidence, rather than merely as an indication, that the succession is more extensive. In this respect see section 6.1.2.

(d)   Preliminary labelling of structures

To summarize the content of the preceding discussion it will be clear that the manner in which fold sets are named is important and is something that should be considered carefully.

For this reason the type of nomenclature used to label different sets of structures should be chosen deliberately at the very beginning of the study as one which intentionally has **no** chronological connotation (cf. Table 1, Hopgood, 1984). Although the labels selected for particular structural sets do not convey an impression of succession, they should serve to identify structures as belonging unambiguously to that particular set (or sets) which formed at the same relative time in the deformational sequence. (Sometimes, however, because the number of sets of structures recognized may be considerable, each set, in addition to its symbol, may be given a number to show its relative position in the structural succession. This facilitates reference to, and comparison between it and the others. But, if this is done the reasons for using the numerical labelling must be given) Hence, the structures can be described in chronological order and while relative age is implicit in this order, this does not necessarily mean that the sets or structures described follow one another immediately. As has been shown already, it is conceivable that one or more structural sets not expressed locally, nor recognized consistently throughout the area, could be observed later as intervening between one or more successive pairs of the structural sets described earlier. A comparable situation arose in the course of a structural study in the Jussarö region of southern Finland where other structures, not identified earlier in the preliminary study (Hopgood, *et al.*, 1976) were recognized as the investigation progressed (Hopgood, 1984). The structures designated F_a in earlier published work (Hopgood *et al.*, 1976) include both F_{aa} and F_{ab} of the later (1984) work, while F_b of the earlier work, although largely representing the 1980 F_{br} structures, also includes some of the F_{bb} and F_B structures subsequently recognized (Hopgood, 1984).

(e)   Labelling structure sets in practice

Although the various forms of nomenclature

(such as a, a′, a″ etc.) proposed earlier in the discussion provide a sound basis for a strictly logical method, in practice such an approach can prove to be unwieldy and could lead to errors arising either from the similarities between some of the labels, or the accidental omission of 'ticks' (′) and inverted commas (″), or both. In practical terms, especially in the initial stages of the study, suffixes indicating fold geometry or style are a very useful means of discriminating between different folds, and a more informative system is that most often used by the author which employs the initial letters of descriptive terms. For example an <u>a</u>symmetrical, <u>i</u>ntrafolial <u>f</u>old can be labelled F_{ai}, a foliation (<u>s</u>urface) is labelled S and a <u>l</u>inear structure, L etc. (see, for example, Hopgood and Bowes, 1995).

Besides being relatively uncomplicated, and therefore less prone to error by the observer or to confusion of interpretation by the reader, this form of nomenclature has the dual advantage not only of providing the reader with some information about the structure (e.g. F_{ai} represents an <u>a</u>symmetrical, <u>i</u>ntrafolial <u>f</u>old; an explanation of which can be given to the reader) but also of avoiding a sequential connotation and the further confusion this could engender in the mind of the reader. It does of course have its limitations in the case of extensive successions where it might not be possible to find enough expressions to label all the structures and then it would be necessary to modify the system, say by incorporating part of one of the labelling systems described earlier, such the addition of superscript 'ticks' (′, ″ etc.).

8.1.7 PARTIAL SUCCESSIONS

At this stage it will be clear that, in most cases, the succession recognized at the first exposure examined during a new field study will be subject to later augmentation by other structures. This is because subsequent identification of other structures from structural relationships seen on other exposures is (almost) inevitable.

If, say, the structural relationships shown in Figure 8.14 are those recognized at the first exposure examined, then the succession F_p, F_q, F_r derived from it, and which represents only a partial succession as opposed to the 'total' succession recognized overall, will be subject to potential modification by the relationships between the sets provisionally labelled F_b and F_x at a second exposure (Figure 8.15) and further by sets labelled provisionally as F_g and F_k at yet another exposure (Figure 8.16). In the earlier stages of the study it might not be possible to add to the information obtained thus far, i.e. that of the three partial successions at different localities such as those of Figures 8.14–8.16, nor therefore to add to and extend (e.g. by correlation) the overall succession. However, as more information is accumulated concerning the characteristic features of the

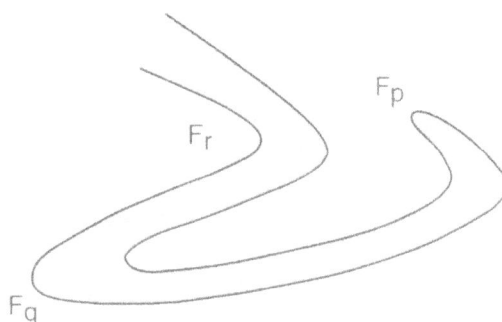

Figure 8.14 Sketch showing the refold relationships discernible between three fold sets (F_r, F_p and F_q).

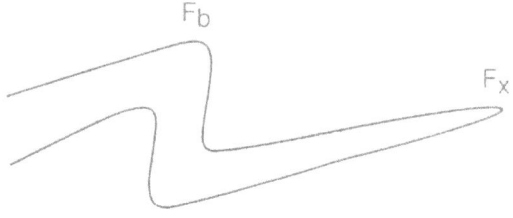

Figure 8.15 Sketch of the refold relationships recognizable between two fold sets (F_b and F_x).

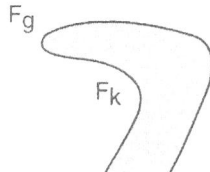

Figure 8.16 Sketch of the refold relationships between two fold sets (F_g and F_k).

structures observed, the potential for correlation increases and with it the prospects of building up the overall succession.

Unless the deformational history has been a simple one, resolving the structural complexity is usually accomplished only by a process of 'trial and error' with continual reappraisal of the field observations as the structures at each new locality are examined. For this reason it is advisable to examine as many exposures as possible (or at least continue until such time as examination of several new exposures ceases to provide new information) so that the succession continues to expand as the study progresses, with the 'complete' structural succession being known only towards the very end of the investigation when the overall picture emerges as the components of this 'three-dimensional jigsaw puzzle' fall into place.

8.1.8   ABSENCE OF STRUCTURES FROM THE SUCCESSION

Implicit in the previous discussions is the fact that the fold succession can differ from one locality to the next, in some cases only slightly and in others considerably. Whether or not this is so will become apparent with the examination of more outcrops as the study progresses. Where the successions are seen to have fold sets in common it is obvious that they are related, even though there are structures apparently missing from them at some localities (see section 6.1.3, and section 8.1.9 below).

For example, at three localities (1, 2 and 3) the successions might be, $F_n + F_p + F_r$ at locality 1, $F_n + F_o$ at locality 2 and $F_o + F_r + F_s$ at locality 3. The total known succession is therefore $F_n + F_o + F_p + F_r + F_s$ with sets F_o, F_s missing at locality 1, sets F_p, F_r, F_s missing at locality 2 and sets F_n, F_p missing at locality 3. The sets present at each locality are shown in bold script below.

$F_n + F_o + F_p + F_r + F_s$ ('TOTAL')

$\mathbf{F_n} + (?F_o) + \mathbf{F_p} + \mathbf{F_r} + (?F_s)$ (LOCALITY 1)

$\mathbf{F_n} + \mathbf{F_o} + (?F_p) + (?F_r) + (?F_s)$ (LOCALITY 2)

$(?F_n) + \mathbf{F_o} + (?F_p) + \mathbf{F_r} + \mathbf{F_s}$ (LOCALITY 3)

The absence of these folds, or their representatives, might be more apparent than real (section 7.7), i.e. although mesoscopic structures might be missing there might be small-scale, microscopic expressions of them, for example in terms of fold-related mineralogy which is not obvious on the outcrop. On the other hand they might at first simply be overlooked in the general complexity of the fold pattern. In any case their absence in no way impairs the efficacy of the method nor does it detract from the validity of the principles involved (see also the discussion in section 8.1.3, section 8.1.9 below, and section 12.6.4.

There is, however, always the possibility that folds (or other structures for that matter) related to a particular deformational episode simply do not exist, or are unrecognizable at all or most localities for one of the following reasons.

1. The fold set was initially only weakly developed and as a result is no longer recognizable because it is overshadowed by later, more intense folding (see section 10.1.2 and Figure 10.1). This may be the reason for the apparently relatively uneventful (early) deformational history of some terranes.
2. The missing fold set and that immediately succeeding it were on the same axial **plane** (Figures 7.3a, 15.1). This condition

does not necessarily arise if the folds are simply **coaxial** (section 3.2.4). See also Figures 3.31, 3.32, 11.13 and 12.4.

3. The missing fold set has been (more or less) completely obliterated by transposition caused by intense later shearing or flow for example (Figure 3.30).

4. The structure was never developed in the rocks exposed at the study locality because of local deformation partitioning (see Figure 6.9), or is missing for the reasons discussed in section 6.1.3 and section 8.1.2.

8.1.9 EFFECTS OF STRAIN PARTITIONING

How does irregular, uneven, distribution of strain (i.e. the effects of strain partitioning) affect the determination of successions and their correlation? This is a question that is frequently asked (see also the discussion of this question and Figure 6.9 towards the end of section 6.1.2.).

One of the greatest sources of misunderstanding (possibly **the** greatest source) that exists with respect to the principle of correlation of structural successions stems from a misconception relating to the concept of **strain partitioning** (Figure 8.17), i.e. the differential response of different rocks to a given stress field, and also the differential response of the same rocks to the same stress field in different crustal (physical) environments (Hatcher, 1992, p. 112).

(a) Effect on structural synthesis in broad terms

Because correlation is based on comparing structural successions derived in terms of overprinting relationships, the fact that strain partitioning can influence the structural expression of a given deformational event in terms of different settings (different depths, temperatures, rock types etc.) has led some observers to question the validity of such correlation. In fact their misgivings are groundless because, in questioning the reliability of the correlation, they have overlooked one factor of fundamental importance, namely that the correlation is not dependent solely on the comparison of only one, or even two, structures in each succession but on several (see Chapter 6, and section 12.5, and

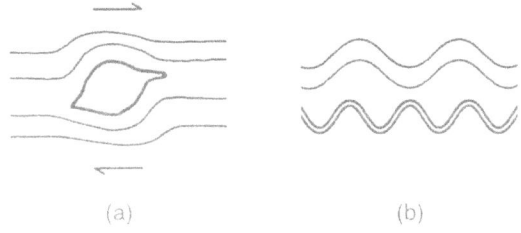

(a) (b)

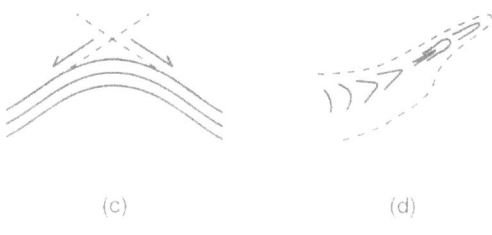

(c) (d)

(e)

Figure 8.17 (a–d) Simple examples of the effects of strain partitioning. The differential response to the stress field of different rocks, different shapes of rock bodies and different layer thicknesses. (e) Fold-ductile shear in a 'transitional' environment to show the effect of differential response of the same rock to different strain rates or to different physical conditions (equivalent to different 'crustal environments'). Pre-Ketilidian migmatite, West Greenland. See also Figures 5.39 and 6.3.

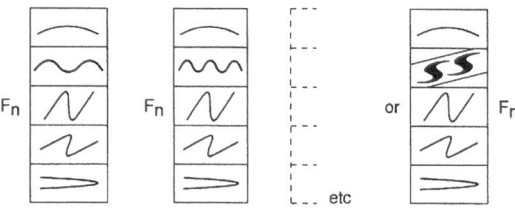

Figure 8.18 Effect of strain partitioning on the structural successions formed at different localities, and shown for three of these, during the same deformational sequence. The structural succession in each case is identical for the most part, except for minor local variations near the top caused by the different responses of the rock units to the imposed stress field.

also the discussion on the immediately preceding pages).

Where strain partitioning has had a significant effect, the structure (F_{n+1}) say, following structure F_n formed during deformational event D_n, will differ from place to place so modifying the structural succession locally (Figure 8.18).

However, even those cases where the difference is so great that a particular D_{n+1} structure is unrecognizable as such, the fact that all, or

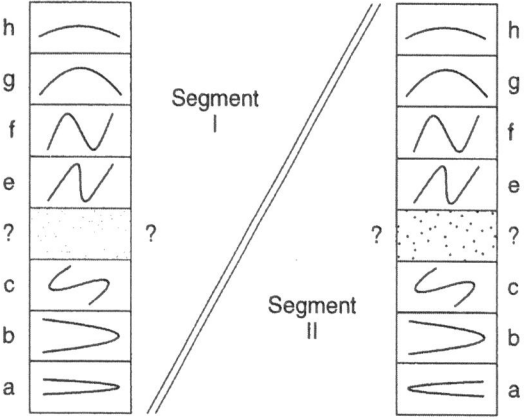

Figure 8.19 Correlation between crustal segments I and II. The validity of this correlation is not compromised by the fact that the structural successions exhibit local differences arising from the effects of strain partitioning.

most, (even many) of the remainder of the structures in the succession (comprising structures 1 to $n + x$) are identifiable and in a particular order, means that correlation of structural successions between crustal segments is valid (Figure 8.19).

This can be graphically illustrated using the simple analogy with a book lacking both cover and title page. The fact that a page, several pages, even a chapter, as well as the cover and title page, is also missing from the book does not render it unrecognizable. The four copies of the book depicted in Figure 8.20 differ in that one is complete while the other three have one chapter missing: a different chapter in each case. The components (chapters, say) of each book can be represented in columnar form (Figure 8.20a) in a way comparable to the representation of a stratigraphic or structure succession (Figure 8.20b) and in both cases the columns (successions) are identifiable and recognizably equivalent. In the one case (a) correlation is 'literal' and in the other (b) it is structural.

(b) Effect on synthesis when integrating local partial successions

Similarly, when the overall succession is being built up by the process of integration of a great many local partial successions, doubts about the identity of a particular structure in a few cases because of strain partitioning may be dismissed. This is because uncertainty about only one structure in a long succession does not affect the correlation and integration so long as the remainder of the structures are identifiable and in the correct order of succession (Figure 8.21). This is not to say that the particular structure is not important, nor that it may not have a significant bearing on the tectonic history of the complex, it is simply to say that the uncertainty of its identity will not affect the identity of the whole succession.

Furthermore, it seems likely that misapprehension regarding the validity of structural correlation also arises because it has not been

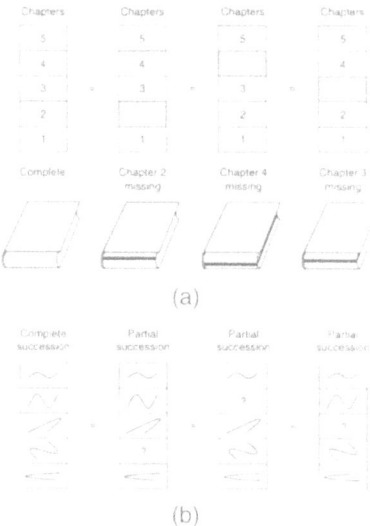

Figure 8.20 Sketches showing 'literal' and structural correlation compared. In (**a**) four copies of the same book are shown, all but one incomplete. Although they differ from each other in terms of completeness, they are still quite recognizably (parts of) the same book by virtue of the remaining pages, etc. Similarly in (**b**), although some of the successions shown are incomplete they are still recognizably part of the same succession despite the missing structure sets. See also section 8.1.8.

realized that a structural succession is based on overprinting relationships at not just one or two exposures, but at several. Ideally it is confirmed only after being substantiated at a great many localities. See Figure 8.22 where the reliability of the integrated structural succession based on data collection at many localities (a) is compared with that where the data collection is limited (b).

8.1.10   APPARENTLY INTRACTABLE STRUCTURAL RELATIONSHIPS

Sometimes, when systematic attempts to establish a succession purely on the basis of observation appear to be making little progress, it may be necessary to consider adopting a radically different approach, one which is hypothesis-led. In such a case considerable care

must be taken to avoid being influenced by preconceptions relating to the initial premise (see section 8.1.1).

In such cases it is advisable to establish a working hypothesis by resorting to intuition in order to make a reasonable guess as to the likely order of the succession, basing this on factors such as the change in style of structures with progressive deformation discussed under 'General features of fold successions', section 7.1. Having set up such a provisional succession it then becomes a matter of seeking new evidence to prove or disprove its validity, and during this time the significance of structures previously not identified may come to be recognized as a result of critical reappraisal of the earlier observations. In adopting this 'last resort' approach the mind is likely to become concentrated on the relationships to the proposed succession of certain structural features which hitherto were not obvious, or were overlooked. This is simply because, having established the succession, one is then faced with having to decide on the relationships of every feature observed. For example, the breakthrough needed might come initially with determination of the relationship between just two fold sets and this could lead on to the clarification of the relationships between the remaining sets.

In any event the determination of fold successions in these circumstances consists in practice of erecting a series of hypotheses and then setting out to prove, or disprove wholly or in part, each of these in turn. Specifically it involves careful and accurate observation of structural data followed by continual modification and refinement of these observations and their interpretation. Depending on the terrain (sic), degree of exposure and structural complexity, this process, for which there appears to be no short cut, is often lengthy and sometimes even tedious unless each of the fold sets is distinctive in some way, e.g. in terms of style, mineral growth, pegmatite veining or lineation etc., such as in the case of key structures (section 6.3). But this is gener-

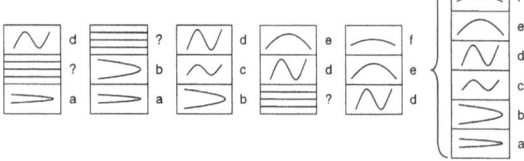

Figure 8.21 Integration of the overall structural succession (**a–f** in the column on the right) from local partial successions, some of which have been influenced by strain partitioning effects at different sites in the succession. The effect at site **c** (or elsewhere) does not invalidate the synthesis of the total succession.

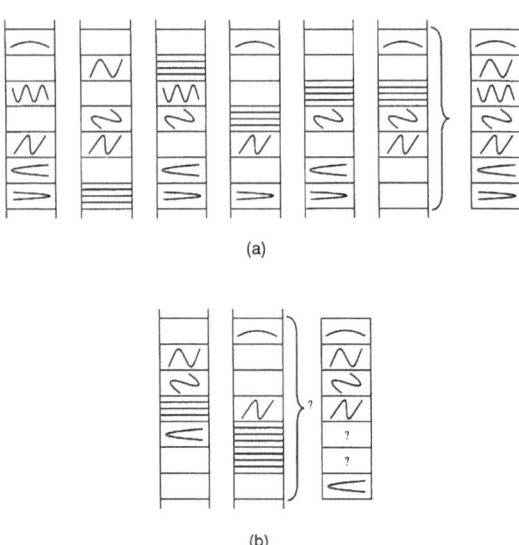

(a)

(b)

Figure 8.22 Comparison between the reliability of integrating structural successions based on **several** observations to produce a 'total' observed succession, on the right in (**a**), with the case where observations have been **few** and insufficient to overcome the uncertainty introduced by strain partitioning effects (**b**). In (**a**) gaps in the succession are of little importance in terms of the reliability of integration whereas in (**b**) these could have considerable significance and the degree of certainty of the synthesized succession being representative of the total succession is likely to be low.

ally the case only in those terranes where there are several distinct rock types including those (e.g. carbonate and shale etc.) sensitive to the effects of metamorphism and deformation. Unfortunately this is not generally the case in gneissic migmatitic terranes. But on the other hand where distinctive marker layers such as continuous amphibolites, anorthosites, marbles etc. are present, the determination of the structural geometry and the establishment of the structural succession might be less difficult because mappable large-scale structures are likely to be identifiable without too much difficulty.

8.1.11 'WILDNESS' OF STRUCTURAL PATTERN

Initially an impression may be given in some cases that the folds and 'wrinkles' affecting the foliation are totally without order, being more or less random or 'wild', and that any regularity is merely apparent and completely fortuitous, with at times one or two directions being common only by chance. The reality of the regularity as opposed to simply coincidence, i.e. 'wildness', will be proved (or otherwise) by collecting as many orientation data as possible at the outcrop, especially if refolding relationships are apparent. The consistence in pattern of orientation relationships will emerge when the data are plotted, thus demonstrating a degree of coincidence which is unreasonably (if not impossibly) high. In this, in fact, lies the proof of the method: the probability that coincidence of so many structural relationships exists solely by chance is too great to be reasonably acceptable. (Refer again to section 1.1.1 and the footnote relating to scientific 'proof' in section 8.1.4).

Finally, while discussion in this chapter has concentrated largely on structural aspects, it is neither possible nor desirable to separate consideration of the structural development of migmatites entirely from the metamorphic and igneous events which have played a part in their development. In fact it is to the investigator's advantage to incorporate observations on these aspects in the overall study of the terrane, not only as an aid to unravelling the structural complexity of the rocks but also because they comprise an integral part of the tectonic history of the terrane. Of course the same applies to sedimentological data as well, if this has contributed to the developmental history of the rocks because it too can assist in resolving structural relationships as well as providing evidence of their sedimentological history.

Using the criteria just discussed one can expect, with a high degree of certainty, to determine structural successions for gneiss and migmatite terranes of considerable complexity. Besides relative ages of structural sets, and igneous, metamorphic and deformational features, successions may include information such as the isotopic ages of the events responsible for these features, the style and geometry of the structures, their syntectonic mineralogy and the igneous and metamorphic petrography of the deformed rocks. All this contributes to the knowledge and understanding of the tectonic history of the rocks and the physical conditions that existed during their development.

9.1 INTRODUCTION

The basic theory of establishing structural successions in complexly deformed rocks has been covered in the previous chapters. In this chapter we will consider what is entailed when it comes to putting the theory into practice and discuss aspects that are especially relevant to fieldwork in highly deformed rocks. In the following chapters we will examine examples from complex terranes that have been studied in detail using this approach.

While the reader may now feel confident of having a good grasp of the principles, it is very likely, depending on such factors as exposure, structural complexity and experience, that when it comes to putting the theory into practice in the field the geologist will feel rather less confident. Obviously it is in the field where most difficulty is likely to be experienced, and it is essential therefore to adhere to a careful and systematic approach. This cannot be emphasized too strongly: without accurate observation and systematic recording of data any study could run into difficulty.

It will be useful therefore at this stage to briefly consider aspects of the field study particularly relevant to multiple deformation.

9.2 FIELD PROCEDURES

The broader aspects of field methods important in the mapping and geometric analysis of structures in deformed rocks are discussed in structural geology text books (Ghosh, 1993; Lahee, 1941; McClay, 1987; Ragan, 1985; Ramsay, 1967; Turner and Weiss, 1963; Whitten, 1966). However, where they bear specifically on the particular problems of structural relationships in gneisses and migmatites, some of those aspects will be treated briefly here.

The approach used in the examination of an unknown terrane will depend to some degree on the time available for the study and also on whether or not the area to be studied has predetermined boundaries. Ultimately the extent of the area (i.e. the proportion of the total area) examined, may depend entirely on the time available.

Ideally at the outset, a reconnaissance should be made of the whole tract with only the briefest time being spent on localities which do not appear to be potentially rewarding. Experience may of course lead to revision of such early impressions.

9.2.1 RECONNAISSANCE

Initial reconnaissance of the ground is preferable to starting at one 'boundary' and working systematically in detail across the ground with the objective of completing the project on arrival at the 'far boundary' at the end of the allotted time. The disadvantage of the latter approach is that much valuable time may be spent on localities which experience proves to be less informative than others examined later in the study where the time would have been spent much more profitably. Also, with experience gained as the work proceeds in the new terrane, the observer becomes more perceptive and is likely to see much more in comparable ground than would be the case earlier in the investigation. For example, supposing that the area being investigated happened to be more or less uniform in terms of exposure and structural expression (although this is unlikely to be the case in reality) then much more in the way of structural relationships would be recognized towards the end of the study than at the start, particularly as regards finer detail and the less obvious structural features, simply as a result of experience. Therefore in the usual circumstances where there is considerable variation in exposure and in structural expression across the ground, it is likely that a good deal of valuable information could be missed in the early stages of the study, simply because the significance of certain features was not appreciated at the outset.

Preliminary reconnaissance means that nearly all, or many of the important instructive localities are likely to be recognized early on in the study. Consequently most of the observer's time and effort can be concentrated on studying the structure in these localities during the main part of the investigation,

rather than being dissipated in the 'routine' collection of data from exposures that are not particularly revealing. Such exposures might be those on the limb sections of large folds rather than in the hinges where refolding relationships are normally better shown (Figure 9.1). When carrying out a field reconnaissance the observer should try to establish whether or not there are large-scale structures such as folds (of the order say of hundreds of square metres in outcrop) as well as smaller-scale (mesoscopic) structures.

(a)   Large-scale structures

If there are recognizable large folds and the exposure is reasonably good, treat each as an area to be studied separately. It is advantageous to distinguish the hinge zones from the limb areas (Figure 9.1) and examine these smaller areas to see whether or not the foliations present are deformed by later folds or whether (e.g. in the hinge zones) the large folds deform earlier structures such as folds, foliations, or lineations. (Also see section 8.1.5, and Turner and Weiss, 1963, pp. 175–180, Figures 5–23, 24, and 25.) Conversely, the trends of the axial planar traces of the large folds will show whether or not they have been

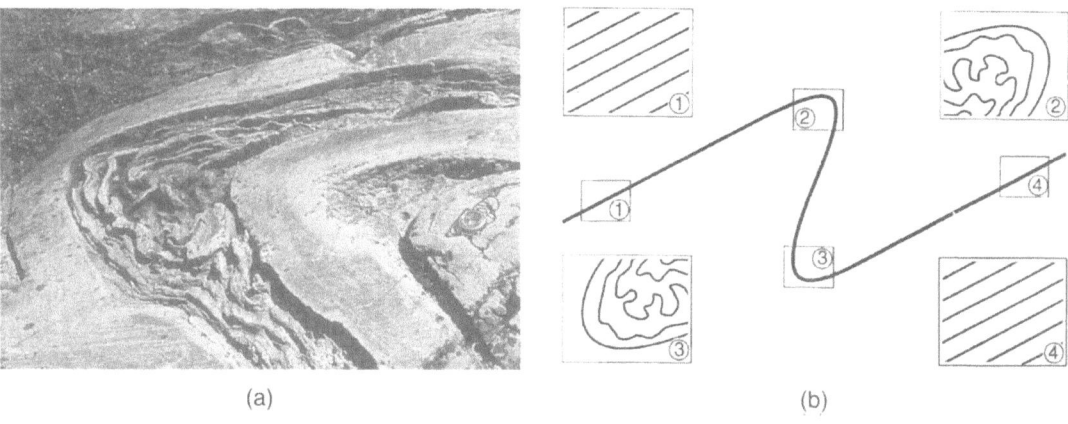

(a)   (b)

Figure 9.1   (a) Complex folding in the hinge of a fold. Kvarnskär, Åland Islands, Finland. (b) Sketch to show the structural complexity in a fold hinge zone (**a** and (**b**2, **b**3) contrasted with the structural simplicity of the limb zone (**b**1, **b**4). Compare Figures 8.4 and 8.5.

consistently deformed by later (including larger scale) structures.

(b) Systematic statistical structural analysis

Although in reasonably well exposed terranes it may be possible to map the form of major structures (Turner and Weiss, 1963, pp. 175–180), this is less likely in gneiss where there is a lack of lithological variation and where distinctive marker horizons useful for lithological mapping are uncommon (section 1.1.4 and Figure 1.18). Also the lithological complexity of migmatite is likely to preclude the mapping of major structures (based on the tracing of distinctive 'marker' units). In both gneiss and migmatite the sheer structural complexity coupled with intermittence of exposure often means that major structures are difficult or impossible to map so that the structure must be resolved in terms of relationships between small-scale structures.

Nevertheless, indications of the presence of large-scale structures should be looked for and it should be borne in mind that the frequency and distribution of exposures necessary to establish the outcrop forms of large structures depends on the scale of these and on their outcrop pattern. Considerable 'danger' exists in sampling with insufficient resolution whether from choice or because of paucity of exposure (cf. Turner and Weiss, 1963 p. 158, Figures 5–7). Remember also that, should it be decided that regular statistical sampling (within the limits dictated by the distribution of exposure) is to be carried out, slope, if it is steep, can have a significant influence on the spacing of sample localities and allowance should be made for this. Similarly, in those cases where the structure plunges, conversion of the form of the outcrop pattern on steep slopes to its appearance on a horizontal surface should be made.

If large-scale folds cannot be recognized, as is commonly the situation because of the absence of continuous marker units, then it is most convenient to subdivide the region into smaller areas where there is reasonably good exposure. Depending on the region, these may be stretches of river valley, ridges, coastal stretches, shoreline, islands etc. (see section 1.1.4, 'Influence of exposure type on the structural study of migmatites').

Note. The ideal condition, where for statistical purposes the region is divided up regularly into smaller areas of the same size or regularly distributed stations, can seldom be realized in practice, and the departure from the ideal controlled by the irregularity of exposure distribution and the attendant subjectivity which this might entail is something that the geologist must learn to cope with. By its very nature any investigation that seeks to resolve a problem is bound to be subjective to some extent, in that the investigator ultimately has to evaluate the evidence, otherwise the investigation would not succeed in accomplishing anything more than the accumulation of masses of raw data. In fact it is the subjective application of the researcher's experience and ability that enables a problem to be solved.

(c) High and low strain zones

Similarly, if a true picture of the deformation is to be obtained, sample spacing is critical where zones of relatively high and relatively low strain outcrop occur (Figure 9.2). This always presents a problem and probably is an effect that can never be satisfactorily eliminated, especially in poorly exposed ground. However, awareness of the problem is important, and where necessary the situation can be rectified later by further examination involving drilling etc. In areas where differences in strain are extreme, the effects of insufficient sampling on the ultimate interpretation of the structure can be significant. The consequences range from forming incorrect conclusions on the amount of total finite strain to ending up with a structural succession from which one or more sets is missing because they have been masked (or obliterated) by intense strain in the zones that happened to be

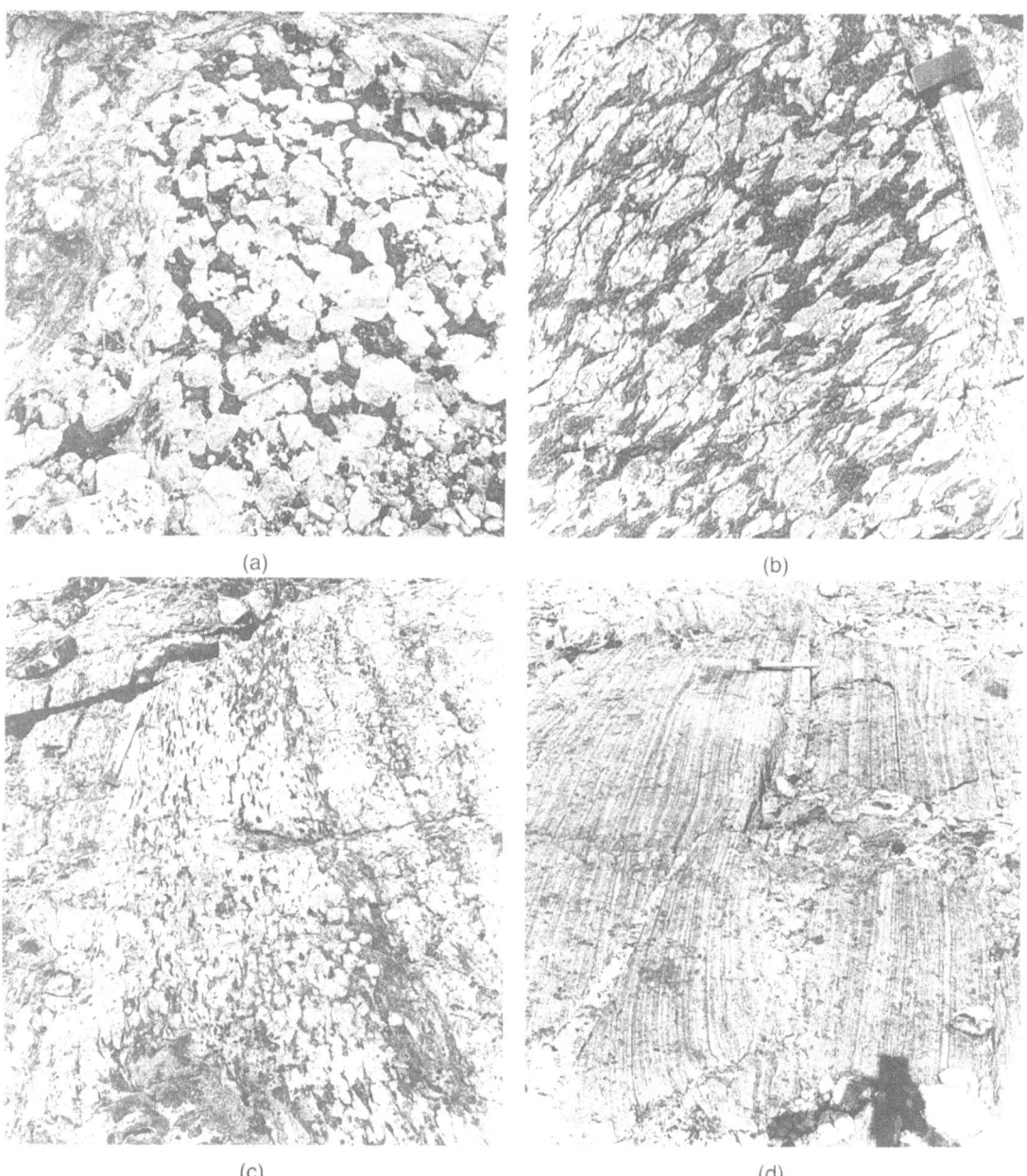

(a)

(b)

(c)

(d)

Figure 9.2   Photographs showing wide differences in strain over a comparatively small area, exemplified by the deformation of anorthosite megacrysts from undeformed (**a**), to moderate (**b**), through strong (**c**), to extreme, where the megacrysts are no longer recognizable as such, having been deformed into thin laminae (**d**). Pre-Ketilidian complex, Nunatak, West Greenland.

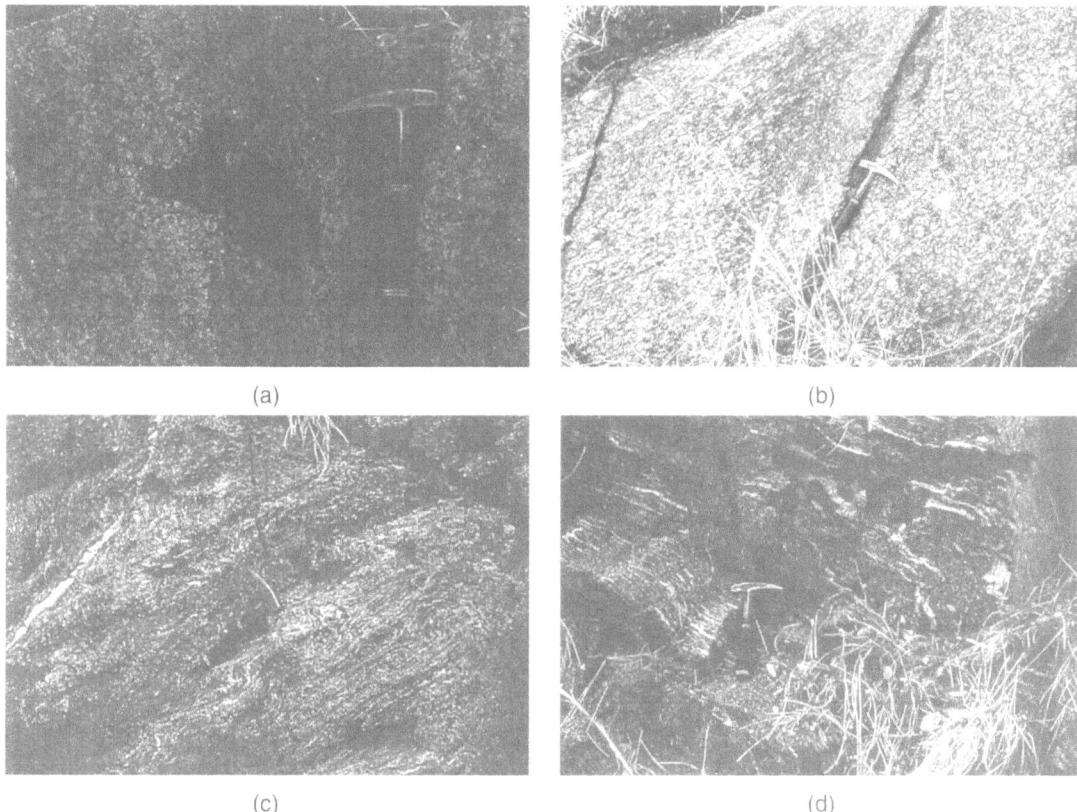

Figure 9.3 Photographs showing successive stages in the development of foliation from a porphyritic enclave-bearing igneous rock (**a**), through the formation of ellipsoidal porphyroclasts (**b**) to foliated rock (**c**) and intensely folded foliated rock (**d**). Kariba Dam, Zimbabwe. See also Figure 9.2.

sampled. The potential for this is illustrated by the examples from near the Kariba Dam, Zimbabwe (Figure 9.3) which show stages in the development of a strongly foliated rock from a porphyritic igneous rock. If, say, the structure shown in either Figure 9.3b or Figure 9.3c was taken to be representative, the impression of the state of strain would be very different from that gained from either Figure 9.3a or Figure 9.3d. See also Figure 9.2.

(d) Potential marker horizons

During reconnaissance look for pelites, carbonates or calc-silicate layers. These are rare in gneisses and migmatite but are invaluable in terms of (i) marker horizons for mapping large-scale structures and (ii) for fine structural detail such as grading, bedding/ cleavage relationships, 'way-up' criteria etc. and (iii) mineralogical-structural relationships and evidence of syntectonic metamorphic conditions. Not only do such rock types tend to be uncommon but differential weathering and erosion means that they are also less likely to be preserved than quartzofeldspathic gneisses. On the other hand, differential weathering often results in accentuation of fine structural detail, even if the mineralogy is strongly altered (Figures 1.25 and 1.27). However, their distinctive weathering and erosion also means that they are more prominent

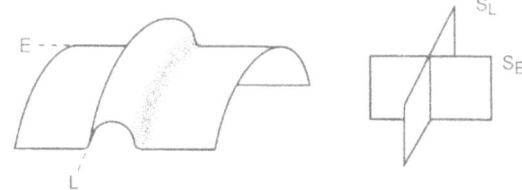

Figure 9.5 Sketches to show the different attitudes of the hinge of a superimposed fold. S_E and S_L are the axial planar attitudes, of the earlier (E), and of the later (L) fold axes. Compare Figure 7.12.

Figure 9.4 Foliation in quartzofeldspathic gneiss crossed by the rectilinear trace of a planar structure (parallel to pencil). Etching by differential abrasion from wind-blown sand has enhanced the expression of the structure. Lewisian complex. Beach at Tangusdale, Isle of Barra, Outer Hebrides, Scotland.

when they do survive and their erosional remnants, colour, staining etc., may be sufficient to allow them to be used in large-scale mapping even if there is insufficient rock left to show fine structure relationships.

(e) Effects of differential abrasion

Note particularly those locations where differential abrasion, e.g. by wind-blown sand on or near beaches, is likely to enhance finer structural detail (Figure 9.4).

(f) Angular discordance between structures

Look for distinct and distinctive angular discordance between successive structures, bearing in mind that of two successive deformational events, the second event will produce fold structures clearly distinguishable from the earlier structures (F_e) only if the orientation of the later potential folds (F_l) have a marked difference in orientation from the earlier set (see Odonne and Vialon, 1987 on hinge migration in superimposed folding and discussion

of this here, (Figure 3.32); also Figure 2.6 and Figure 2.5 from Ghosh and Ramberg, 1968). This can lead to the apparently anomalous situation where in some exposures the later fold set (F_l) is present whereas in others it is absent. The condition can arise as a result of the earlier set (F_e) having different orientations in different places and would depend on there being folds prior to (F_e) on which they (F_e) were superimposed. Fold hinge variation caused by overprinting is shown in Figure 9.5 where F_L superimposed on S_E varies in plunge. If S_L was inclined, the range in attitude of F_E would be even greater, varying in both trend and plunge (see also section 6.2.2 and Figure 6.42). It is important therefore to look at all (or as many as possible) localities where refolding relationships can be seen. They may all be important in unravelling the overprinting relationships between successive fold sets which otherwise can be distinguished only in terms of events intervening between F_E and F_L (i.e. overprinting) such as post-F_E tabular intrusions which are themselves deformed with the formation of F_L (see Chapter 4).

(g) Distinctive structures

At all times look for characteristic features in the folds (structures) seen and note if any of these features appear to be distinctive and sufficiently consistent to contribute to their identity overall and therefore comprise useful aids to the recognition of structures of particular sets. This is particularly important where parallelism or near parallelism exists between different sets of structures whether this is a primary feature or due to rotation of folds into near-parallelism during subsequent deformation, or deflection of a developing fold by a pre-existing structure during its deformation (see Figure 3.32; also Odonne and Vialon, 1987). Also see (l) below.

(h) 'Cryptic' structures

Look also for indications of the existence of other fold sets that are not as clearly defined as the others ('cryptic' structures), and so might have been overlooked during reconnaissance. Evidence to confirm or disprove the existence of such possible folds will then be sought during the main part of the study (as well, of course, as during the remainder of the reconnaissance).

(i) Variation of structural expression

Look for indications of, or evidence for, **different** expressions of the same set of structures in different rock types, e.g. closer, tighter

Figure 9.6 Fold hinge showing differential axial planar cleavage expression. Cleavage is strongly expressed in the leucocratic layers of the fold hinge, whereas the dark (amphibolite) layers appear to be unaffected. Svecofennian migmatites, Jussarö area, Finland.

folding in some rock types, open in others; cleavage development in some rock types and not in others (Figure 9.6) and of course changes between 'brittle' and 'ductile' structural expression (cf. Figures 5.39, 6.53, 8.1, 8.17).

(j) Introduced structures

Look for evidence of emplacement (intrusion, segregation, 'growth-in-place' etc.) of different lithological types within pre-existing layering and foliation either concordant (this may be difficult to recognize) or discordant (Figure 2.1 and cf. Figures 4.8, 4.9 and 4.10). This includes veins, effects of partial melting and mineral growth. The significance of these structures is discussed in Chapter 4.

(k) Orientated specimens

Note localities (carefully, accurately and unambiguously) for collection of (orientated) specimens and for sketching and photographing structural features and relationships. This is not only because this information forms the basis of subsequent geometrical analysis: it must also be borne in mind that subsequent events may decree that further opportunity to visit some localities is restricted (even impossible), particularly those that are remote or where access is difficult e.g. controlled by weather and sea conditions if on islands etc. Furthermore, the data must be adequate for subsequent convincing presentation. Note also that it is worth comparing exposures in both wet and dry states. Often fine structural relationships show up better on wet surfaces but the converse can also be true (Figure 9.7).

(l) Key structures

Be on the lookout for a commonly occurring, consistently distinctive structure set (with characteristic cleavage, perhaps with a consistent axial planar trend, mineralogy etc.) recognizable throughout the field. If such a set can be found then attempt to determine which of

Figure 9.7 (**a**) Absence of visible structural detail on dry rock surface (extreme background), whereas structural detail is enhanced on the wet surface. (**b**) In contrast, with this rock type, the structural detail is more clearly displayed on the dry surface and wetting the surface causes the structural detail to be lost (dark exposure on right and foreground). (**c**) Where there is no local water source the exposure can be wetted by using a portable spray.

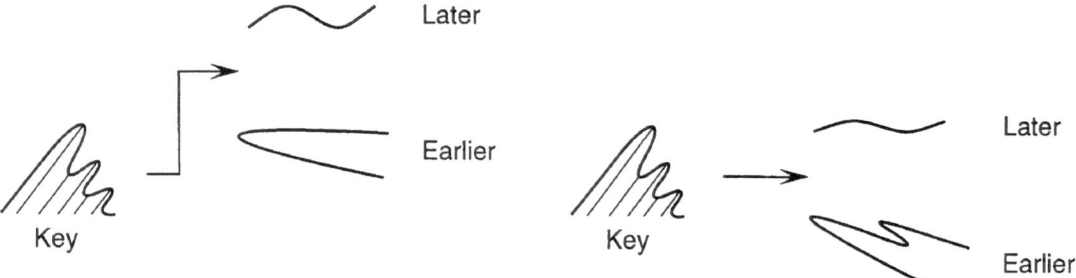

Figure 9.8 Sketches to show relationships between the key structure and different earlier and later structures of the same succession at two localities.

these features are consistently present in the set. These can be used as a **datum** set of folds or **key structures** (section 6.3 and Figure 6.43).

(m)   Relationships to key structures

During (and after) the reconnaissance note the relationships between the set(s) of key structures and those structure sets that are earlier and those that are later than the key structure(s), so provisionally classifying them (Figure 9.8).

(n)   Asymmetry of folds relative to large structures: vergence

When examining asymmetrical folds that appear to belong to the same set it is useful to note the sense of movement implied by the vergence of the folds to see whether it is consistent or variable (Figure 9.9). Vergence of structures is variously defined in textbooks and relates to the sense of rotation implied by the fold geometry (Figure 9.11). If it is consistent this suggests the structure is fairly late. If it is not consistent this could mean that (a) more than one fold set has been included with folds grouped as belonging to a single set or that (b) the folds have been reorientated strongly or that (c) they are due to a local rather than a regional asymmetrical movement, e.g. 'drag' on the limbs of a larger struc-

ture (Figure 9.9) or that (d) they belong to a conjugate set of folds (Figure 9.10) in which case they are another case of (c) above, i.e. they show the effects of local asymmetrical movement.

Hence, when using the principle of vergence to establish the presence of large-scale structures on the basis of small-scale asymmetrical folds it is important to ensure that all the small-scale structures used are **contemporaneous**, that is belonging to the same set (Figures 9.9, 9.12). This may not always be easy if the fold pattern is complex (see also section 10.1.3).

(o)   Record of exposures showing clear structural relationships

During reconnaissance, exposures that look as if they will yield useful information readily by

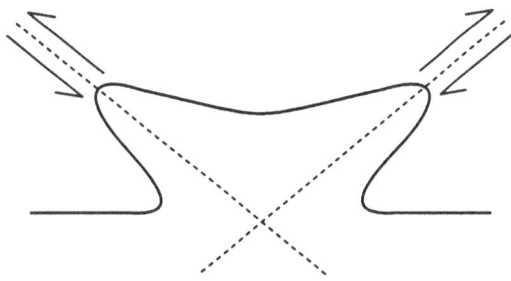

Figure 9.10   Sketch showing movement sense associated with the development of a conjugate fold.

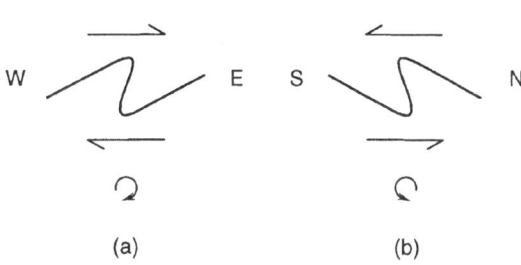

Figure 9.11   Sketches of the vertical section ('elevation') of folds showing (**a**) easterly vergence and (**b**) southerly vergence (from the sense of rotation of the **upper** part of the stress couple).

Figure 9.9   Sketch showing the sense of vergence of minor folds on the limbs of a larger fold. Structures on both the upper and lower limbs verge towards the hinge of the large ('host') fold, those on the upper limb having a clockwise rotation whereas the sense of rotation for those on the lower limb is anticlockwise.

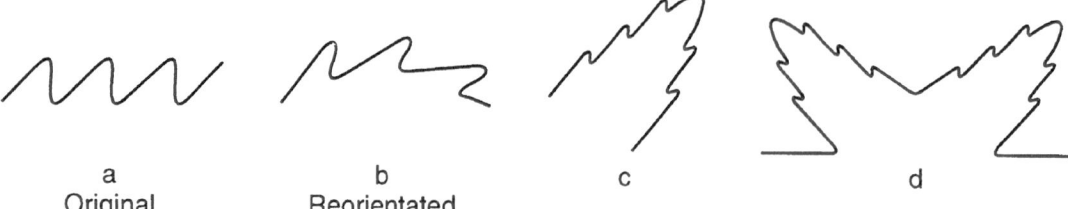

| a | b | c | d |

Original   Reorientated

Figure 9.12 Sketches of asymmetrical folds in different structural settings to show that the significance of their vergence sense differs from case to case. It depends on their structural relationship to other folds with which they are associated in the field. (**a** and **b**) Development of a compound fold by refolding of an asymmetrically folded layer. (**c**) Fold with contemporaneous asymmetrical folds ('drag' folds) on its limbs. (**d**) Conjugate folds with contemporaneous asymmetrical folds ('drag' folds) on their limbs. The attitudes and vergence sense in (**b**) does not have the same significance as a means of determining the relationship of small-scale folds to a larger one as it does in cases (**c**) and (**d**). See also Figures 6.41 and 10.10.

detailed examination should be noted and these revisited when the detailed study is begun. (It is important that these are noted and recorded unambiguously so that they are easily found later during the main part of the study). Examples are those where complex structural relationships can be observed, where fine structure detail is well displayed, where sketches may be made of good examples of structure which can be photographed for later study and for illustrating the report subsequently. Incidentally, when marking out an area for sketching or indicating a particular structure do not use paint or other 'permanent' marks. Chalk or water-soluble paint is acceptable, but if possible use pebbles, twigs, etc. which can be removed afterwards to avoid defacing the exposure. A comparable policy should be adopted when taking drill core samples. If sampling is necessary, then, if possible, select places that do not spoil the appearance of the outcrop or obliterate or remove useful or critical (structural) relationships which could be useful for subsequent investigation by you or others.

Later, when detailed examination has begun, it is worth taking some time to look again, if only briefly, at the other exposures selected for close examination during reconnaissance because these may prove to show certain structural features more clearly than the one currently being studied. With experience gained as the work progresses, one can expect to see considerably more in the way of structural relationships than at the beginning of the work. This heightened perception is likely to lead to revision of earlier impressions and the process of revision will continue as the study progresses.

(p) Ending reconnaissance

Reconnaissance should be continued until it is felt with reasonable confidence that an outline has been obtained of at least some or all of what appear to be the 'main events' in the sequence. At the very least one should have an idea of the problems that must be solved. It is important to recognize these problems. While one is unlikely to be certain about **all** the events, it is important to focus attention on unresolved aspects of the study so that, unconsciously at least, they are continually under active consideration and review (see also the following sections: 2.3, 5.1.1, 8.1.1, 8.1.7, 13.3.1 and 15.1). In this way, and also provided one goes into the field without prejudice and preconceived ideas, one is unlikely to be looking only for limited evidence to support a picture that is determined solely on superficial observation and is therefore almost certainly over-simplified. Conversely it should **never** be concluded that the structure is incapable of resolution and is merely an example of so-called 'wild folding': in effect avoid taking the 'easy way out' and surrendering to the problem. It is most important to keep an open and receptive mind throughout the investigation.

A willingness to modify conclusions in the light of new evidence is important, even when (in spite of, or because of doing so) it 'spoils' the picture already built up. The consequences for the structural interpretation of failing to change a standpoint based on earlier observations when new evidence shows that the interpretation should be otherwise, are likely to be serious. Even worse is a lack of awareness that bias is preventing the change of conclusion necessitated by new evidence. Therefore it is necessary to be wary of consciously (or unconsciously) looking selectively for evidence that supports a particular hypothesis and ignoring evidence that 'doesn't fit'. If some features cannot be explained at first, note them and accept that they seem to be anomalous. There is a reason for them. In this respect refer again to Cloo's comment, quoted in section 1.1.2. At some stage evidence is likely to be uncovered that explains them, perhaps completely changing the picture (cf. the discussion on the apparently anomalous behaviour of igneous contacts and folded cooling joints, Hopgood and Bowes, 1980).

Be self critical of all hypotheses. If possible test them on an independent third party who is prepared to be critical. In any case ensure that hypotheses can be defended and will stand against any objections that you or anyone else can raise.

Subjectivity can to some extent pose an impediment to the resolution of the structure because, paradoxically, it is an essential part of the approach. The final analysis requires reasoned **interpretation** by the observer and cannot be done by 'rule-of-thumb' techniques. Therefore the investigation is, by definition, subjective, and must remain so because it is dependent on the application of experience, observational powers and interpretative skill, something that has not yet been supplanted by the computer (see again Figure P.1).

It is important, particularly in those terranes where the structure is complex, not to be overawed by the complexity and become discouraged at an early stage in the study even if the picture appears at first sight hopelessly complex and beyond resolution. The problem is almost certain to appear difficult at some stage or other, if it is tackled in a straightforward manner. That is, unless the structure is relatively simple. However, even 'simple' structures can yield a great deal of information and may turn out to be more complicated than first appearances suggest. The amount of information derived in any case depends on the degree to which the problem is resolved and that almost inevitably depends to a large extent on the amount of effort that is applied to the study.

9.2.2 DETAILED EXAMINATION

(a) Introduction

When satisfied that the initial reconnaissance of the terrane is adequate, begin the detailed investigation. If there is a strict limit to the time available for completing the field work a balance must be struck between (a) covering the whole area adequately in a reconnaissance fashion and (b) leaving sufficient time to complete the whole field study. Concentrate initially on the special areas selected because they show clear structural relationships, and endeavour to demonstrate the relationships between those sets of structures so far recognized as making up the sequence. Some less instructive locations may never be looked at in detail, depending on the time and the nature of the project, e.g. the work may involve the preparation of a map or it might simply be a structural investigation to provide supplementary evidence for particular structures, etc.

(b) Geometrical analysis using stereographic projection

If inspection suggests either that the deformation sequence is a comparatively simple one involving no more than three or four fold sets or, where the sequence is more complicated, the latest three or four sets can be distinguished, then attempts to resolve the structural succession using stereographic analysis may prove successful. The success or otherwise of this approach and indeed of the initial assumption (viz. that only three or four fold sets are involved) is one that only trial and error will substantiate. See Figures 2.10, 2.11 and 2.12 in section 2.2.5; and section 10.1.1.

Apart from the circumstances discussed in the preceding paragraphs, the relationships of those structures that are in doubt can be checked by plotting their attitudes stereographically over a small area to see if they vary in a regular manner consistent with systematic refolding between the three or four sets distinguished.

Distinction between similar-looking folds by stereographic analysis

If there is reason to suspect that similar-looking folds belong to more than one set, their geometrical relationships can sometimes be checked using the following simple procedure.

Plot the attitudes of the axes of these folds in stereographic projection.

A. (i) If they lie on a single point maximum then they are likely to be the latest, or a very late, set of folds and must have developed on a more or less planar (i.e. unfolded) surface (but see effects of relative scale of refolding (Figure 2.12).

(ii) If they lie on a great circle girdle maximum then the folds are likely to be the latest, or a very late set superimposed on pre-existing folds (Figures 2.11, 5.14, 5.15b).

B. If there is considerable dispersion of the plotted points this may be due to one or more factors.

(i) It may not have been possible to measure and record the orientation of the axial attitudes accurately because the topography is such that measurements have been taken on exposed surfaces with an 'infinite' (or great) number of angular relationships to the fold axes. Dispersion will be more or less random.

(ii) If the dispersion is polarized in one or more directions this may be due to the fact that measurements have been made on surfaces inclined at one or more directions to the fold axial direction (which is the same throughout the field).

(iii) The dispersion may be due to the effects of reorientation during refolding, so demonstrating that the folds are not the latest set. Evidence must be sought on the outcrop for refolding relationships (cf. Figures 2.10, 2.11, 5.14b).

(c) Orientated specimen collection

In those cases where refolding relationships can be seen between structures that otherwise appear to be identical, look for distinctive features which can be used to distinguish them from each other. Collect orientation specimens for thin-section work.

Note that structural (or textual) specimens collected should **always** be orientated if possible and their position and attitude with respect to other (especially major) structures carefully recorded (with a sketch to show this), because (a) they show the relationship between minor (and thin section) structures and others in the field and (b) they may show hitherto unsuspected, or unrecognized relationships or (c) confirm the presence and nature of structures identified later elsewhere.

Orientated specimens should be collected whether or not it is intended that thin sections are to be made. It could later prove to be the case that the particular specimen was the only one (or one of only a few) to show some significant structural or mineralogical relationship, in which case orientation is essential.

(a) Break the specimen off the outcrop first, in order to ensure that a reasonable, intact sample can be collected before spending

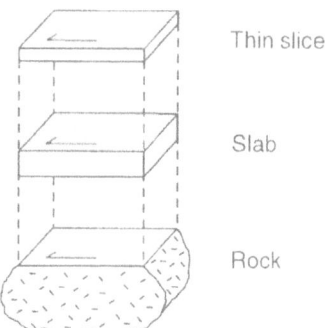

Figure 9.14 Sketch to show the preparation of a thin rock slice cut from a block that was orientated in the field. This shows the use of the single-barbed arrow to prevent ambiguity of orientation of slice and subsequent thin section.

Figure 9.13 Marking the orientation of a specimen prior to its removal, after it has been broken free from the outcrop (the fracture shows above the hand).

time marking and labelling it (Figure 9.13). Otherwise there is always a risk that, having spent time marking the specimen first, it disintegrates when the attempt is made to break it off the outcrop. If on the other hand there is only one possible collectable specimen, it is important to mark the specimen thoroughly **first** because the marks will aid in the reassembly of the fragments if the specimen shatters during collection.

(b) Wedge the specimen back in place and mark (1) at least one horizontal line with a single barbed arrow at one end plus the bearing, and (2) the attitude of at least one surface, noting whether or not it overhangs.

The preparation of orientated thin sections from orientated specimens must be undertaken with care to ensure both that the section is related to the correct cut face on the specimen and that the section has the correct orientation on the face and also is not inverted. The risk of inversion is avoided if the cut slice, the rock face and the glass mount are all marked with a single barbed arrow (Figure 9.14).

1. In order to relate the orientation of the image of small structures and movements seen under the microscope with respect to the same structures on the original rock specimen, examine the thin section after rotating it corner-to-corner about a vertical axis (Figure 9.15).

For most purposes the mount glass should not be thick.

2. Or correct for optical inversion by placing the thin section upside down on the stage (Figure 9.16) **with the arrow in the correct attitude**. This can be useful when working at high magnification.

(d)   Sketches of structural relationships

Sketch and label relationships, if only approximately (although correctly) to start with, from good clear exposures. The act of sketching has the effect of drawing attention to structural features and relationships which might otherwise have been unnoticed, perhaps because of visual distraction (Figure 9.17) by some other more prominent feature (not necessarily geological) or because one's attention was already focused on some feature that has special significance at that stage of the study. As an example of subjectivity in observation consider the photograph of a road cut exposing

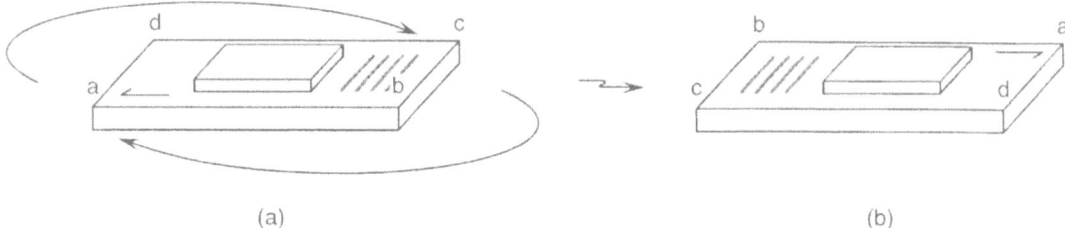

(a)   (b)

Figure 9.15 Sketches of an orientated thin section (a) rotated 180° about a vertical axis (b) in order to correct for optical inversion of the image by the microscope so as to view the structure in the correct orientation with respect to the original rock slab.

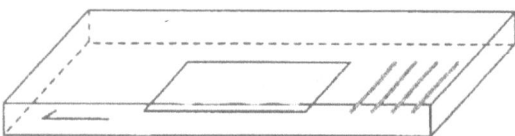

Figure 9.16 Sketch of a thin section placed upside-down on microscope stage, i.e. with the label underneath.

ptygmatic structures in migmatite (Figure 9.17a). Prominent on the rock face are the marks left from a series of parallel drill holes. These tend to dominate the picture, visually distracting from the geology and in some circumstances (particularly in bright sunlight parallel to the rock face) obscuring the structure almost entirely. It is important to ignore such visual distractions and develop the habit of attempting to 'see through' them to the structure underneath.

For detailed sketching, grid the area and measure and mark the grid intersections with pebbles etc. rather than ink or paint. As noted in section 9.2.1(k), try to avoid painting, or otherwise permanently marking the rock surface, for besides spoiling the appearance of the outcrop for yourself and others, some important structural information, not at first recognized, could be obscured by the paint.

In making a sketch, for example to illustrate the style of a fold, it might become evident in the course of the sketching that some feature such as the differential development of cleavage (Figures 9.18 and 6.47) is characteristic of the structure. Or, in another case,

the axial plane of a fold provisionally distinguished from other folds of similar style because it appeared to be undeformed (Figure 9.19a) is after all seen to have been affected by a slight warp, which was unnoticed prior to sketching. The recognition of the later warp demonstrates that the classification and style of this fold are consistent with the other folds (Figures 9.19b, 9.19c).

The kinds of relationships suitable for sketching that demonstrate relative timing of events, i.e. overprinting relationships, include the following: folded foliation, crenulation of pre-existing foliation, refolded folds, discordant tabular vein or discordant foliation, deformed originally tabular intrusive vein or planar foliation, etc. (see section 3.1.2 for a fuller discussion).

(e) Orientated photographs

When taking orientated photographs, note the direction of view, the date and the time (for sun direction and consequently, shadowed features). Sketch the structure, at least simply, if only to pinpoint the details and relationships of the features and mark on it the orientation of the labelled features. It is possible that later on the photograph could reveal other structural features not noticed at the time of the photography (see again Figures 9.17, 9.19), and knowledge of the attitudes of these structures could be important. Knowing the time, and the direction of the sun, will provide information about the attitude of

(a)

(b)

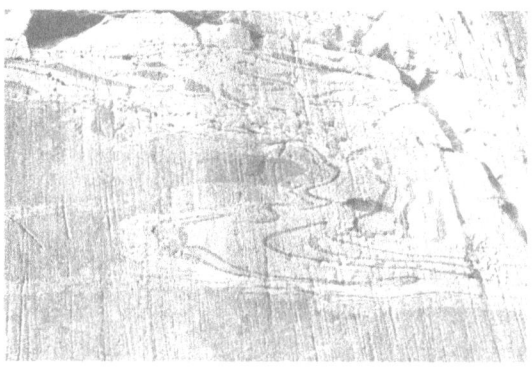

(c)

◄ **Figure 9.17** Visual distraction. Prominent features on exposures such as the drill marks catch the eye more easily and tend to distract the observer from geological features such as joints, fractures and folds. Every effort should be made to ignore their effect, and to 'see though' them to the structures in the rock. (**a**) Ptygmatic structures exposed on a roadside rock face crossed by prominent drill marks. Ekenäs–Helsinki road, Finland. (**b**) The narrow isoclinal fold hinge at road level 2m to the right of the figure is not nearly so obvious as the drill marks on the rock face. Road cut at Hyypiänmäki, (**c**) Clarity of structural detail reduced because of the distraction caused by shadowing from glacial striae as a result of oblique sunlight. Note particularly that the refolded isoclinal fold hinge to the left of the pencil (and nearer the camera) might easily be overlooked during cursory observation. Svecofennian migmatite, Jussarö region, southern Finland. See also Figure 10.12 for other examples of visual distraction adversely affecting the observation of geological structural features.

◄ **Figure 9.18** Asymmetrical differential development of cleavage in a fold. The cleavage on the right of the hinge is expressed as a prominent planar fracturing whereas on the left the cleavage development is more subtle. Hammaslahti, Finland.

shadows arising from structures such as fracture cleavage that might have been overlooked in the field. If there is no opportunity to revisit the site this could the only available information about the cleavage at that locality (see also section 9.2.2(c) above).

(f) Field notes

Make notes each day as a matter of course, but attempt immediately during reconnaissance to build up some sort of sequence of events and update this **daily**, depending on the acquisition of new information. In doing so, one's attention is drawn to discrepancies and the need for (further) specific information to relate certain features etc., i.e. one sees anomalies and aspects of the study for which further evidence or confirmation is required.

Beware of accepting some aspects of the structure without acknowledging that their exact relationships are still to be clarified, especially when the structure is complicated.

Notes provide a means of checking what needs to be collected, photographed, sketched, corrected, and what other deficiencies need to be remedied (e.g. poor labelling, lack of data such as missing dips and strikes, etc.).

It is essential too to read through daily, and correct, or rewrite and amplify, field notes before important detail is forgotten and while it is still possible to remember enough to allow rewriting of notes that are so poorly legible (e.g. because of rain) they can no longer be deciphered.

In some circumstances it is desirable to keep loose leaf notes or copies of notes as insurance against loss of the notebook (with the complete

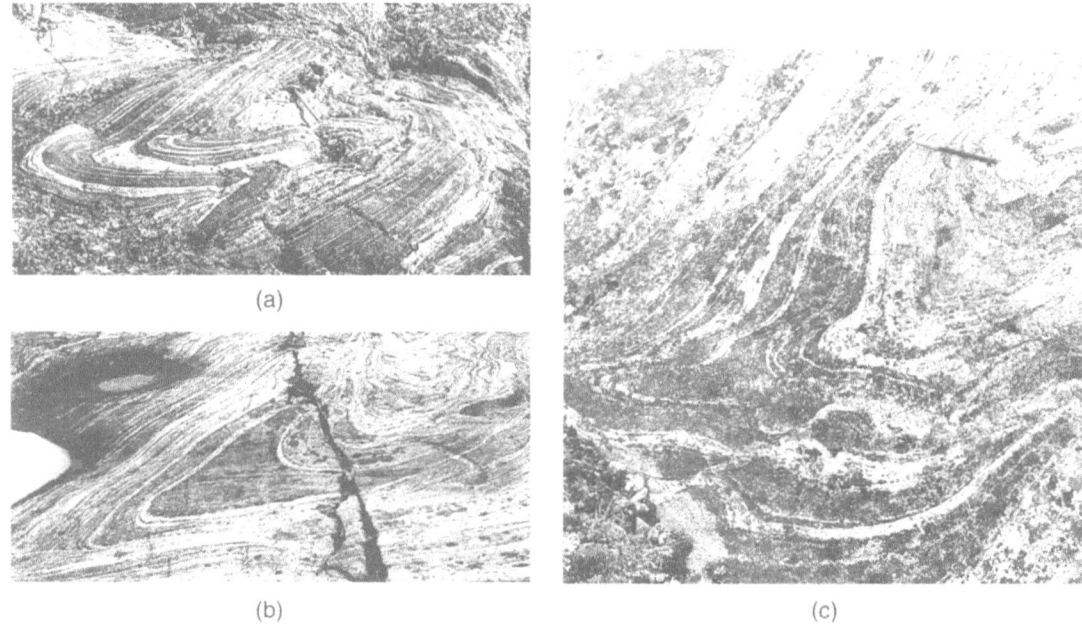

(a)

(b)

(c)

Figure 9.19   Three examples of folds belonging to the same set, but with different expressions. (**a**) Subtle warp of the axial plane in a tight fold. The very slight curvature of the axial planar trace of the tight fold, although not immediately obvious, is the type of feature that might remain unrecognized until a sketch is made of the structure. Pre-Ketilidian gneiss, Fiskenaesset area, southwest Greenland. (**b**) Simple curve of a fold axial plane. Pre-Ketilidian gneiss, Fiskenaesset area, southwest Greenland. (**c**) Sinuous curve of fold axial planar trace in a tight fold by later folding. Pre-Ketilidian gneiss, Fiskenaesset area, southwest Greenland.

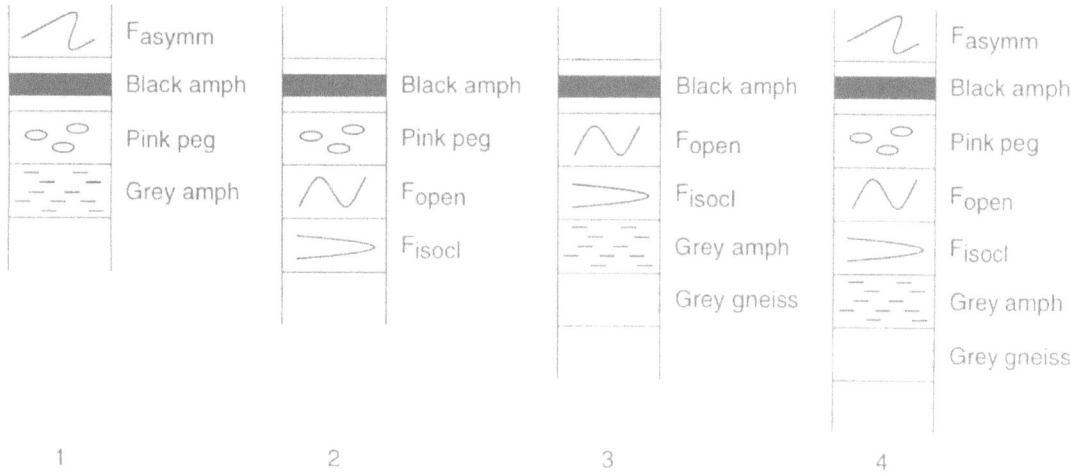

Figure 9.20 Structural succession (column 4) compiled from local successions observed at three localities (1–3).

set of notes), especially when working from a boat.

(g) Structural relationships check

Check those relationships requiring confirmation in small areas by plotting them stereographically where appropriate, e.g. the relationship between pairs of folds. Refer back to 9.2.2(b) above.

(h) Consistence of successions check

For example, a succession found in one place might be:

$F_{asymmetrical}$
Black amphibolite
Pink pegmatite
Grey amphibolite

At another place it might be:

Black amphibolite
Pink pegmatite + flattened feldspar
. F_{open}
$F_{isoclinal}$
etc.

While at a third outcrop it might be:

Black amphibolite
F_{open}
$F_{isoclinal}$
Grey amphibolite
Grey gneiss

Ensure that the order of such successions is consistent (as it is here) before building up the 'total' succession (Figure 9.20).

(i) Good field practice

In general observe the procedures of good field practice (see text books for discussions on field practice e.g. Ghosh, 1993; Lahee, 1941; McClay, 1987; Ragan, 1985; Ramsay, 1967; Turner and Weiss, 1963; Whitten, 1966.)

(j) Summary

In summary, the determination of a deformational sequence in highly deformed gneissic terranes depends principally on careful, objective direct field observation of refolding (and other structural) relationships, commonly in areas isolated from one another. Naturally it depends on there being sufficiently good exposure to build up a complete sequence of fold (structure) sets.

10.1 OBSERVATION OF STRUCTURES AND STRUCTURAL RELATIONSHIPS

Before beginning the discussion on the systematic observation of structures and their relationships to one another, six topics of particular significance in the observation of structures in migmatites and other complexes will be considered. These are, further comment on the role of stereographic projection and its limitations, problems associated with overprinting relationships with open folds, the use of vergence of small-scale folds in the identification of large folds, the need for objectivity and accuracy in observation, the limitations of observation from strictly two-dimensional exposure and the use of 'apparent dips' as a criterion for structural discrimination.

10.1.1 MIGMATITE STRUCTURAL ANALYSIS: LIMITATIONS OF STEREOGRAPHIC PROJECTION

It has already been seen that, apart from those instances where only two or three (or four) sets of folds deform the rocks (or where only the latest two or three (or four) sets are being considered), establishment of the structural succession and determination of the nature of the folding and the deformational sequence cannot be achieved by analysis of non-selective stereographic plots of structural data (Turner and Weiss, 1963, p.219, *Selective and Collective Diagrams*) (See also section 9.2.2(b)). Even if, after having established a number of fold sets which have been distinguished on

the basis of criteria such as characteristic features (style, orientation, etc.), stereographic plotting of fold axial plunges results in a distinct maximum for each set, it still might not be possible to establish the age relationships of the sets on the basis of the types of the maxima so obtained. In fact it is more than likely that no discrete maxima will result because it is seldom possible to recognize and record a sufficient number of attitudes from each set to form clear, strong maxima except where there is a very good exposure (e.g. West Greenland, the Finnish Archipelago, islands in the Baltic Sea etc. – See Figures 1.23 and 1.24). Even so, recognition of individual maxima is likely to depend on their being clearly separated. However, as often as not, a single, irregular, diffuse scatter diagram is likely to result which fails to suggest any definite maxima (cf. Figures 2.11, 5.15) so that stereographic plotting of data is unlikely to be constructive and a different procedure must therefore be adopted to analyse the relationships between multiple fold sets in order (i) to determine their sequential relationships and sometimes (ii) to determine beforehand, the number of fold sets in those cases where preliminary examination has failed to reveal all or any of them.

In those cases noted above (in fact almost always when more than three or four sets of folds are involved) the identification of fold sets and the determination of their age relationships must be carried out primarily on the basis of direct field observations, viz. of fold and refold relationships established on the exposure. While this type of study is systematic and involves a series of progressive stages, it does not resolve itself into anything quite so straightforward as say the procedure entailed in inorganic chemical analysis where the use of a flow sheet enables the analyst, by a process of elimination, to establish the nature of the unknown compound systematically in a series of steps. Instead it can be regarded as a two-stage process: firstly that of determining overprinted relationships and secondly the extension of these relationships by correlation involving extrapolation of, and interpolation within, the structural succession. In the simplest case it is necessary to be able to equate the overprinted relationships between two structures, x and y say, and two other structures such as folds a and b where a and x belong to one and the same family of folds or fold set (F_p say) and b and y to another fold set (F_q say). These same or equivalent relationships have also to be recognizable between other folds of the same two fold sets F_p and F_q. To achieve this correlation between folds of the same set

and hence confirm the refold relationships between them requires some means of identification, some criteria that will allow the distinction of the folds of one fold set from those of another. In this matter the refolded relationships between the sets and their identity (criteria) are inextricably linked: they are inseparable, the one goes with the other.

10.1.2   OPEN FOLDS

Where the structure is very complicated, the resolution of overprinting relationships between late open folds and the earlier structure generally poses special problems.

Accepting that later fold sets will be less tight than earlier ones, because they have been subjected to a shorter deformational history and hence to fewer deformational 'pulses' which could have tightened their inter-limb angle (section 2.3, Assumption 3), these should theoretically be distinguishable from earlier folds. (A cautionary note: Always bear in mind that the assumption that late folds are more

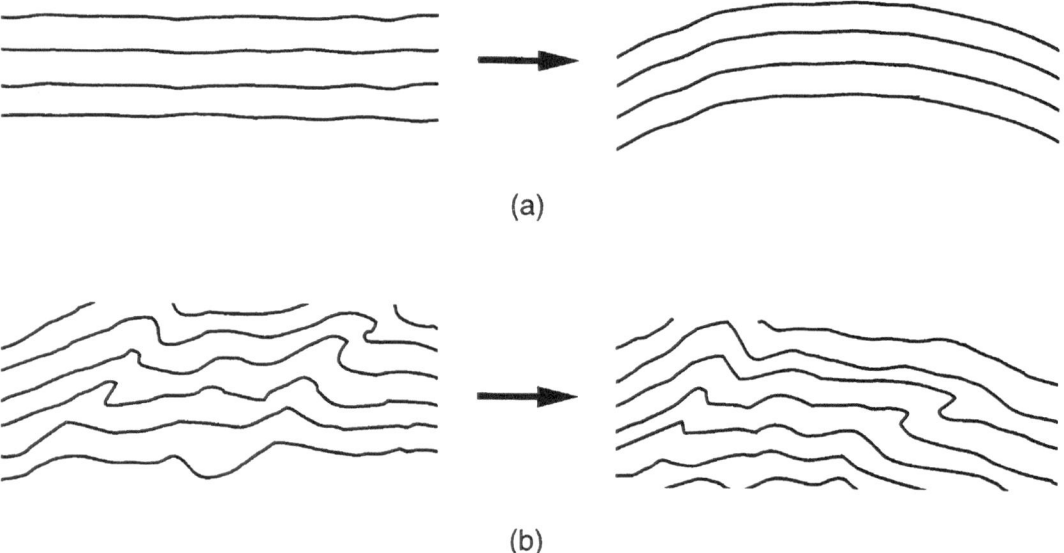

(a)

(b)

Figure 10.1   Sketches showing open folding. The slight curvature caused by open folding of a smooth planar surface is easily recognized (**a**) whereas the same slight curvature of an already-folded surface is unrecognizable (**b**). The structures on the left in both (**a**) and (**b**) have been affected by the same amount of curvature, although this curvature is not discernible in (**b**).

open than earlier ones is based on **observation only** and, as has already been seen, observation is unlikely to reveal early-formed **open** structures which have been modified later by tight folding.)

Furthermore, factors such as 'work hardening' by repeated deformation will have affected the response to deformation of earlier structures, thus influencing their style and so accentuating the difference between them and later folds. Recall that the folding of a layered unit serves to increase its rigidity, as is the case with corrugated sheeting, so altering its

response to applied stress so that it behaves in a more brittle fashion (Figure 7.11 in section 7.1.1). Later structures such as folds formed in previously folded layers will tend to be accompanied by fractures, have straighter limbs and angular hinges etc. typical of structures formed in response to higher strain rates and those late in the succession (e.g. Figures 7.16 and 7.18, in section 7.1.3).

However, it will be appreciated that even though they are late, very open folds in complexly folded terranes may be difficult to recognize at all because the simplicity of their

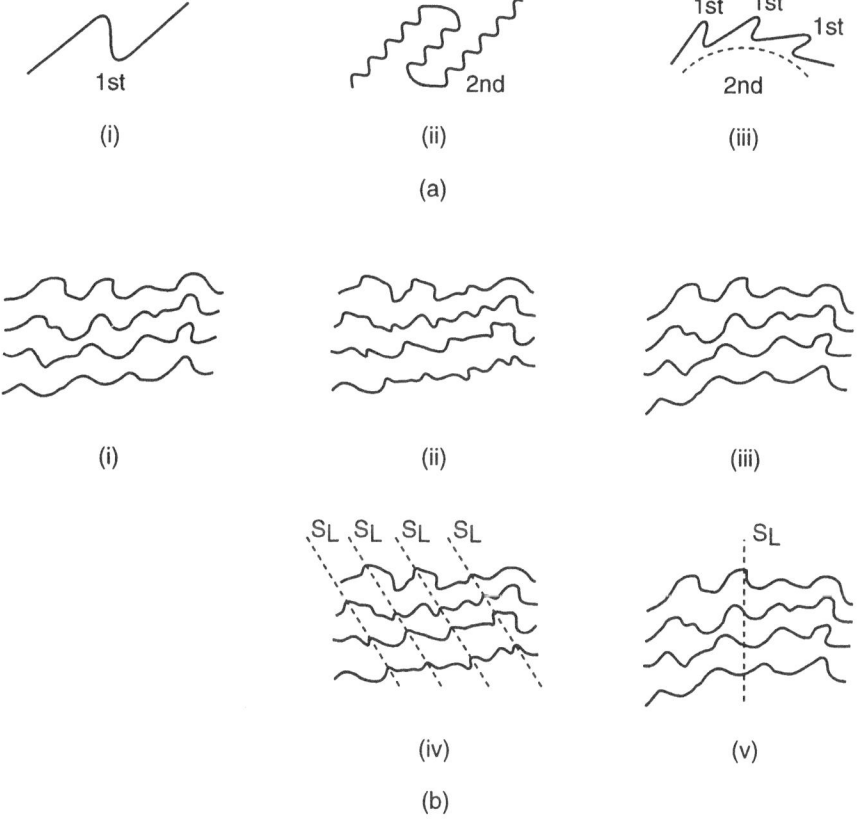

Figure 10.2 Sketches to show how scale might influence the ease of recognition of later folds in foliation already affected by multiple folding. (**a**) Where the pre-existing folds are comparatively simple (**i**), the relationship between relatively small-scale folds superimposed on existing large-scale folds (**ii**), and that of relatively large-scale folds superimposed on small-scale (**iii**) is clear. (**b**) Where the foliation is already affected by multiple folding (**i**), the relationships between smaller-scale later folds (**ii**) and larger-scale later folds (**iii**) is less clear until demonstrated by the later axial planar traces, (S_L) in (**iv**) and (**v**).

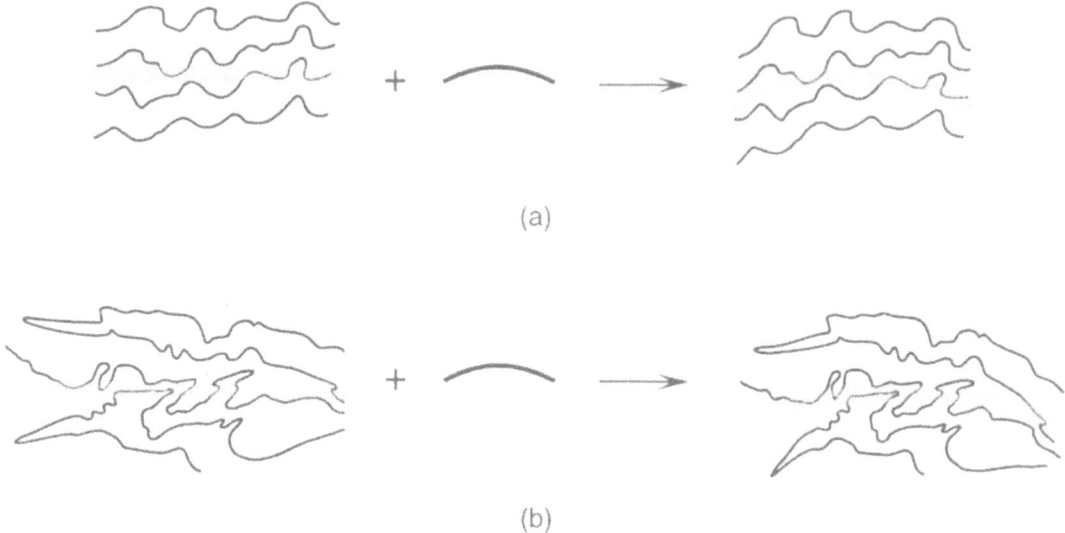

(a)

(b)

Figure 10.3 Sketches showing two cases of open refolding of complex structure. Although the later curvature of the left side of (**a**), as shown by the grey bands in both **a** and **b**, might be recognized if the observer was already alerted to the possibility of its existence, there is no prospect of recognizing it in the structure of (**b**) even if it was known to exist. Sketches traced from a photograph.

shape is lost in the intricacy of the overall structural pattern. This is especially so where there is more than one set of folds in near-parallel orientation so that the form of an earlier fold is partly dependent on the effects of any later ones. Although, as late structures, open folds can be expected to have relatively regular orientation and trend, provided their axial planes are upright as they often appear to be, the very fact that they are open structures makes this orientation, and hence its regularity, difficult to assess because their limbs are not likely to be simple curves stemming from a single folding of initially planar layers (as in Figure 10.9) but compound (Figure 10.1a), viz. the multicurved surfaces of the fold envelopes of all the preceding fold sets (Figure 10.1b). The chance that these later folds would be recognized is more likely when they are either (a) very **small**-scale, i.e. penetrative; or (b) very **large**-scale relative to those on which they are superimposed (Figure 10.2). But even in these cases ease of recognition would depend very much on individual circumstances and where

the structure existing prior to the later folding is complex, recognition of the effects of the later folding is not at all likely (Figure 10.3a, b). An example of structural complexity where superimposed open folding would be unrecognizable is shown in Figure 10.4.

Therefore it might be difficult to decide whether the observed structure is (i) simply exhibiting a slight departure from the predicted orientation of fold sets arising from mutual interference or local slight irregularity (inhomogeneous deformation) or whether it is in fact (ii) the response to a late stress field superimposed on several earlier fold sets. In practice, however, it is often possible to sense the structural regularity associated with these folds intuitively but the subjectivity inherent in such intuitive recognition of fold sets should always be acknowledged and the existence of the late fold set must be confirmed, even though the quantitative establishment of open fold sets based on recording the necessary structural data can be difficult. For this reason some other means of identifying the existence

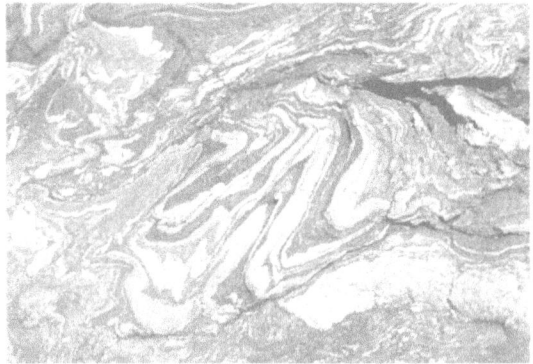

Figure 10.4 Photograph of structure exhibiting the degree of complexity represented in Figures 10.1–10.3. Svecofennian migmatite, Djupkobbarna, Jussarö area, southern Finland.

If exposure allows it (and if measurement is not precluded by the structural complexity) the best way to deal with this problem is to plot several poles to foliation and thus establish the axial attitudes of the later folds. This often means attempting to record the attitudes of the **envelopes** of the earlier fold sets, or measuring the poles to the same set of earlier fold axial planes (Figure 10.5a) and then plotting them stereographically (Figure 10.5b). The pole to the resultant scatter diagram (Figure 10.5c) provides the attitude of the hinge of the later fold. Although attempts to measure and record axial planar traces or axial attitudes directly from three-dimensional exposures may produce a fairly wide spread in orientation, a reasonably close approximation should be possible if a sufficient number of measurements is made. As Figures 10.5b and 10.5c show, these, when plotted on a scatter diagram, should produce a recognizable, if weak great circle girdle maximum normal to the fold axis comparable to that derived from the measurement of poles to foliation in a simple open fold (Figure 10.6). The poor resolution of such a maximum, in spite of the regularity of orientation of the structure because of its late position in the sequence, will not be serious

of open folds might be necessary, such as recognizing the curvature of simple structures imposed part way through the succession, structures comparable to the grey bands of Figure 10.3 (see also Chapter 4, and Figures 4.19, 4.20). Of course such simple structures may not always be present in the succession and even if they are, they must first be recognized as such.

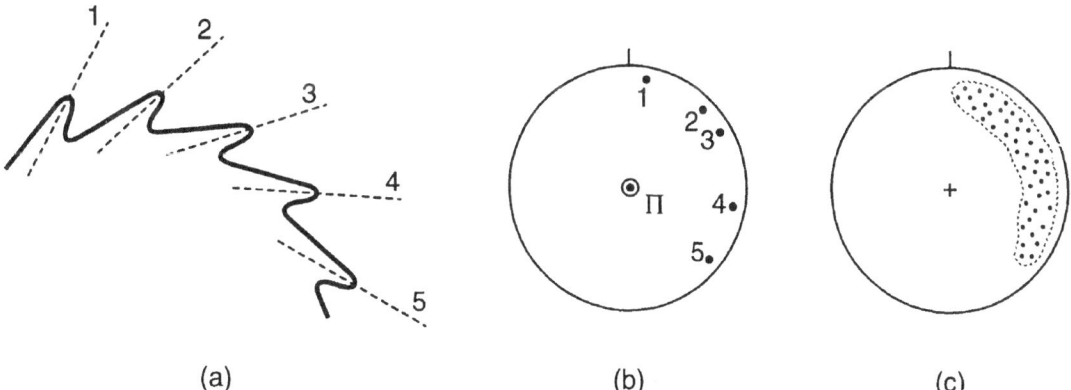

(a) (b) (c)

Figure 10.5 Sketch of a compound fold comprising tight folds plus a later open fold (a) Axial planar traces of the earlier folds dispersed by refolding on the open fold. (b) Stereo plot of the poles of the earlier axial planar traces. (c) Scatter diagram of these poles showing dispersion on a great circle girdle, a plane. The pole (II) to this great circle girdle is parallel to the (later) open fold axis. Compare (a) with the dispersion of folds shown in the photograph of Figure 6.41c. Also Figures 6.41b, 9.12b and 10.10b.

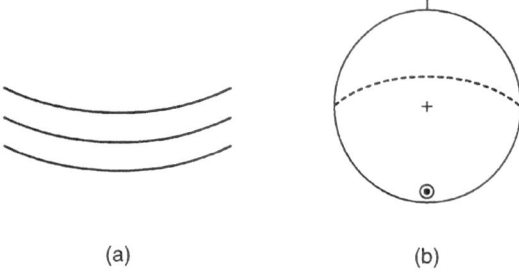

(a)

(b)

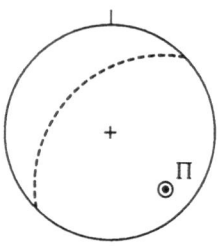

Figure 10.6 (a) Sketch showing the traces of slightly warped sub-horizontal foliation defining a simple open fold such as that shown in the photograph of Figure 10.9. (b) Plot showing the great circle distribution of poles to the foliation of the warp in (a). The pole to this great circle defines the axis of the warp.

Figure 10.8 Stereo plot showing a great circle girdle representing the poles to folded foliation and the pole to this girdle (II) which is parallel to the axis of the fold. A clearly defined, consistently obtainable pattern on stereo plots implies the existence of a distinct fold set.

unless one is unfortunate enough to find oneself in the position of having to distinguish between more than one set of such structures whose orientations are relatively close to one another. In that case three-dimensional exposure allowing the measurement of poles to foliation to enable the plotting of a II-diagram is essential, and several plots must be made where the existence of more than one set of late open structures is suspected.

Note that with the exception of instances such as those when the succession is short, the existence of open structures formed early in the sequence is most unlikely to be recognized. At most their influence on the overall structure would be to cause a slight, and in fact probably imperceptible, departure from the predicted distribution or orientational behaviour of other folds (Figure 10.7). Such structures will not be discerned in a complex

Figure 10.9 A late, very open fold (warp) recognizable because it is (apparently) unaffected by other folds. Lewisian gneiss, Isle of Lewis, Outer Hebrides, Scotland.

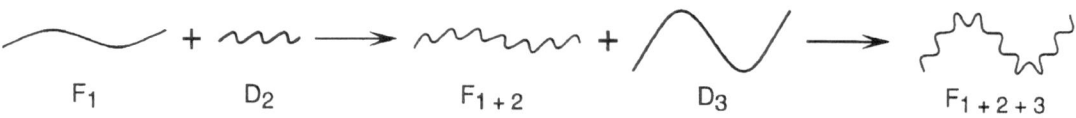

F_1 D_2 F_{1+2} D_3 F_{1+2+3}

Figure 10.7 Sketches showing how an early, very open fold form (F_1) becomes disguised by later, overprinting structures formed during deformations D_2 and D_3 so that ultimately there is no (obvious) trace of F_1.

deformational succession and will not affect the orientation of data of the other folds on a stereo plot. Hence if a consistent pattern suggesting the existence of another structure does emerge when the structural data are plotted, it is normally safe to accept this as evidence of the presence of a distinct fold generation (Figure 10.8).

In summary then, late open structures are most likely to be identified where the effect of preceding deformational events has been minimal, such as in the limb zones of earlier large structures (particularly if these are iso-clinal), so that the structural pattern observed is relatively uncomplicated and hence does not obscure the simple geometrical form of the overprinted fold (Figure 10.9).

10.1.3   VERGENCE AND SYMMETRY ASPECTS OF LARGE FOLDS

When examining asymmetrical folds it is use-ful to note the sense of movement implied by the vergence of the folds to see whether it is consistent or whether it varies (Figure 10.10). If it is consistent, it suggests the structure is fairly late. If it is not consistent this could mean that (a) more than one fold set has been included with folds grouped as belonging to a single set or that (b) the folds have been reorientated strongly or that (c) they are due to a local rather than a regional asymmetrical movement, e.g. drag on the limbs of a larger structure, or that (d) they belong to a conjugate set of folds in which case they are another case of (b) above i.e. showing the effects of local asymmetrical movement.

Therefore, when using the principle of ver-gence to establish the existence of large-scale structures from the presence of small-scale asymmetrical folds, it is important to ensure that all the small-scale structures used are con-temporaneous, i.e. belonging to the same set (see Figures 9.12a to 9.12d in section 9.2.1(n) where this subject is dealt with more fully), as well as coeval with the inferred large-scale

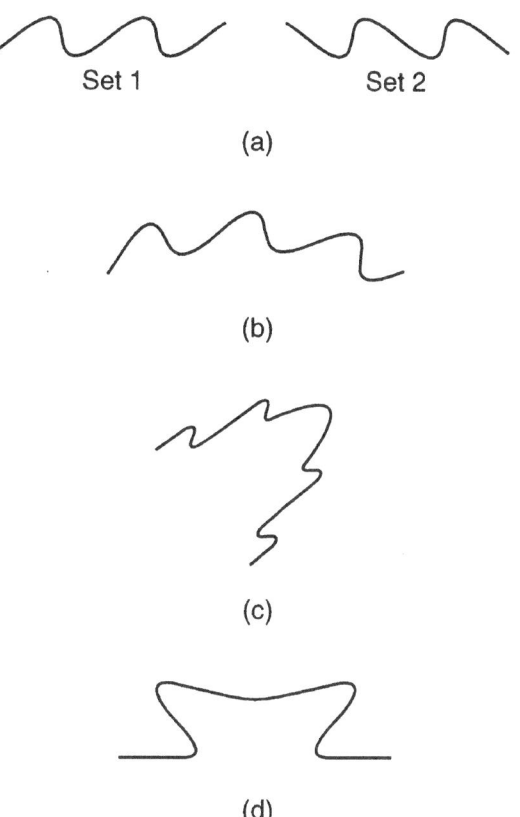

Figure 10.10   Sketches showing, (**a**) Two sets of folds with opposite vergence sense. (**b**) A single set of folds reorientated by later deformation (e.g. by a fold of another set). (**c**) Folds of the same set ('drag' folds) with opposite vergence sense on opposite limbs of a single larger fold. (**d**) Conjugate folds of the same set with opposite symmetry sense. Compare Figures 9.9, 9.10, 9.11, 9.12.

fold. This may not always be easy if the fold pattern is complex.

10.1.4   IMPORTANCE OF METICULOUS OBSERVATION

If the analysis of complex structures can be said to depend on any single factor then that factor has to be the ability to observe objec-tively. This means that in assigning structures to sets, there is a need not only for observa-tional ability (something that develops with

Figure 10.11 The dominant strong cleavage shown in this photograph has almost completely obliterated the bedding which is only faintly discernible as a rough-weathering band traversing the field of view from top centre to lower left. The bedding in this case might easily be overlooked in poor exposures. Falmouth Beds, St Mawes, Cornwall, England.

experience) and objectivity, but also a need to exercise meticulous attention to detail when examining structural relationships, if important and significant detail is not to be overlooked (section 10.1.7 below).

One of the dangers to be guarded against is the effect of visual distraction caused by other, strongly-expressed features which may or may not be of structural significance but which nevertheless distract the observer from structures that are important (Figure 9.17).

An example of this effect is shown in strongly cleaved slates in Figure 10.11. The dominant structure is the cleavage but the original bedding is also visible, albeit faintly, and shows as an irregular, narrow, rough band traversing the picture from top centre to lower left. In this example the bedding might easily be overlooked in poor light.

Another case of the potential for oversight or misinterpretation when significant structural detail is overlooked is shown in Figure 10.12a which shows the trace of an open warp at the contact between quartzofeldspathic gneiss and amphibolite. The ice-planed exposure is crossed by rectilinear glacial striae

trending SSE. A cleavage trace perpendicular to the contact is visible in the amphibolite (upper right and above the pencil at lower left) and on close inspection this can be seen to 'fan' around the open fold. However, this structure, its changing trend and the significance of this, might easily be ignored because of the dominant visual effect of the glacial striae which is almost parallel to the cleavage trace on this outcrop.

A further example of this effect is illustrated in a striking fashion by the comparison of two photographs of the same structure taken at different times (Figures 10.12b and 10.12c). Both are of a fold associated with axial planar alignment of garnet ellipsoids exposed on an ice-planed surface. In Figure 10.12c the garnets and their alignment parallel to the fold axial plane are clearly discernible and are unlikely to be missed by the observer, even during casual inspection, whereas in Figure 10.12b neither the garnets nor their alignment is obvious because of the visual distraction caused by the striae.

The examples described above are just a few of many likely to be encountered and demonstrate the importance of careful observation, something that cannot be emphasized too strongly. Attention to detail is essential for the succession interpretation of the structure of highly deformed rocks. Superficial examination on the other hand is likely only to lead to erroneous conclusions and is unlikely to result in the succession elucidation of the structural succession.

10.1.5 OBSERVATIONS FROM
TWO-DIMENSIONAL EXPOSURES

In gneisses and other foliated rocks (especially quartzofeldspathic rocks) which do not readily weather to give three-dimensional outcrop surfaces it is usually unwise to attempt to infer even the **direction** of dip of the foliation to the outcrop surface and certainly not the **angle** of dip. Experience shows that where clear confirmation of the true angle can be obtained from

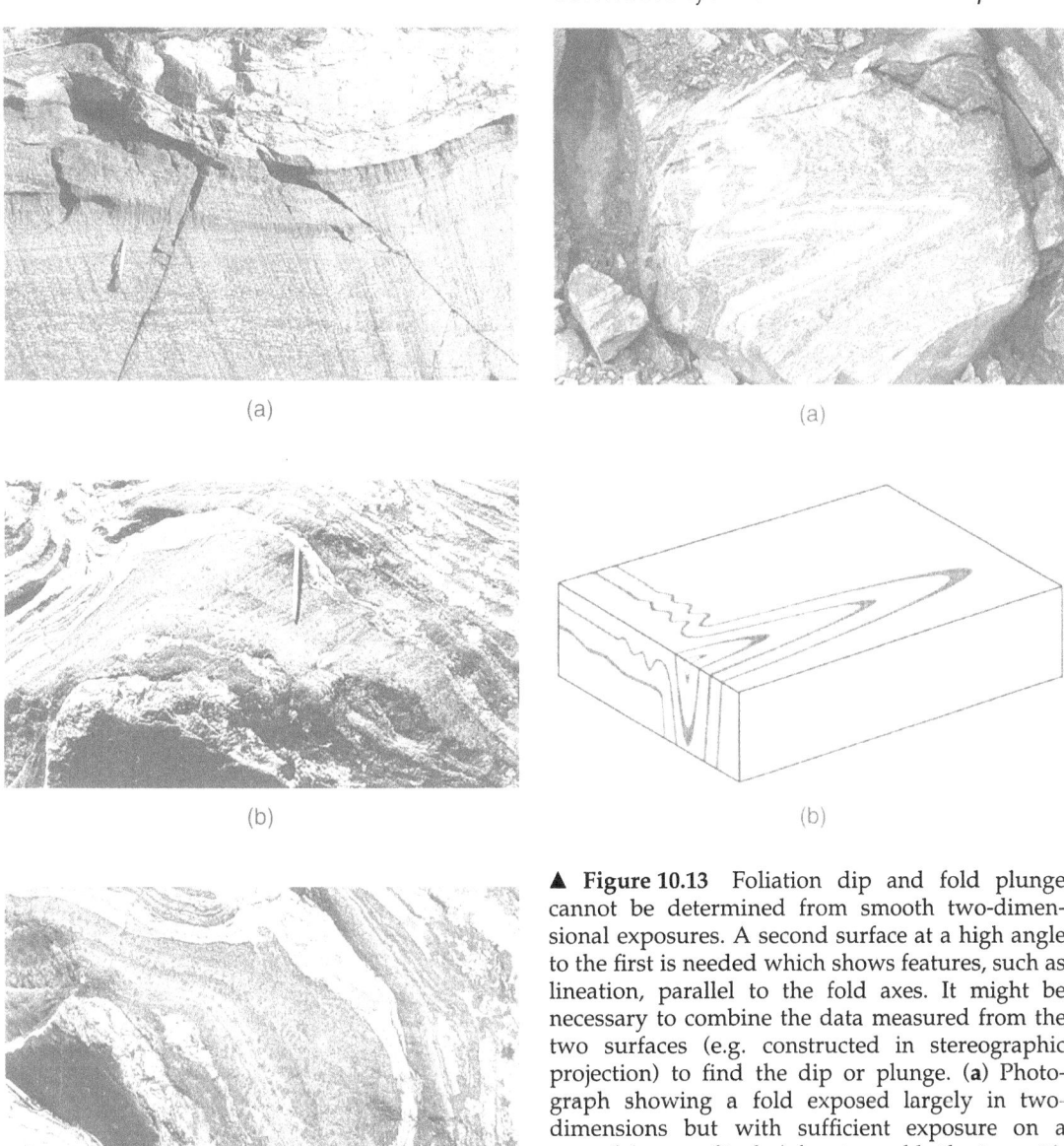

(a)

(a)

(b)

(b)

(c)

▲ **Figure 10.13** Foliation dip and fold plunge cannot be determined from smooth two-dimensional exposures. A second surface at a high angle to the first is needed which shows features, such as lineation, parallel to the fold axes. It might be necessary to combine the data measured from the two surfaces (e.g. constructed in stereographic projection) to find the dip or plunge. (**a**) Photograph showing a fold exposed largely in two-dimensions but with sufficient exposure on a second (perpendicular) face to enable the true attitude to be determined as shown in (**b**). Lewisian complex, Scourie, Northwest Scotland.

▲ **Figure 10.12** (**a**) Warped contact between gneiss and amphibolite crossed by rectilinear glacial striae. Cleavage traces in the amphibolite fanning around the open fold are almost obscured by the striae and could easily be overlooked in poor light. Svecofennian migmatite, North side of Skåldö, southern Finland. From Figure 10g, Hopgood, 1984. Reproduced with the permission of the Royal Society of Edinburgh. (**b**) Folded amphibolite with axial planar ellipsoidal garnets. The garnets, although prominent on the outcrop as dark ellipses, are barely discernible here because of visual 'interference' caused by prominent glacial striae highlighted by the oblique illumination of the outcrop. Compare this view with that of the same fold shown in (**c**) when illumination of the outcrop did not enhance the striae and where the garnets are much more obvious. Svecofennian migmatites, Pohja, southern Finland.

other surfaces (joint faces etc.) inclined to the exposure it is often very different from what it appears to be on the outcrop. Correspondingly, attempts to guess at the angle of plunge of folds from smooth two-dimensional exposures in quartzofeldspathic foliated rocks should also be resisted (Figure 10.13) and interim approximations ('guesstimations') should be recorded **only** as approximations, and should always be confirmed whenever direct measurements can be made. In both cases the strike of the foliation or fold axial trace, or the trend of a linear structure, are the only certain measurements that can be recorded.

10.1.6 'APPARENT DIPS' AS DISCRIMINATORY CRITERIA

In highly deformed terranes, especially when the folds of several sets are of roughly the same order of size, the resultant interference structures will often preclude confirming the identification of folds as members of particular sets solely on the basis of orientation of geometrical features such as axial planes, except in the case of the latest folds. This is because reorientation by later structures means that even large initial differences in angle of dip which might have served as factors in the discrimination between early structures may be reduced or even eliminated during the subsequent deformation so that even opposing senses of axial planar dip may no longer be valid criteria for separating two sets of folds whose axial planes were inclined to one another prior to later deformation. (See the discussion of the significance of dip in the classification of structures (especially fold axial planes) in section 5.2.4). This of course can also apply to any attempt to use attitudes of fold axes as a means of discriminating fold sets, except in the case of the latest structures. Furthermore, acceptance of **apparent dips** as an approximation of the true dip can be very misleading. 'Apparent dip' – a term sometimes used for the angle of plunge of the trace of a dipping surface viewed on the outcrop – is not a satisfactory expression

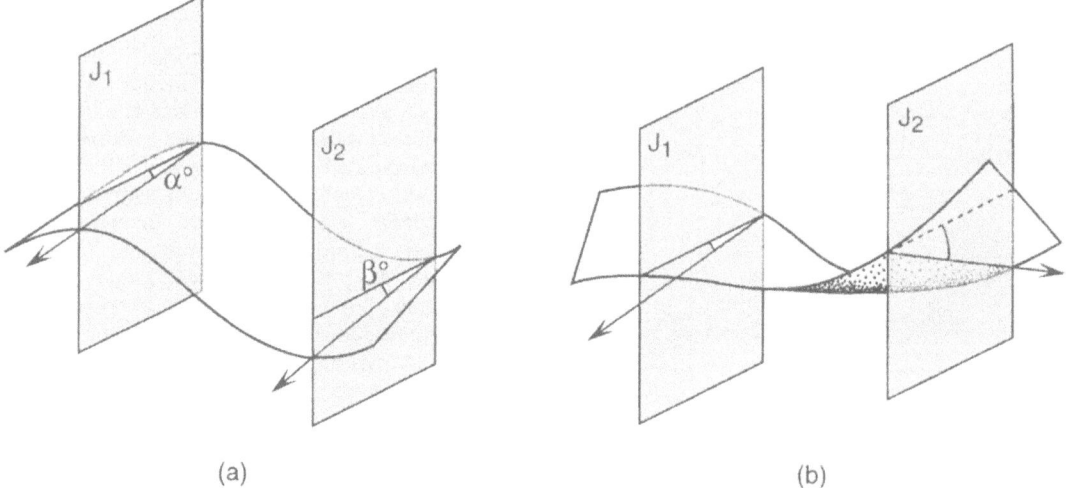

(a)   (b)

Figure 10.14   Sketches showing 'apparent dips' (strictly angles of plunge of the traces of the planar structures). (**a**) Different 'apparent dips', $\alpha°$ and $\beta°$, of the same, deformed structural surface (e.g. foliation or fold axial plane which was originally planar) exposed on successive parallel exposures such as joint faces, J_1 and J_2. (**b**) Opposite 'apparent' dip senses of the same structural surface (e.g. fold axial plane) exposed on parallel joint faces. The differing plunges of the intersections of the surface with the joint planes (in **a**) and the opposed plunge senses (in **b**) are the result of deformation of the originally planar surface.

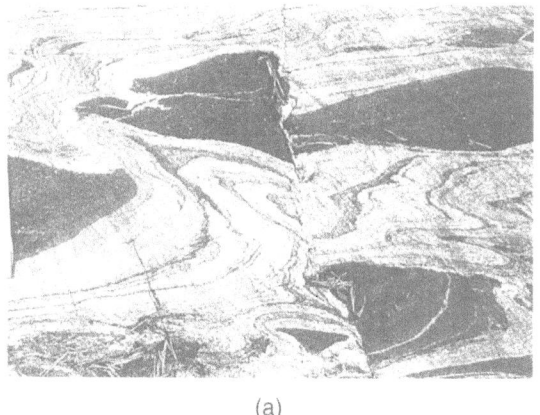

(a)

Figure 10.15 'Scar' folds formed by flow of leuco-cratic neosome into lower pressure zones between darker amphibolite palaeosome blocks. Sveco-fennian migmatite, (**a**) Flackholmen, (**b**) Orrholmen, Jussarö region, southern Finland.

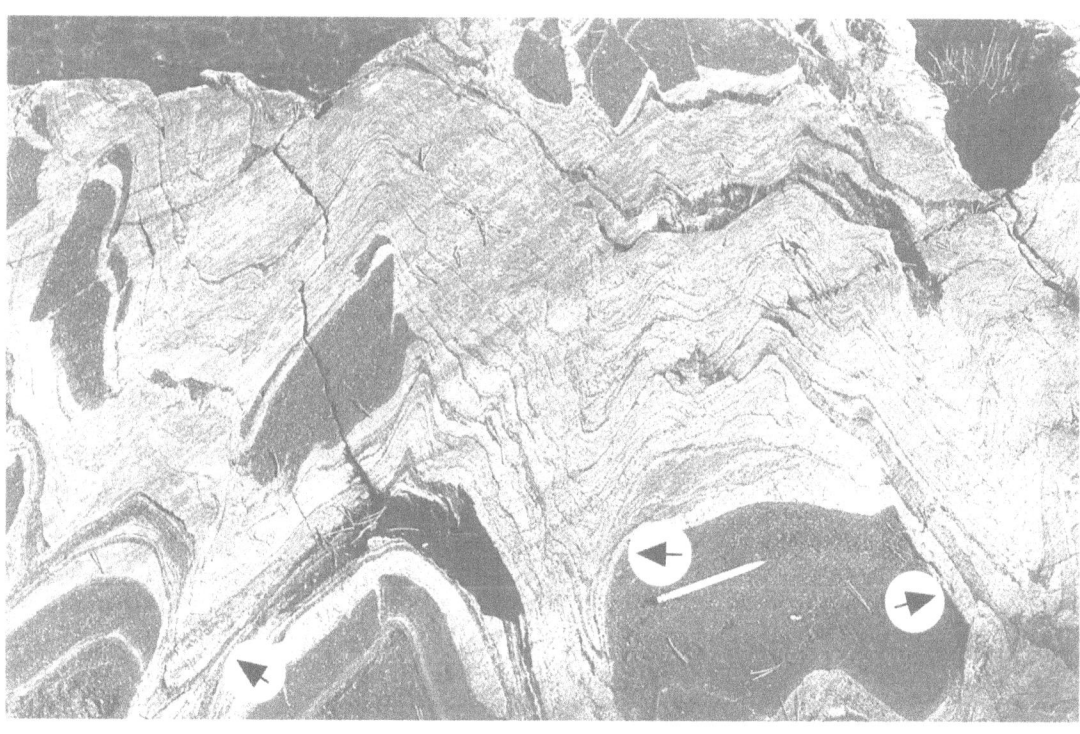

(b)

in that it mixes terminology for the attitude of both linear (one-dimensional) and planar (two-dimensional) structures. This is not only because the angle of dip suggested by the plunging traces of the intercepts of fold axial planes (or other planar structures) with successive parallel rock faces or parallel joint faces can be incorrect, but also because the plunges observed may vary considerably if the struc-

ture is deformed (by folding, say), so compounding the inaccuracy of the observations (Figure 10.14).

10.1.7   EXAMPLES OF SIGNIFICANT OBSERVATIONAL DETAIL

The following examples of sets of structural relationships are used to show what is meant

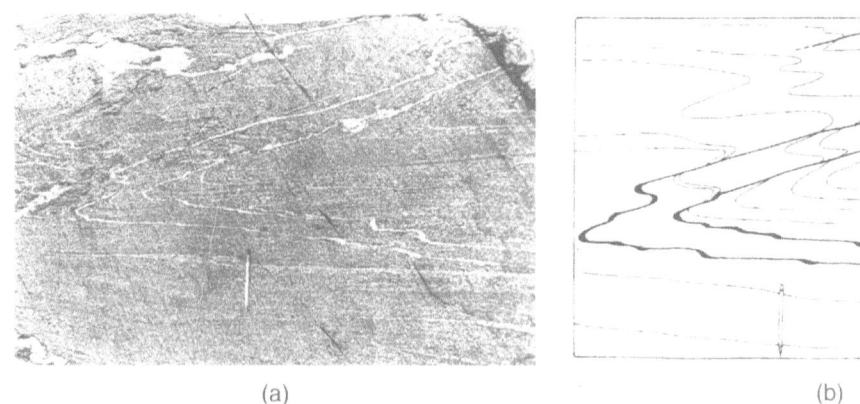

(a)   (b)

Figure 10.16 (a) Incongruent folds. Two sets of folds with parallel axial planar traces, one set of folded leucocratic veins and the other, fainter set, in the folded grey 'host' gneiss. (b) Sketch of the relationships between the two fold sets. Svecofennian migmatites, Angholmen, Skåldö. southern Finland. From Figure 17g, h Hopgood, 1984. Reproduced with the permission of the Royal Society of Edinburgh.

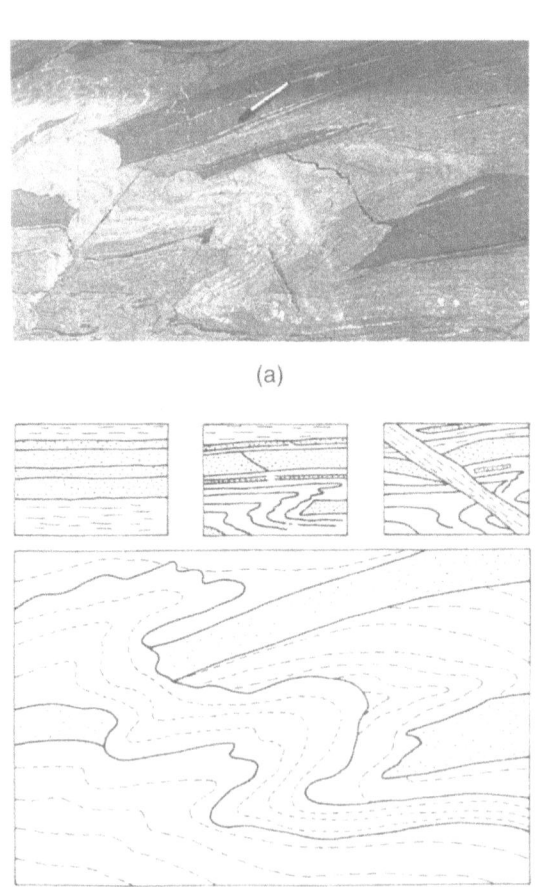

(a)

(b)

◀ **Figure 10.17** Photograph (a) and explanatory sketch (b) of the stages in the development of the relationships shown in (a) between a fold, a folded leucocratic vein and agmatite. Svecofennian migmatite, southern Finland. From Figure 17a, b Hopgood, 1984. Reproduced with the permission of the Royal Society of Edinburgh.

by 'significant' detail and the kind of structural features that contribute to it, and also to further demonstrate the need for careful observation. These examples illustrate structural features and relationships of the kind that might easily be overlooked during a cursory inspection of the exposure and yet could well be of considerable significance in the structural succession and deformational sequence.

(a) Figures 10.15a and 10.15b show agmatite comprising a palaeosome of amphibolite blocks in a neosome of leucocratic gneiss. Note (i) that Figure 10.15b shows that the agmatite is derived from folded migmatite with fold segments preserved in the palaeosome blocks. The train of amphibolite blocks is disposed in an open fold form whose axial planar trace trends perpendicular to the pencil. Note (ii) that following break-up of the amphibolite layers to form the agmatite, there was further development of folds ('scar' folds) in the leuco-cratic neosome (clearly shown in the detail of 10.15a) caused by flow into the 'space' (lower pressure zones) between the amphibolite blocks. Note that in Figure 10.15b the axial traces (and possibly the axial planes) of the two sets of folds are parallel, although they were formed at distinctly different times and by decidedly different mechanisms.

(b) Figure 10.16a shows thin, folded leuco-cratic veins in folded grey gneiss. Superficially the 'vein-folds' and the 'host gneiss-folds' appear to be concordant, but close exami-nation, confirmed by tracing the form of the 'vein-folds', shows that the veins transect the foliation and the grey gneiss. A possible cause of this structural relationship is that the veins were originally planar and transected the gneiss foliation in an original direction equivalent to the present lower right to upper left and they, together with the foliation, were subjected to dextral slip on planes trending from bottom left to top right resulting in folds in both the veins and the gneiss. It is also possible that the fold forms were modified by dextral slip on planes trending left to right.

(c) Figure 10.17a shows a folded, rimmed leucocratic vein crossing grey migmatitic gneisses containing blocks of foliated amphi-bolite. The vein appears to have been folded together with the migmatites after agmatiza-tion of the amphibolite layers(s). However, closer inspection shows that the vein was intruded into the migmatites which were already folded and that the whole, including the vein, was then subjected to folding again as shown in the sequence sketched in Figure 10.17b (cf. section 4.1).

(d) Figure 10.18a shows what, on cursory examination, appears to be a single asymmet-rical fold with well developed axial planar structure and associated minor folds with sharply peaked hinges. But closer inspection of the relationships in the ringed area shows that there are at least two sets of folds. The (curved) axial planar trace of the small, nearly isoclinal fold on the left of the ringed areas is

deformed across the trace of the SE (see compass) axial plane of the dominant larger, more open, angular fold. Although the photo-graph shows other small minor folds on the left to have axial planar trends that are slightly oblique to the trend of the main (SE) fold, these curve into parallelism with it, rather than cutting across it. Therefore these minor folds are likely to be the same age as the main folds and in this respect their rela-tionship differs from that between the (ringed) near-isoclinal fold and the larger fold. Here the NE axial planar traces cross the axial planar trace of the right fold which lies **within** the lithological banding. This alerts the observer to the fact that at least two sets of structures coexist here. Further inspection could show that the attitudes of the minor folds on the left (i.e. not parallel to the main fold) are the result of reorientation by post-SE deformation of both major and minor folds, and support for this is provided by a strong, fine mineral trace oblique to the SE trend and parallel to the axial traces of the minor folds on the left.

10.1.8 FOLD/CLEAVAGE RELATIONSHIPS

Before proceeding with the discussion of fold relationships, some simple and some appar-ently simple cleavage relationships to folds will be considered. It is important when deal–ing with folds and cleavage to first ensure that, even when the deformational history appears to be simple (e.g. there are only two folds), 'the' cleavage observed does not belong to more than one generation, i.e. be aware that there could be more than one set of cleavages (see Figure 10.20, adapted from Rutland and Etheridge, 1975). Therefore look for the following:

(i) consistent transecting relationships be-tween cleavages, if these are not parallel;
(ii) inconsistencies in symmetry relationships and vergence, e.g. an earlier cleavage cutting lithological layering and itself cut

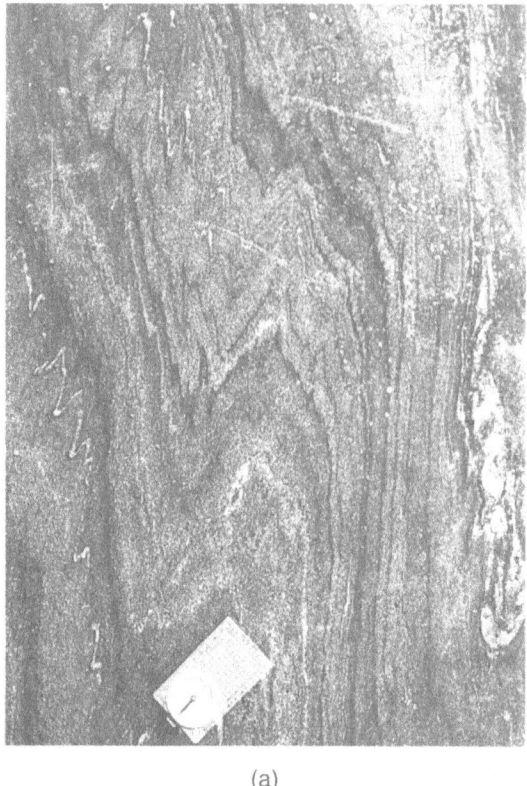

(a)

(b)

Figure 10.18 (**a**) Photograph showing what appears at first sight to be a single fold set comprising a larger fold with associated minor folds. (**b**) Closer inspection of the area circled in (**b**) shows that the structure comprises at least two sets, a tighter fold with curved axial planar trace crossed by a later set (the 'dominant' set in the photograph) with rectilinear axial planar traces. Lipasvaara, Karelia, Finland. Photograph by courtesy of G. Gaál.

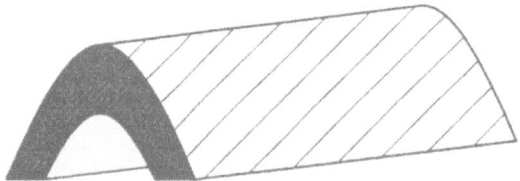

Figure 10.19 Sketch of the general case of a fold associated with planar cleavage, shown here simply as the cleavage **trace** on the folded surface. The cleavage is planar, but not axial planar.

by a later cleavage such as one that is axial planar to a later fold.

Consider the case where there is apparently a single cleavage which is non-axial planar to the fold as shown in Figure 10.19.

The following four possibilities can be considered.

(a) The cleavage is non-parallel, in which case
    (i) it might have been folded;
    (ii) it might be related to the fold (i.e. fanned cleavage if fanning intersects in the axial direction), or
    (iii) if the cleavage is fanned but the inter-section of the fan is not parallel to the axial direction it might be related to the fold but 'transected' (see Stringer and Treagus, 1980; section 3.2.2 and Figure 3.24).
(b) The cleavage is parallel, in which case
    (iv) it is (likely to be) later than the fold.

Examples

(a) An example showing a range of relationships between folds and cleavage (schistosity) is shown in Figure 10.20. The figure is based on relationships between folds associated with schistosity in the Broken Hill region, New South Wales, established by Rutland and Etheridge (1975, Figure 4) using the basic principles of overprinting (*Überprägung* of Sander, 1970). See also sections 3.1.1, 3.2.2 and 5.3.) This followed published work by a number of observers, including Hobbs (1966) and Williams (1967) as a result of which it was considered that the folds could be classified by style into two groups each representing folds belonging to two generations. The folds, while of similar geometry and shape, were originally separated into two 'style groups' ('Group 1' and 'Group 2') in terms of the presence or absence of axial planar schistosity and of folded schistosity. Although the establishment of the two style groups was in part based on the relationship of the folds to schistosity, what had not been appreciated at the time but was subsequently demonstrated by Rutland and Etheridge (1975) was the fact that there were at least **two** periods of (high-grade) schistosity-formation so that the criterion of 'presence or absence' alone of high-grade schistosity was not in itself diagnostic of folds of one or other of the particular two generations. According to Rutland and Etheridge (1975), Group 1 folds had been defined on the basis of possessing a lineation parallel to their axes and an axial planar schistosity, whereas Group 2 folds folded that schistosity but lacked an axial planar schistosity of their own. The two groups were considered to represent two generations of folds.

However, Rutland and Etheridge showed that, contrary to earlier observations, axial planar cleavage existed for both first and second generation folds. Therefore in the case where the axial planar schistosity of a group of folds happened to be the later schistosity, then folds classified as the 'earlier' set on the basis of possessing an axial planar schistosity (i.e. 'Group 1' folds) would in fact belong to the **second** generation and not the first. Rutland and Etheridge distinguished a number of relationships between folds and S_{1-2} cleavage and illustrated these. Three of these are shown stylized in Figure 10.20.

Because S_1 is not distinguished from S_2 all that may be said of the folds shown here is that (a) and (b) are post-S_{1-2} because they fold the (originally axial planar) cleavage, whereas (c) is likely to be contemporaneous with S_1, which (with S_2) is axial planar to the fold (S_1 is unfolded). S_2 is fortuitously parallel to S_1 in the fold hinge because of parallelism of F_1 and F_2 axial planes (Figure 10.21). Another, but

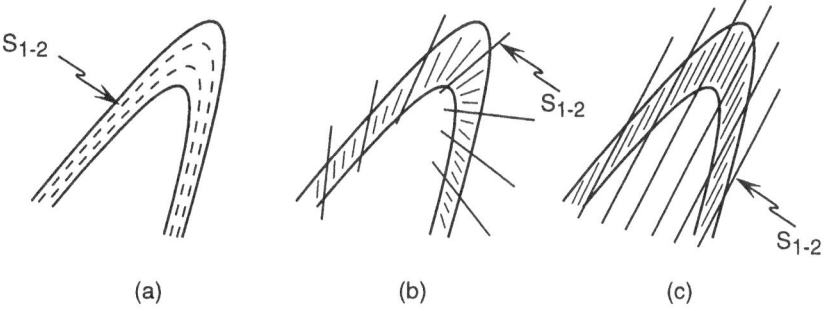

(a)   (b)   (c)

Figure 10.20   (a,b,c) Sketches of different relationships between axial planar cleavage (S_{1-2}) and folds in three cases. Derived from Rutland and Etheridge, 1975.

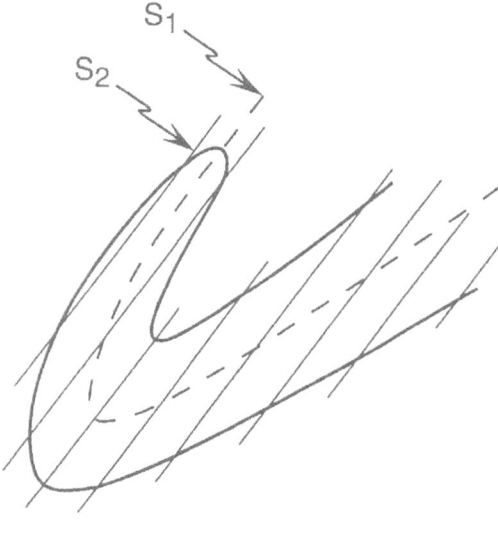

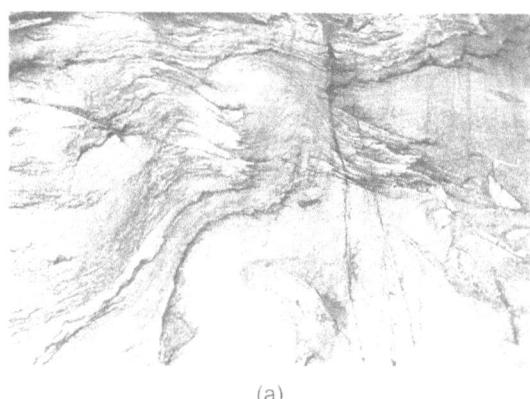

◀ **Figure 10.21** Sketch of the structure representing the pattern of overprinting between two folds (F_1 and F_2) with axial planar traces, S_1 and S_2. The earlier (F_1) hinge, fortuitously parallel to S_2, is equivalent to the hinge shown in (**c**) of Figure 10.20 above. Modified from Rutland and Etheridge, 1975.

(a)

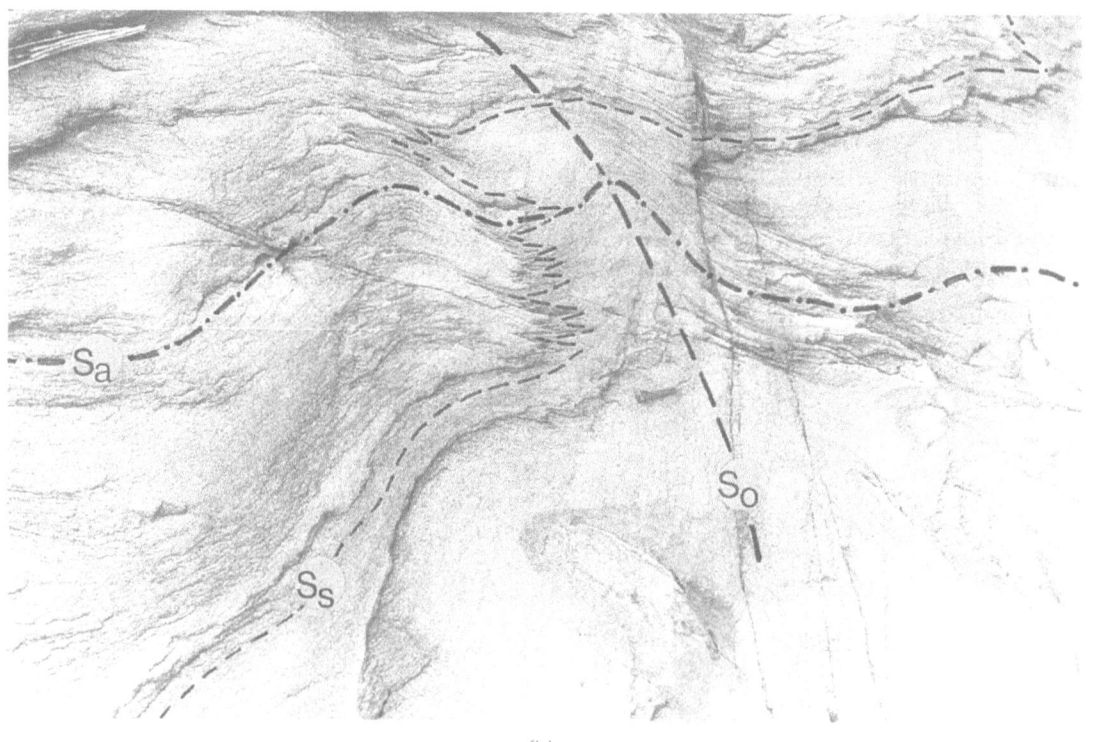

(b)

less likely, theoretical possibility is that (c) is an even earlier fold (F_0), prior to the development of S_1 and S_2, that had both S_1 and S_2 superimposed on it parallel to its axial plane.

This example is intended to highlight the need for care in the observation of structural relationships and the necessity to guard against accepting unconfirmed conclusions regarding the identity of structures (in this case the conclusion that there are two fold generations and only one schistosity). It also shows the need to exercise caution when interpretations have to be made on the basis of limited observational evidence.

(b) Usually where exposure is good enough to permit detailed observation it will be possible to demonstrate the overprinting of early schistosity by later folding (or the converse). Figure 10.22 shows clearly the relationship between schistosity and fold sets in schistose metasediments. Early, tight to isoclinal folds (F_a) formed in a surface (S_S) have a strongly developed axial planar schistosity (S_a) which is predominately parallel to S_S, except in the hinges of F_a. S_a (and S_S) have been affected by later open upright folds with a curved axial planar trace (S_o) at a high angle to the trace of S_a.

(c) While cleavage and lineation is often clearly expressed, as in the example shown in Figure 10.23, in many instances, however, cleavage, schistosity or lineation may be so weakly developed that it can easily be overlooked. In such cases the angle of the light incident on the exposure can have a very important bearing on whether or not the cleavage trace is discerned (cf. Figure 10.12). This means that the time at which observations are made can be critical because not only

is the presence or absence of sunlight important, but so is the angular relationship between the cleavage or lineation traces and the direction of the sun's rays.

Figure 10.23 shows a clearly developed alignment of coplanar mineral aggregates readily recognized as intermittent traces almost

Figure 10.23 Steeply plunging coplanar mineral lineation in migmatite, visible as vertical, short white traces, slightly discordant to the leucocratic veins parallel to the pencil. Dabie complex, southern China.

◀ **Figure 10.22**   (a) Folded metasedimentary layering in Proterozoic quartz-mica schist. (b) Explanation of (a). Folds (F_a) in the metasedimentary layering (S_s) have horizontal secondary foliation (S_a) parallel to the F_a fold axial plane. Both S_s and S_a are deformed by open folds (F_o) with axial plane (S_o) which is itself curved by later deformation. North Karelia, Finland. From Figure 16, Hopgood, 1980. Reproduced with the permission of the Royal Society of Edinburgh.

(a)

(b)

Figure 10.24 (a) Fine cleavage visible intermittently across the exposure as faint traces parallel to the pencil. (b) Detail of (a) showing pervasive cleavage traces affecting the whole surface and expressed as lighter and darker fine lines and streaks parallel to the pencil.

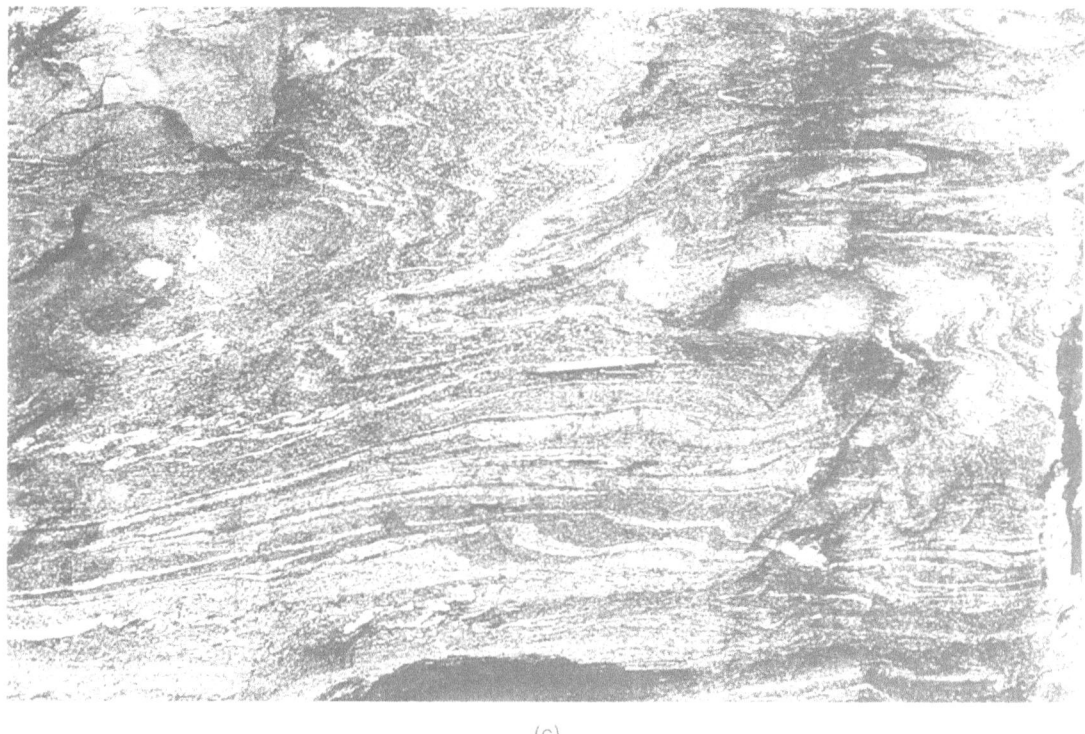

(c)

Figure 10.24 *(cont.)* (c) Fine cleavage recognizable largely because of parallel fractures shown by shadowing in the lower right of the photograph. The NE-trending fractures are 'highlighted' by the obliquity of the illumination of the surface. Migmatite, Dabie complex, Feng Huang Guan Dam, southern China.

parallel to the pencil. The contrast between the grey host rock and the pale colour of the mineral aggregates ensures that their alignment is plainly visible, and even in dull light as in this photograph it can also be seen that there is an angular discordance between the lineation and the leucocratic veins. Although the mineral aggregates shown are small they are nevertheless still recognizable, even when the light is poor.

In contrast, the cleavage traces in the following three figures (Figures 10.24a, b and c) from the migmatites of the Dabie complex in Southern China (Hopgood *et al.*, 1989) are more difficult to recognize. Although the structural relationships in Figures 10.23 and 10.24 from the same complex are comparable, in Figure

10.24 the structure is not so easily seen as it is in Figure 10.23.

Compare Figures 10.24a, b and c which are all from the same exposure. In Figure 10.24c the linear traces are clearer because of shadowing of the fractures, purely because of the quality of the light. At the time the photograph was taken the light was oblique to the rock face. In contrast, at the time of the photographs of Figures 10.24a and 10.24b, the weather was cloudy and here the linear traces are less clear.

It can be seen, therefore, that fine structures such as schistosity and cleavage show best when the light is oblique to the outcrop and also when the illumination is at a high angle to their trace on the outcrop. This is because they are associated with fine fractures which produce faint shadowing in oblique light, so causing the cleavage traces to be highlighted. While bright sunlight increases this effect it should be remembered that it will also accentuate surface irregularities and this could detract from the appearance of the structure. In such cases a balance must be struck between these lighting conditions (see the comments on the effects of visual distraction and Figure 10.12).

10.1.9 STRUCTURES ASSOCIATED WITH ONE OR TWO FOLDS

The following examples show structural relationships associated with one or two folds.

Structural relationships and structures likely to be associated with folds in highly deformed terranes may include the following, although the number of these is likely to be considerably less in high-grade metamorphic rocks: primary structures; bedding (graded cross-, younging – useful for structural facing); joints (e.g. joints perpendicular to the fold hinge); mineral (elongation – 'stretching') lineations;

intersection lineations; cleavages (fracture, crenulation, slaty) and their different relationships (cross-cutting, axial planar, fanning, transected etc.); fold axial plunge, azimuth (from hinge); enveloping surface of folded surface and relationship to minor folds; vergence and location of minor folds with respect to the major hinge; symmetry/asymmetry, axial planar attitude etc. of folds.

Examples

(a) An example of two sets of similar-looking small-scale folds or crenulations which are sporadically developed because of lithological variations and hence seldom seen together on the outcrop is shown in Figure 10.25.

Although the two fold sets are superficially similar and often show comparable wavelengths in the same rock type, the two are distinguishable because of their different sense of asymmetry (cf. Figures 6.33 and 6.34) and ultimately their overprinting relationships when seen together on the outcrop. Where the folded foliation of one or both of such struc-

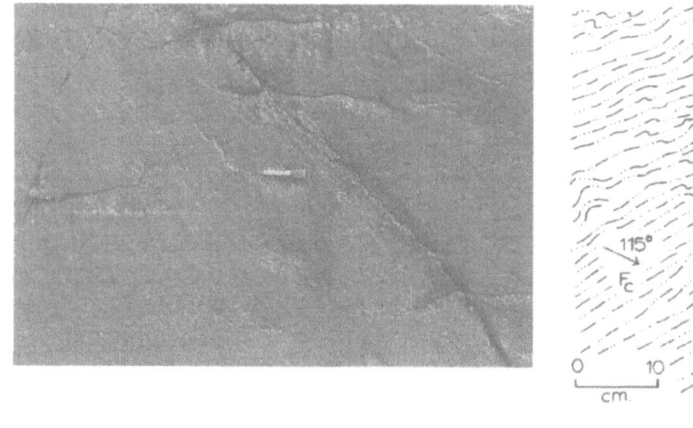

(a)

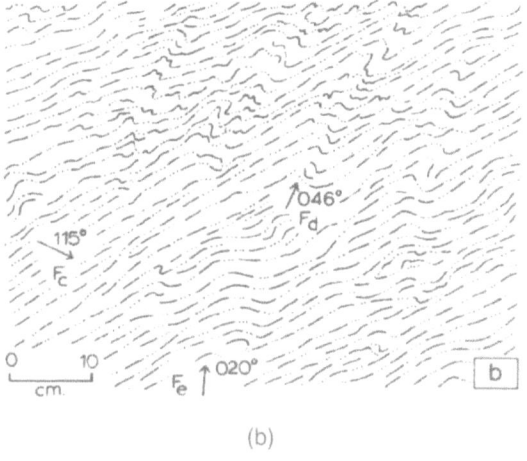

(b)

Figure 10.25 (a) Photograph of the interference pattern effectively resulting from two sets of small-scale folds (two asymmetrical sets on inclined axial planes). The pattern is insignificantly modified by a third fold set with more or less vertical axial planes. (b) Labelled sketch of (a) showing the relationships between the asymmetrical folds (F_c and F_d) and the later, more open fold (F_e) which causes barely discernible curvature of the axial traces of F_c and F_d (cf. Figures 10.1–10.4). Svecofennian migmatite, Skåldö, southern Finland. (Compare Figure 6.37.)

tures is associated with cleavage, schistosity or other planar structure, such as (thin) veins, this can further aid their discrimination.

(b) It is important to ensure that folds that appear to be of the same set are in fact congruent, i.e. harmonic, in all cases, because it is possible that there could be at least two sets of folds with similar geometry and parallel (but not coincident) axial planar traces, such as those in the example shown in Figure 10.26.

(c) Distinguish between similar or different types of lineation(s) and linear structures, which could be present even on single folds (Figure 10.27), e.g. extension ('stretching') lineation or non-penetrative linear structures such as intersection 'lineations', e.g. bedding/ cleavage intersection. As in the case of cleavages discussed earlier (Figure 10.20), it is important not to assume that all lineation observed is the same or even related.

In the example of Figure 10.27, $L_i \wedge L_m$ is not 90°. Therefore one can conclude that:

(i) if the original foliation was horizontal, there was folding on a horizontal axis F_1 and the maximum extension during deformation was not vertical;

(ii) if $L_i \wedge F_m$ was originally 90° then there has been differential rotation of L_1 and L_m, because of heterogeneous strain.

Then either or both L and F have been rotated to their present attitudes.

(iii) The original foliation was not horizontal etc.

(d) Distinguish between the hinges of folds that are similar-looking, but which belong to different sets (cf. (b) above). Where refolding (F_c) of an isoclinal fold (F_i) has been coaxial, or nearly so, as in Figures 10.28a to 10.28d, distinction between the first and second fold sets can be difficult and F_i and F_c are not so easily distinguished if F_c is also **isoclinal**. An illustration of the importance of being able to make such a distinction for economic reasons is seen in the work of Koistinen (1981). Here it was important to be able to differentiate between the earliest folds (F_1 and F_2) in the ore bodies at Outokumpu, Finland (Figures 10.28d and 10.28e).

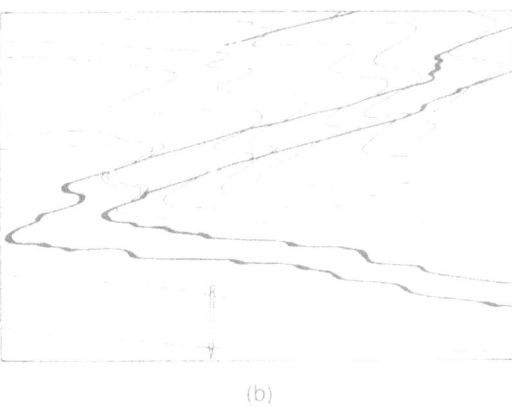

(a) (b)

Figure 10.26 Photograph (**a**) and explanatory sketch (**b**) of incongruent relationships between two fold sets. These, which have parallel axial planar traces, are folded leucocratic veins and folds with similar geometry in the host gneiss. Seen before as Figure 10.16. Old iron prospect. North shore, Ängholmen, Skåldö, southern Finland. (From Figure 17g, h Hopgood, 1984. Reproduced with the permission of the Royal Society of Edinburgh.)

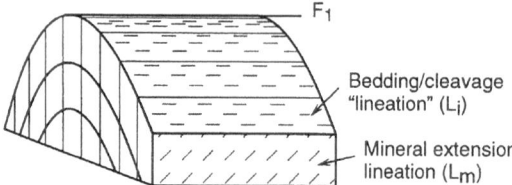

Figure 10.27 Sketch of folded bedding with two linear structures, L_i (bedding/cleavage intersection – a non-penetrative linear structure) and L_m (mineral extension – a penetrative lineation).

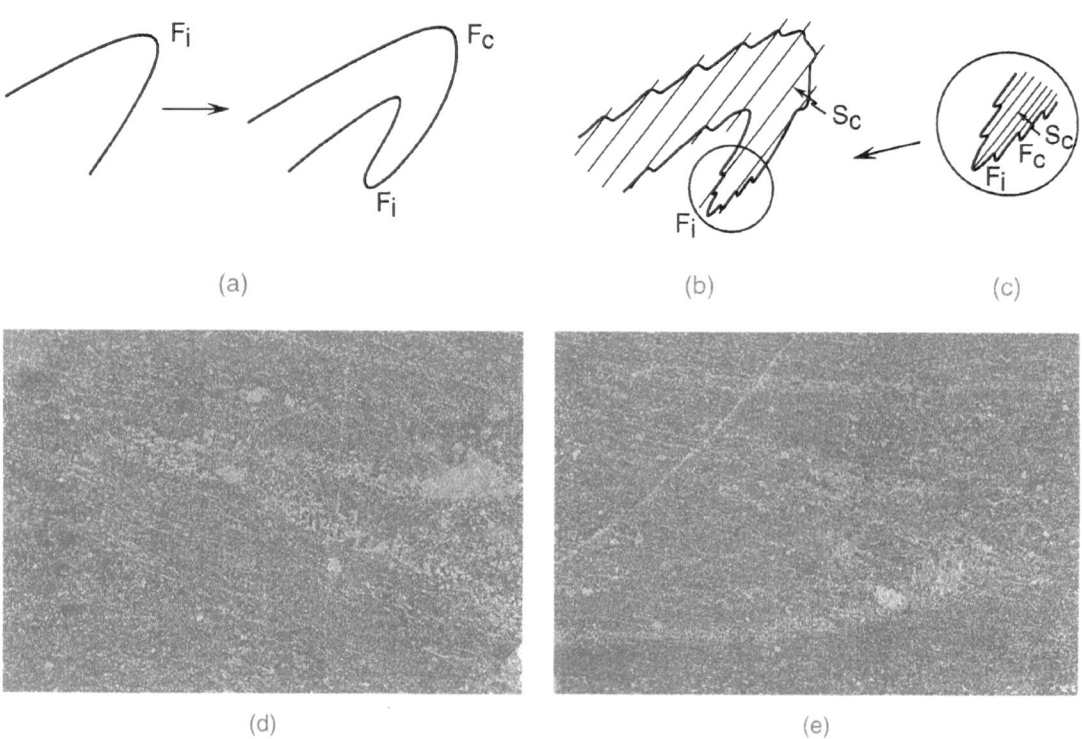

(a)   (b)   (c)

(d)   (e)

Figure 10.28 (**a, b** and **c**) Sketches of the relationships between two coaxial planar isoclinal fold sets F_i and F_c showing the difficulty of distinguishing between them, especially when cleavage forms with the second folding. (**a**) Refolding relationships where cleavage is not involved. (**b**) If F_c is associated with axial planar cleavage (S_c) which will also be imposed on F_i hinges, then these and the F_c hinges will not easily (if at all) be distinguishable from F_c so that in some instances (**c**) the true status of the earlier fold (F_i rather than F_c) might not be appreciated. As Figure 5.4. Also compare Figures 6.29 and 10.18. (**d**) Early isoclinal folds in the Outokumpu ore body showing the characteristic sharply angular hinge of the first (F_1) fold set. (**e**) Characteristic round hinge of the second fold set (F_2) in the Outokumpu ore body. Outokumpu, Finland.

◀ **Figure 10.29**   Overprinted relationships between three fold sets in quartzofeldspathic gneiss. An upright chevron fold which folds an intrafolial isoclinal fold (in the hinge and on the left limb) is folded by an inclined open fold (gently curving the right limb of the chevron) associated with an axial planar fracture cleavage (rectilinear trace plunging to the left). Lewisian complex, Harris, Outer Hebrides, Scotland.

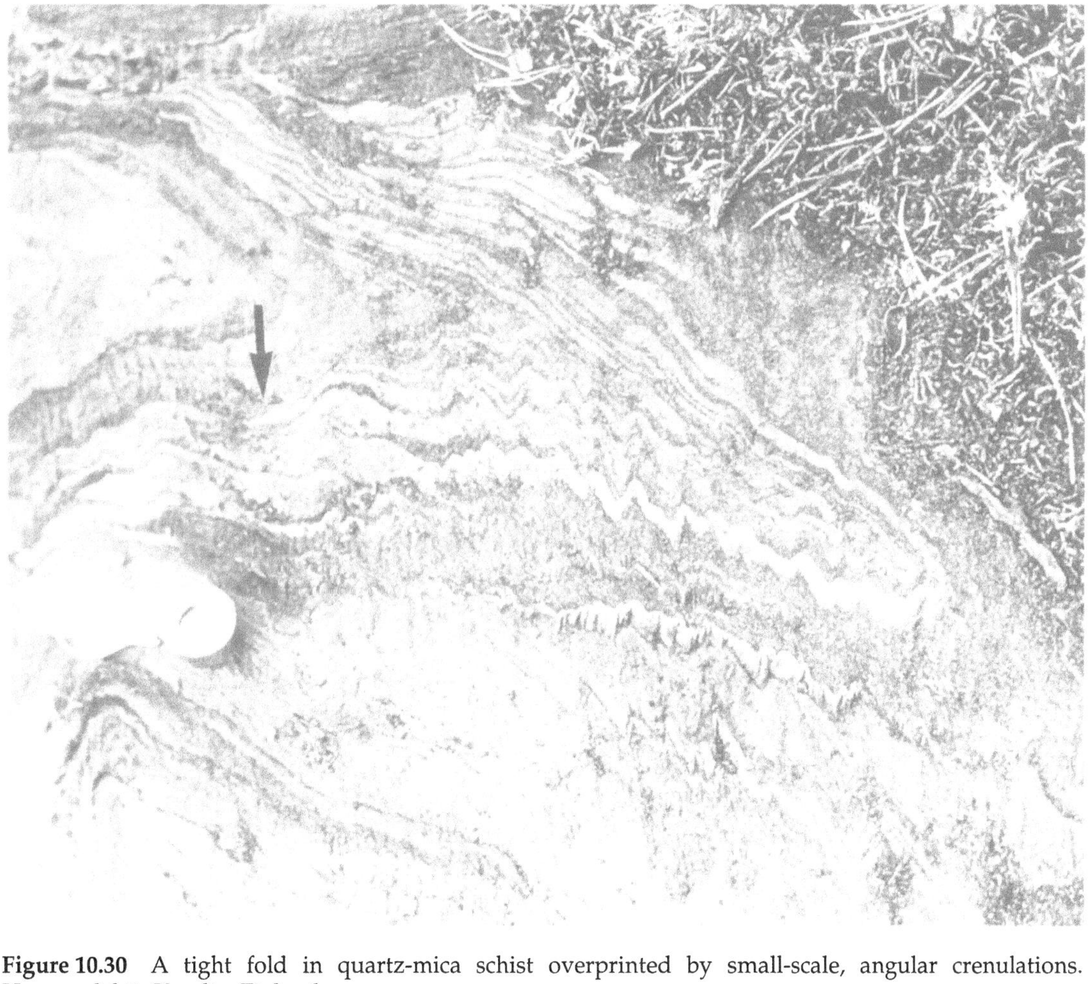

Figure 10.30   A tight fold in quartz-mica schist overprinted by small-scale, angular crenulations. Hammaslahti, Karelia, Finland.

10.1.10 MULTIPLE FOLD RELATIONSHIPS

(a) Figure 10.29 shows simple refold relationships between three folds (F_i, F_u and F_o with an associated cleavage S_o) in gneiss of the Lewisian complex in the Outer Hebrides, Scotland. An isoclinal fold (F_i) can be seen to be refolded by an upright fold F_u, and small intrafolial 'eye' interference patterns also appear in the upper right of the photograph. These patterns result from the intersection of the F_i hinge with a later, open fold set (F_o), which curves the right limb of F_u. An axial planar cleavage (S_o) associated with F_o dips to the left at a slightly steeper angle than the leucocratic vein trending from the centre to the top right of the photograph. The fact that this cleavage trace clearly cuts across F_i and F_u, and is rectilinear, demonstrates that it is later than both fold sets and that F_o is one of the latest fold sets.

The sequence of deformation was: F_i followed by F_u, followed by F_o/S_o.

(b) Where an earlier and a later fold set have distinctly different expressions (in terms of geometry and style) and also have their axial planes inclined at high angles to one another, then their mutual age relationships are generally easy to recognize. In Figure 10.30 the tight fold (F_t, say) with the curved axial planar trace trending E–W is clearly overprinted by a set of crenulations (F_{cr}, say) with an associated N–S axial planar cleavage inclined at a high angle to the axial trace of the earlier fold. Convergence, and bifurcation of the foliation in places (e.g. at the arrow and above the hand lens on the limb of the larger fold) indicates that this fold refolds earlier intrafolial isoclinal folds (F_{ii}, say). Furthermore, the curvature of the large fold, which accords with a slight fanning of its superimposed crenulation traces, suggests the existence of an open fold (F_o, say), the latest of four fold sets observed here.

The sequence of events responsible for the structure was: F_{ii} followed by F_t, followed by F_{cr} and then by F_o.

(c) Figure 10.31a (upper) shows the mutual interference between folds of two sets in migmatites of the Lewisian complex, Scotland. These structures are labelled in Figure 10.31a (lower) and explained in Figure 10.31b. The hinge of an intrafolial isoclinal fold hinge (F_i) is visible at the lower right. Curvature of this fold, together with the trace of its axial plane (S_i), can be seen to define an asymmetrical fold (F_a) with its hinges below the hammer head at the far left of the photograph. The trace of the axial plane (S_a) of the asymmetrical fold is inclined to the left through the centre of the photograph. The sequence of events (i to iii), shown in Figure 10.31b was, (i) development of dominant foliation (S_o), (ii) isoclinal folding (F_i) and (iii) asymmetrical folding (F_a).

(d) The structural relationships shown in Figure 10.32 are less commonly seen. In this case clear overprinted relationships between several sets of folds (four in this case) can be observed on a single exposure. Here the axial planar traces of a set (F_i) of intrafolial isoclinal folds (the sharp hinges visible to the right of the compass in the centre) are associated with a tight fold (F_t) whose hinge can be seen below and to the left of the compass, in the lower left corner. Both of these folds are deformed across the axial plane of a more open, recumbent fold (F_r). The axial planar trace trends across the middle of the picture and recumbent fold hinges are visible to both left and right of the compass. Upright angular folds (F_u), expressed by the small hinges visible in the lower right of the photograph, affect coarse horizontal cleavage in dark amphibolite. The cleavage, which is parallel to the axial plane of the recumbent fold (i.e. it is axial planar to the recumbent fold), is only weakly developed at the centre and to the left of the photograph. The compass is resting on a cleavage surface.

The refold relationships can be demonstrated by considering the disposition of fold hinges and axial planar traces. If a line is traced through successive hinges (viz. those on the left and on the right of the compass) of the recumbent fold (F_r), it will be seen that this line is sub-horizontal and gently curved.

Figure 10.31 (a) Two photographs (the lower showing fold axial planar traces) of intersecting axial planar directions of two different fold sets. Amphibolite facies migmatite, Lewisian complex, South Harris, Outer Hebrides, Scotland. (b) Explanatory sketch. The foliation (S_o) has been folded isoclinally and the isoclinal fold (F_i) with its axial plane (S_i) has been refolded by the later cross-fold (F_a) with axial plane (S_a). From Figure 4, Hopgood, 1980. Reproduced with the permission of the Royal Society of Edinburgh.

It forms the profile of an open upright (F_u) fold which must be later than F_r. Similarly the trace of the axial plane (S_t) of F_t, which passes 'through' the compass, can be seen to have been folded by F_r (whose axial plane S_r intersects S_t at the compass). F_r is therefore later than F_t. It can also be seen that the axial planar traces of F_i, when extrapolated, must have been folded around F_t.

The deformation sequence, shown in Figure 10.32b, was: (i) isoclinal folding (F_i), (ii) tight folding (F_t), (iii) recumbent folding (F_r) and (iv) upright folding (F_u).

(e) Figure 10.33 shows two examples, each of two sets of folds with almost parallel axial planar traces, in Svecofennian migmatites in southern Finland.

In 10.33a, the axial planar trace of the isoclinal fold (F_i), to the left of the diagonal fracture, is comparatively easy to distinguish from that of F_a, to the right of the fracture. Also F_i is more or less symmetrical and its axial

(a)   (b)

Figure 10.32   (a) Photograph of refolding in amphibolite showing the relationships between four fold sets. (b) Sketch of the fold sets comprising the structure of (a) and showing its development. Isoclinal fold (F_i), tight fold (F_t), recumbent fold (F_r), and upright fold (F_u). Lewisian complex, Rona, Inner Hebrides, Scotland.

planar trace can be seen to have been deformed by the asymmetrical fold F_a to the right of the centre of the picture. On the other hand, the distinction between the two axial planar traces in 10.33b is less obvious. In this case the tight to isoclinal folds (F_i), at lower right and also below and to the right of the pencil, are asymmetrical also. Their deformation by the asymmetrical folds (F_a) whose hinges can be seen above and to the left of the pencil and also at the pencil point is not so readily discernible. Although there is a superficial similarity between the two fold sets (F_i and F_a) in this case (they are both asymmetrical), their distinction immediately becomes clear when the attitudes of their axial planar traces are compared. The axial planar trace of F_i is parallel to the main foliation trend (vertical in this instance) whereas that of the asymmetrical set (parallel to the pencil) is inclined to

the foliation trend and cuts across it, so confirming its later age.

The sequence of events responsible for the structure in both examples was: F_i followed by F_a.

(f) In many instances where the attitudes of two sets of folds are almost the same, the attitude of the later fold axial plane will be constrained by that of the earlier one such that one may merge with the other, with the later axial plane curving into the earlier. Recognition of this condition, and of the existence of two separate fold sets, might be difficult, particularly if exposure of the structure is discontinuous (see Odonne and Vialon, 1987; Figure 3.32, and the discussion of control exerted on an overprinting fold by the pre-existing fold trend in section 3.2.4). In the example shown in Figure 10.34 the axial planar trace of the asymmetrical fold (S_c) curves asymptotically

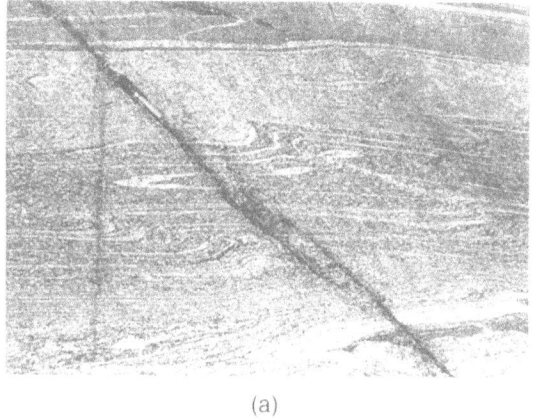

(a)

(b)

Figure 10.33 Two examples of near-parallelism between different fold sets (F_i and F_a). In both cases the axial planes (S_a) of close asymmetrical folds with a sinistral sense of asymmetry diverge by only a few degrees from the axial planes (S_i) of isoclinal folds. In (**a**) the divergence of S_a from S_i is more easily recognizable but in (**b**) the difference between the trace of S_a and S_i is more subtle, partly because F_i is also asymmetrical. S_a is parallel to the pencil and oblique to, and crossing, the vertical broad foliation trend which shows on the left of the photograph. S_i is vertical and parallel to the broad foliation trend. Svecofennian migmatite, southern Finland.

into that of the more or less symmetrical tight to isoclinal fold (S_{br}) forming a composite axial plane ($S_{c:br}$).

The deformation sequence responsible for the structure in this example was: S_{br} followed by S_c to produce $S_{c:br}$.

(g) Figure 10.35a shows the effects of agmatization of a folded amphibolite layer explained in Figure 10.35b. Fracturing parallel to early dextral asymmetrical folds (F_e) in the amphibolite layer (dark grey) was followed by conditions favouring mobility of leucocratic neosome allowing separation of blocks of the more refractory amphibolite (to form an agmatite locally). The leucosome (white in the sketch) has flowed into lower pressure zones developed between the basic agmatite blocks to form 'scar' folds (Hobbs, Means and Williams 1976, Figure 6.10) in the leucosome (arrows in the photograph). The axial planes of both the early folds (F_e) and the later scar folds (F_S) are parallel, but the two sets are easily distinguished in terms of the sequence of events described. Note that the thin leucocratic vein (V) parallel to S_e shows clear discordance not only with the F_e fold but also with the scar

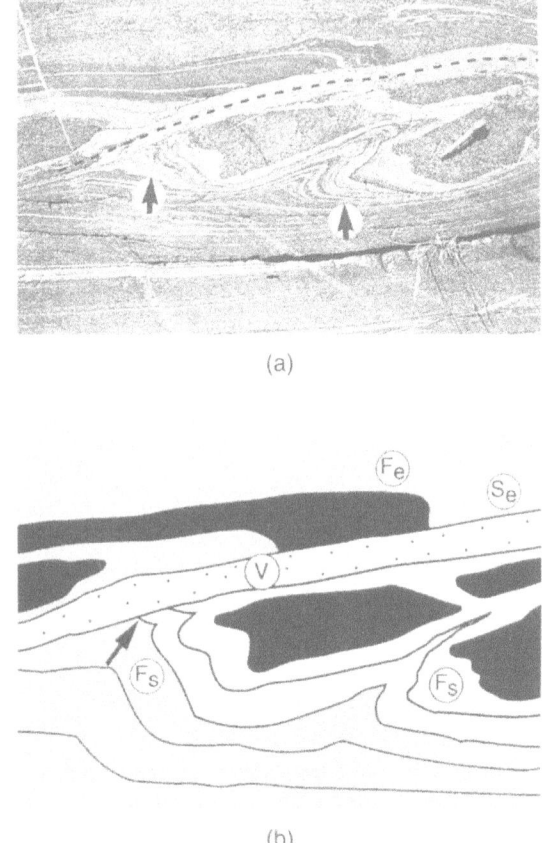

(a)

(b)

Figure 10.34 Photograph showing how the attitude of the axial plane of a later fold is constrained by that of a pre-existing axial plane. The axial plane of the asymmetrical later fold (S_c) merges with that of the isoclinal earlier fold (S_{br}) to form a composite axial plane ($S_{c:br}$). Svecofennian migmatite, Jussarö area, southern Finland. From Figure 14h, Hopgood, 1984. Reproduced with the permission of the Royal Society of Edinburgh. See also the discussion of refolding on parallel axes and Figure 3.32.

Figure 10.35 (a) Early folds (F_e) in amphibolite (dark) followed by 'scar' folds (F_s, arrow in the photograph) developed in later (post-F_e) leucocratic neosome. Folds F_s (at arrow in sketch) and F_e are cut by the vein (V) parallel to both S_e, the axial plane of F_e, and S_s. Therefore vein V is the latest structure. (b) Explanatory sketch of (a). Svecofennian migmatite, Flackholmen, Skåldö, southern Finland.

fold (F_s) at the arrow on the left. This indicates that it was emplaced even later than the development of the scar fold.

The sequence of events responsible for the structure shown in Figure 10.35a was: foliation followed by F_e, followed by partial melting and agmatization, followed by F_s, followed by V.

(h) Figure 10.36 shows another case of near parallelism of axial planar traces of two different fold sets. In this example asymmetrical folds in Svecofennian migmatites are discordant to folds in thin leucocratic veins (Figure 10.36b). Both host rock foliation and veins appear to have been slip-folded on slip surfaces approximately parallel to the axial

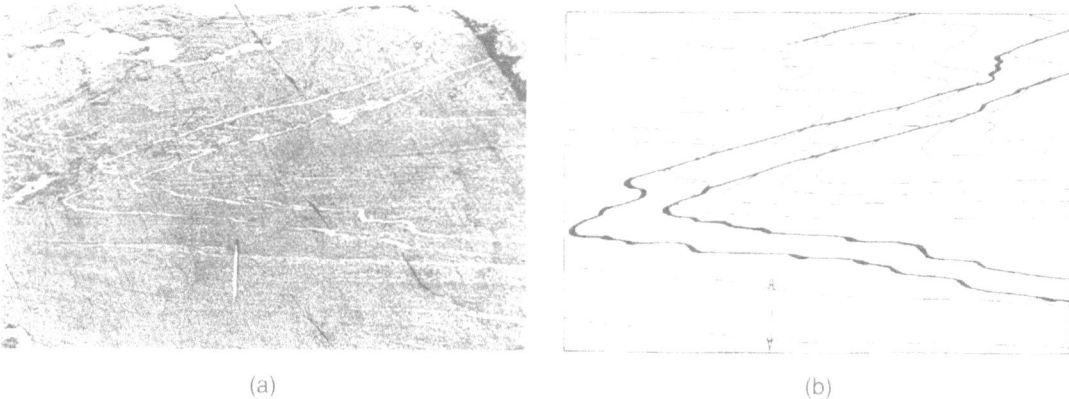

(a)   (b)

Figure 10.36 (**a**) Disharmonic folds with approximately parallel axial planar traces. Folds in the thin leucocratic veins discordantly cross the (less distinct) folds with similar profiles in the host rock. (**b**) Sketch of the relationships between the two fold sets. Svecofennian migmatite, Ängholmen, southern Finland. From Figure 17g, h Hopgood, 1984. Reproduced with the permission of the Royal Society of Edinburgh.

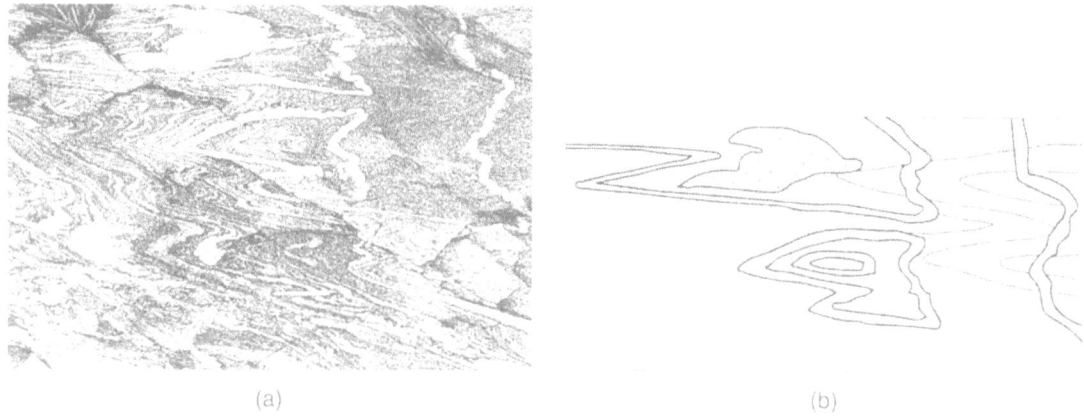

(a)   (b)

Figure 10.37 (**a**) Interference pattern caused by refolding between two fold sets on parallel axial planes (axes). The second set folds the prominent white veins and the first set folds the host rock foliation. (**b**) The second fold set, although not distinguishable as such from the folds of the dominant foliation, can be identified from its relationship to the folded, originally rectilinear trace of the 'N–S', shallow-dipping, leucocratic vein on the right. Although the folds in the veins and the folds in the host rock all appear to be concordant from the centre to the left of the photograph, at the right the veins can be seen to be discordant to the tight folds in the host rock. Compare (**a**) with the sketch (**b**). Therefore there must be at least two fold sets on parallel-trending axial planes. Dabie complex, Feng Huang GuanDam, southern China.

plane of the asymmetrical folds to produce two sets of disharmonic folds with more or less parallel axial traces. There is additional discussion of the sequence of events leading to this structure in relation to Figure 10.16.

The sequence of events causing this struc-

ture appears to have been: foliation, intrusion of thin leucocratic veins, slip-folding.

(i) The pattern shown in Figure 10.37 is comparable to that in Figure 10.36 of example (h) above, and stems from the interference between two sets of folds. In the present example early-formed folds in grey gneiss were crossed by thin, probably essentially tabular, leucocratic veins. The veins (together with the early folds) have apparently been affected by slip to form another, later set of folds which superficially appear to be co-genetic with the pre-existing fold set in the host gneiss. The distinction between the two sets is possible because of the discordance to the pre-existing folds (visible in some places such as the upper right of the photograph) of the leucocratic veins which were involved in the later slip-folding.

The events responsible for the structure shown were: foliation, folding, intrusion of thin leucocratic veins, slip-folding.

(j) More complicated structures, with relationships that are less obvious than those previously described, are shown by the typical migmatite of Figure 10.38.

The principal structural features are labelled in Figure 10.38b. These include: the prominent gneissic foliation which is deformed into upright irregular, round- to angular-hinged folds (F_u) with sinuous axial planar traces (S_u). This axial planar curvature defines broad open recumbent folds (F_r) with gently inclined axial planar traces (S_r). Dark grey amphibolite (surviving as irregular vertically elongate masses on the right) is partly discordant to the foliation and appears to disrupt the upright folds also. The whole was agmatized by leucocratic neosome and the amphibolite, along with blocks of folded gneiss, comprises part of the palaeosome. While for the most part, the neosome is expressed as diffuse patches blurring the structural detail, in places it has coalesced to form irregular discrete bodies (to the left, and on the right enclosing the amphibolite blocks). Individual F_u hinges have been kinked, disrupted and offset relative to one another on vertical slip on surfaces with rectilinear traces (S_S), often associated with narrow, discrete leucocratic veins, producing modified upright fold forms (F_S). The slip surfaces and veins cut across the neosome and appear to be the latest of the structures described.

The events leading to this structure include: foliation, F_u, basic intrusion (amphibolite), partial melting (agmatite), F_r, F_S, vertical veins.

The succession of structures at this exposure (shown again in Figure 11.15), and the sequence of events causing them, are discussed in greater detail in section 11.1.2(l).

(k) The structure of Figure 10.39a records yet another complicated series of events. The photograph shows inter-layered Svecofennian quartzofeldspathic gneiss and amphibolite that has been folded and disrupted, cut by a leucocratic vein and then refolded.

The photograph shows angular amphibolite blocks and grey gneiss (their continuity disrupted by separation and some rotation) in layered leucocratic neosome with scar folds in the neosome between the amphibolite layers. A sinuous banded leucocratic vein trending left to right cuts the amphibolite and the scar folds. Folds in this vein are disharmonic with respect to the scar folds but the axial planes of both fold sets are parallel.

Some of the events contributing to the complex, beginning with the layered succession, are shown in Figure 10.39b and include the following: (i) folding and fracturing in the amphibolite layers, (ii) mobilization of leucocratic neosome and separation of amphibolite blocks to form an agmatite with flow of neosome into low pressure areas between the blocks to form 'scar' folds, (iii) emplacement of a discordant leucocratic vein cutting across the whole structure including the 'scar' fold, (iv) folding of the vein accompanied by refolding of the 'scar' fold on more or less parallel axial planes to produce folds of different ages and with different geometry distinguishable in terms of the events described above.

In the 22 examples discussed above, the relationships are not always obvious at first sight, but are reasonably clear once attention is drawn to them. They serve to emphasize again the need for careful observation and also show the value of making sketches of relationships (relationships that might not have even been

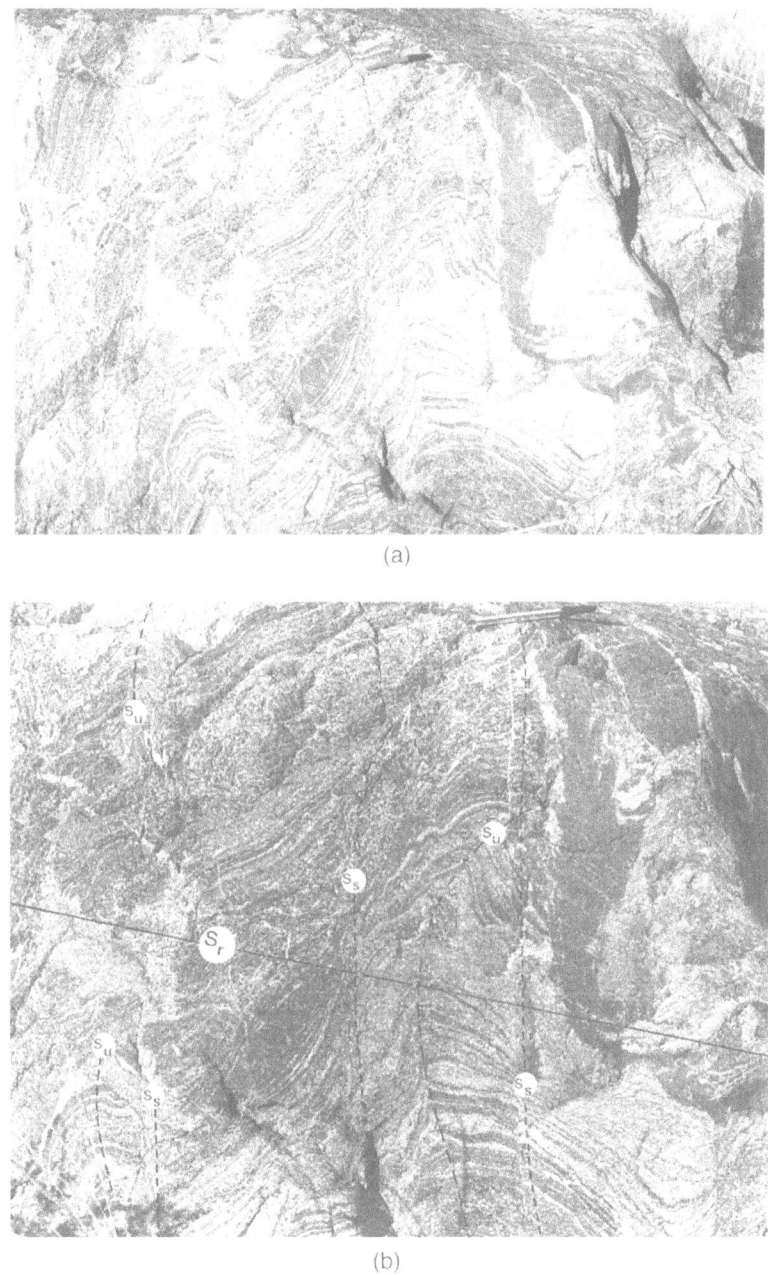

Figure 10.38 (**a**) Migmatites comprising agmatite with irregular amphibolite palaeosome blocks, folds, shears and leucocratic neosome veins and pockets. This shows complex structural relationships (labelled in **b**) between folds (F_u, F_s and F_r) with axial planar traces S_u, S_s and S_r, leucocratic neosome (pale grey patches and thin, vertical rectilinear veins) and basic igneous veins (irregular dark masses on the right). Svecofennian migmatite, Klobban, Jussarö region, southern Finland. Modified from Figure 10a, Hopgood, 1984. Reproduced with the permission of the Royal Society of Edinburgh.

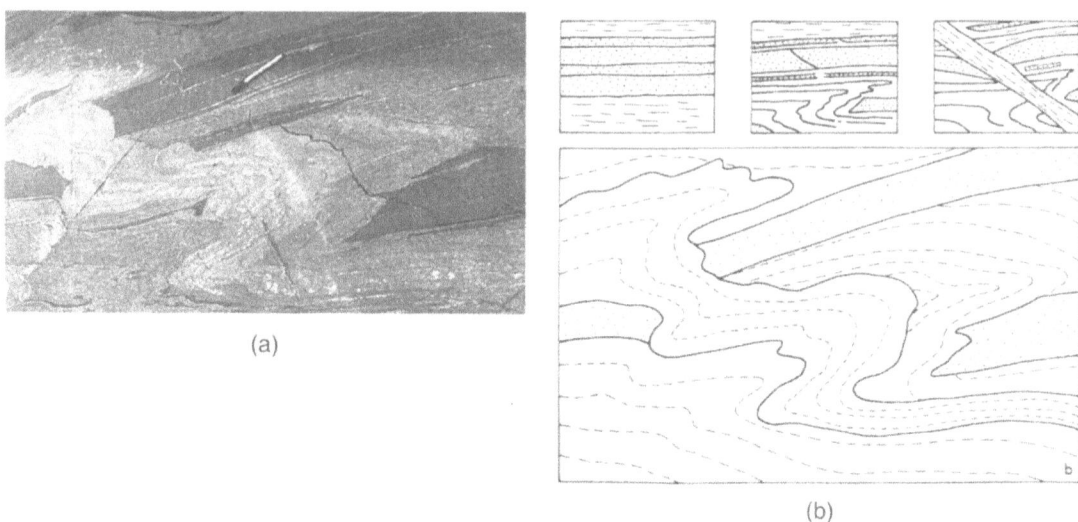

(a)

(b)

Figure 10.39   Photograph (**a**) and sketch (**b**) of interrelationships between folding, neosome development, vein emplacement and agmatite formation. The sketch outlines the stages in the development of the structure shown in the photograph. Svecofennian migmatite, southern Finland. From Figure 17a, b Hopgood, 1984. Reproduced with the permission of the Royal Society of Edinburgh.

noticed prior to sketching). This is because the act of sketching immediately focuses the observer's attention on the relationships between the structural features as they are being drawn. It is commonly the case that the observer 'sees' the features on outcrop but fails to appreciate their significance.

11.1 LOCAL STRUCTURE SUCCESSIONS COMPRISING SEVERAL SETS

11.1.1 INTRODUCTION

Discussion of the procedures for the resolution of overprinted relationships is continued in this chapter by further consideration of structural associations, this time of greater complexity. It begins with a re-examination of some structures already looked at, followed by cases of increasing complexity, and then examines cases of more extensive successions comparable to that comprising the structure seen in Figures 11.1 and 11.2. The approach used in resolving the overprinted relationships is fundamentally the same as before, but the successions determined are 'formalized' here with the structure sets listed for the most part in the form of 'stratigraphical' columns.

Firstly, as a reminder of how daunting first impressions can be when examining exposures of complicated structural relationships, consider the complexity of the interference patterns of Figures 11.1 and 11.2.

11.1.2 PROCEDURE WITH EXAMPLES

In spite of the initial impression of disorder and confusion and of a total absence of regularity, the structural relationships shown in Figures 11.1 and 11.2 can, and have been, resolved. Careful, 'step-by-step' consideration of the relationships between structures when these are examined in detail, inevitably leads to the determination of the structural succes-

sion and hence to the establishment of the deformational sequence. Initially one must avoid attempting to deal with the total complexity of the pattern produced by multiple folding (as shown in the examples above) and concentrate on the study of the relationships between only parts of the whole structure. At any one time the structure in only small areas, each time exposing the relationships between just two or three folds, should be considered (see section 8.1.5).

The example of extreme structural complexity shown in Figure 11.2 is almost certain to suggest the term 'wild folds' (see the discussion in section 1.1.1). However, although they are very complicated, the structural relationships shown at this outcrop have been determined, and the structural succession established using the methods described in Chapter 8. An analysis of the structure in the photograph of Svecofennian (formerly named Svecokarelian) migmatites of Figure 11.2 is shown in the series of sketches comprising Figures 11.3a to 11.3c. These show form lines drawn from the structure shown in the photograph. In Figure 11.3b the hinges of recognizable folds are emphasized and in Figure 11.3c these hinges are labelled in accordance with the fold sets identified in the succession (Hopgood, 1984).

The procedure can best be explained by working systematically through particular cases in a manner similar to that used in Chapter 10. The following 14 examples illustrate the approach used to establish a structural succession. The succession (an objective,

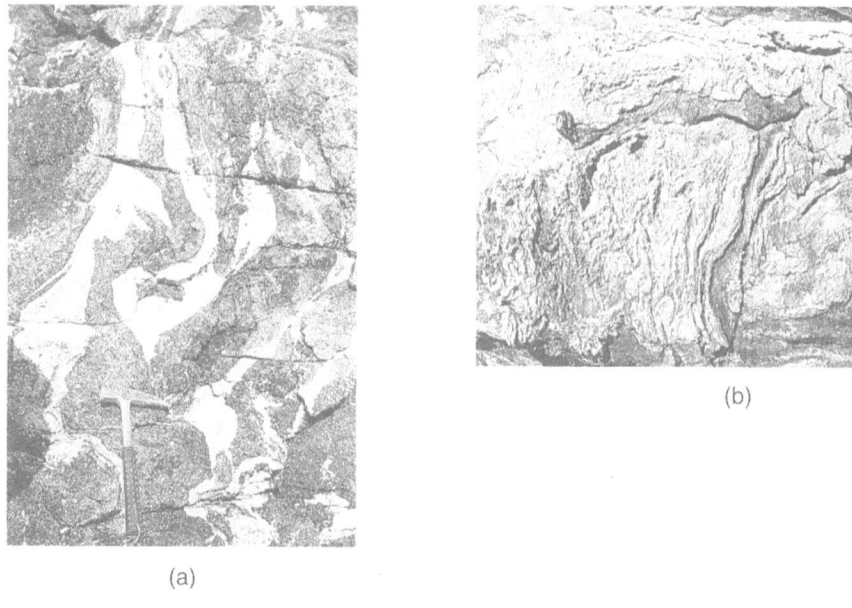

(a)

(b)

Figure 11.1   Patterns resulting from multiple folding in migmatites. (**a**) Interference pattern stemming from overprinting between three fold sets. Lewisian complex, Isle of Barra, Outer Hebrides, Scotland. (**b**) Interference pattern stemming from overprinting between at least five fold sets. Rona, Inner Hebrides.

Figure 11.2   An example of complex refold relationships in Svecofennian migmatites. Structure such as this has given rise to expressions such as 'wild migmatites' and so-called 'wild' folds. South shore of Djupkobbarna, Jussarö area, southern Finland. The 15 cm pencil on the left is aligned E–W. From Figure 21a, Hopgood, 1984. Reproduced with the permission of the Royal Society of Edinburgh. Compare Figure 10.4.

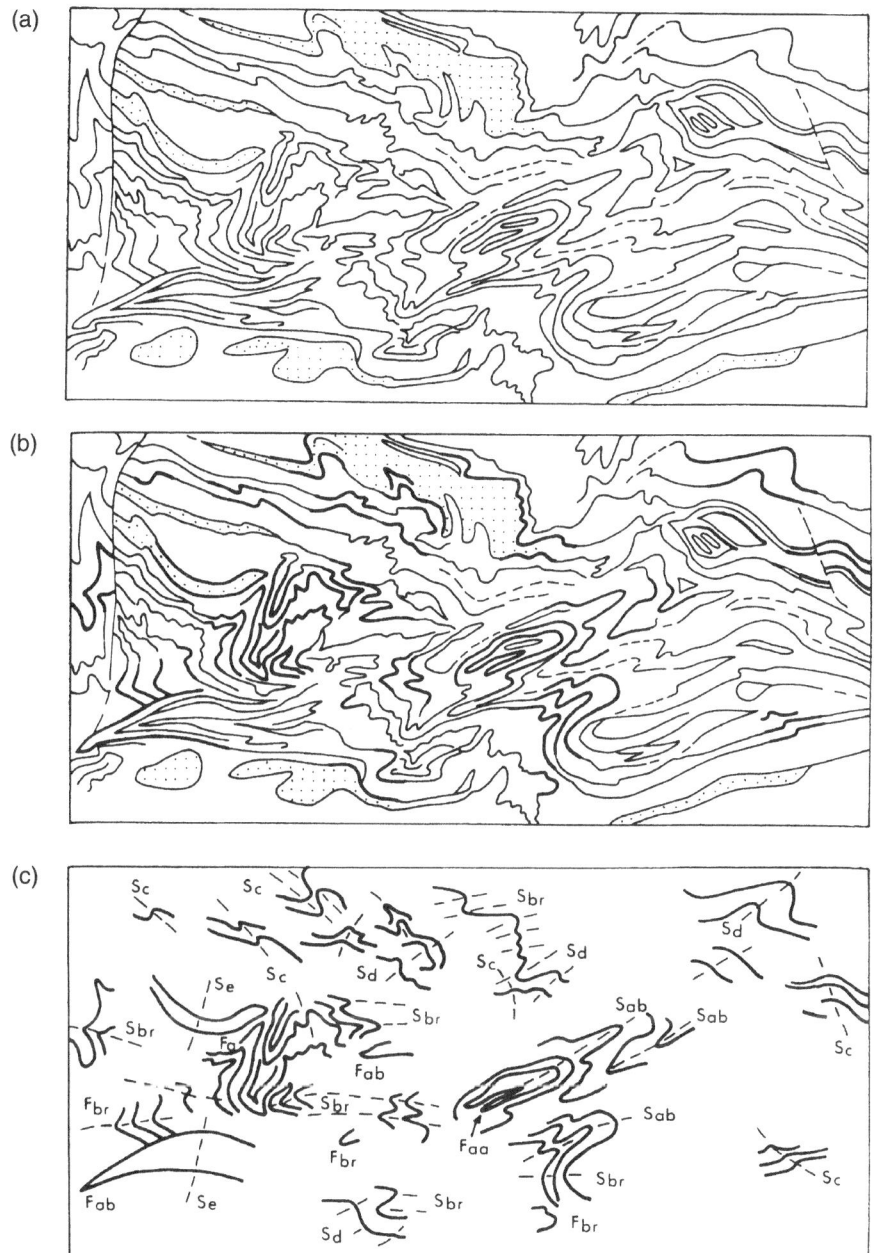

Figure 11.3 (a) Form lines drawn from the photograph of Figure 11.2. (b) Analysis of the complex fold pattern of Figure 11.2; form lines with heavy lines to outline the hinges of the fold-sets causing the structure shown. (c) The fold-sets identified in the area of Figure 11.2; these, in order of development, are F_a, (F_{aa} plus F_{ab} at centre), F_{br} with fold axial traces curved by later folds (e.g. S_d at the bottom centre), F_c at the top left, F_d and S_e, with the effects of F_w shown by the gentle warping of F_{br} axial planar traces (S_{br}) at left and centre; the divergence of the F_{aa} and F_{ab} fold axial traces from the 'normal' easterly trend here stems from D_{br} reorientation because they lie within the hinge zone of a larger F_{br}. From Figures 21b, c, d, Hopgood 1984. Reproduced with the permission of the Royal Society of Edinburgh.

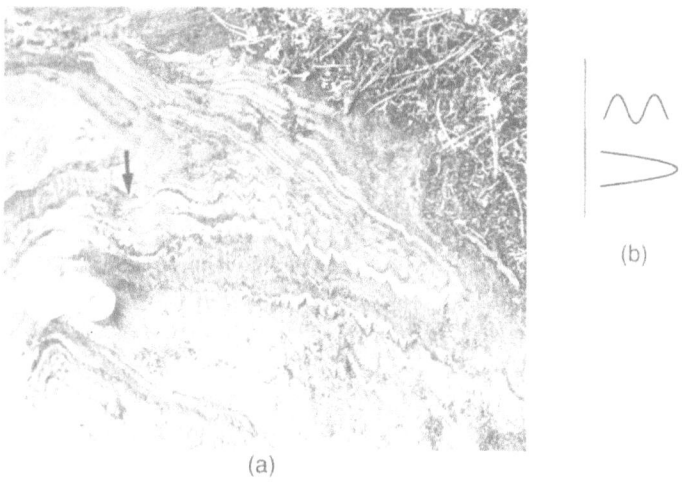

(b)

(a)

Figure 11.4 (**a**) Photograph of a tight fold (F_t) overprinted at a high angle by crenulations (F_c). (**b**) column showing the confirmed succession of structures (other, earlier and later, structures discussed in the text are not included). Svecofennian migmatite, Karelia, Finland.

observational fact) provides the basis for determining the sequence of events leading to the observed structure. The causes of these events can be the subject of individual interpretation.

Examples

(a) First consider some relatively uncomplicated structural relationships, for example the setting of Figure 11.4, previously seen in Figures 6.40 and 10.30. Here, an early tight fold (F_t) with axial plane S_t can be seen to be crossed by a set of superimposed crenulations (F_c) with axial traces (S_c) trending at approximately 90° to the F_t axial plane (succession in column of Figure 11.4b). There is of course the possibility that other fold sets, not observed at this exposure, exist both prior to, and after, the two folds F_t and F_c as well as between them (refer to section 8.1.5 and the discussion relating to Figure 8.12). For example, there appears to be an isoclinal intrafolial fold hinge (F_{ii}) immediately above the hand lens (arrow) and this must have formed prior to F_t (its axial planar trace extrapolated continues around

the hinge of F_t). Further support for the existence of pre-F_t folds can be seen in the intrafolial divergence of the foliation at the centre of the photograph. Furthermore the axial trace of the tight fold is broadly curved (convex down) implying later (post-F_c) very open folding (F_o) and this accords with the slight fanning of S_c (cf. Figure 10.30).

The fold succession is thus likely to be at least: ... + F_{ii} + ... + F_t + ... + F_c + ... + F_o + ..., with + F_t + ... + F_c + certain.

In line with the discussion of the 'The structural succession' (section 2.1.1) relating to the derivation of a sequence of events from a succession of structures, the sequence of events (after the development of the foliation and including the possible F_{ii} and F_o) leading to the structure in Figure 11.4 would have been, (i) isoclinal folding, (ii) tight folding, (iii) crenulation folding, (iv) open folding.

Note: It is important to bear in mind that although the events have been referred to for convenience as 'isoclinal folding' etc., this is not likely to be strictly correct. The folding would not necessarily have been such and the observed 'intrafolial isoclinal' and 'tight'

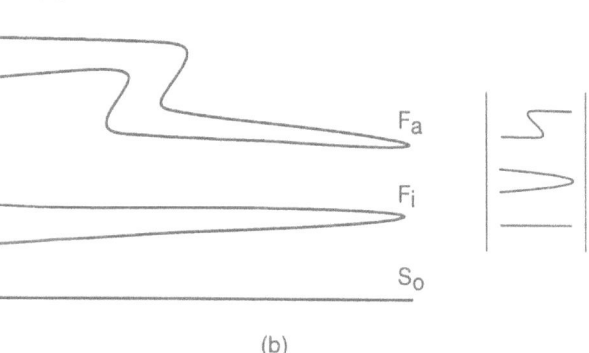

Figure 11.5 (**a**) Refolded gneiss. Lewisian complex, Harris, Outer Hebrides, Scotland. (**b**) Sketch showing the development of the structure shown in (**a**) and the known succession of structures comprising (**a**).

forms of the folds relating to each event are more likely to be the result of the cumulative effects of all deformational events rather than a result of the initial deformation (section 7.4).

(b) In Figure 11.5a the interference pattern resulting from the overprinted relationships between two folds is shown: a close asymmetrical fold F_a with an inclined axial planar trace (S_a) inclined to the left on the outcrop, an

isoclinal fold F_i with an axial planar trace (S_i) that is 'horizontal' at lower left but curves across fold (F_a) on the left.

The fold succession is thus at least: $\ldots + F_i + \ldots + F_a + \ldots$ with the suggested possibility of fold sets before, after or intermediate between, the two sets F_i and F_a, for example within the foliation between the hammer head and the (F_a) hinges there is the suggestion of a

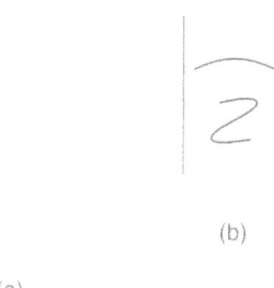

(b)

◀ (a)

Figure 11.6   (a) Asymmetrical fold (F_a) with a dextral vergence sense. The curvature of the trace of the axial plane (concave upwards) results from deformation by a later open upright fold with an axial plane dipping steeply to the right. (b) Succession of structures known to be involved in (a). Svecofennian migmatites, Jussarö area, southern Finland.

thin intrafolial isoclinal fold. If confirmed this would add another structure (?F_{ii}) which is earlier than F_i.

The succession (in the column) and sequence of events responsible for the structure in Figure 11.5, following the development of the foliation, are shown in Figure 11.5b, with the sequence following the development of the foliation (S_0) having been, (i) isoclinal folding, (ii) asymmetrical folding (see the discussion in section 2.1.1 relating to the derivation of a sequence of events from a succession of structures).

(c) The photograph (Figure 11.6) of an asymmetrical fold (F_a) with a curved axial planar trace in Svecofennian migmatites shows structural relationships which might not be immediately obvious to the inexperienced observer.

The axial planar curvature is in the form of a broad open fold (F_{bo}) with a steep, upright axial plane. Note, however, that the apparently simple, upwardly concave curve of the trace formed by joining successive hinges seems to compound, comprising a series of smaller, remarkably regular, second order curves which could imply the presence of another fold set of smaller open folds (F_{so}) whose axial planar traces appear to be dispersed ('fanned') about that of the larger fold (F_{bo}). These would

then have followed the asymmetrical fold but have been prior to the more obvious broad curve (F_{bo}) whose axial planar trace on the exposure face is inclined steeply towards the right, almost parallel to the pencil.

The fold succession (shown in the column) would thus have been at least: ... + F_a + ... + F_{bo} + (with the possibility, to be confirmed, of ... + F_{so} + ...).

Drawing on the discussion in section 2.1.1 relating to the derivation of a sequence of events from a succession of structures, it can be seen that the sequence of events which led to the structure shown in Figure 11.6 would have been, (i) asymmetrical folding, (ii) ?open upright folding, (iii) broad open upright folding.

(d) The axial planar traces of the recumbent folds (F_r) in the Lewisian gneiss shown in Figure 11.7 curve across the axial planes of open upright folds (F_w). Although there is no **tangible** surface (S_r) associated with these folds, because this surface is defined by the disposition of recumbent fold hinges, the warp in the axial plane can be demonstrated by tracing through successive recumbent fold hinges.

The succession is therefore ... + F_r + ... + F_w + ... (shown in the column of Figure 11.7), with the possibilities that other sets, not observed at these two exposures, may be

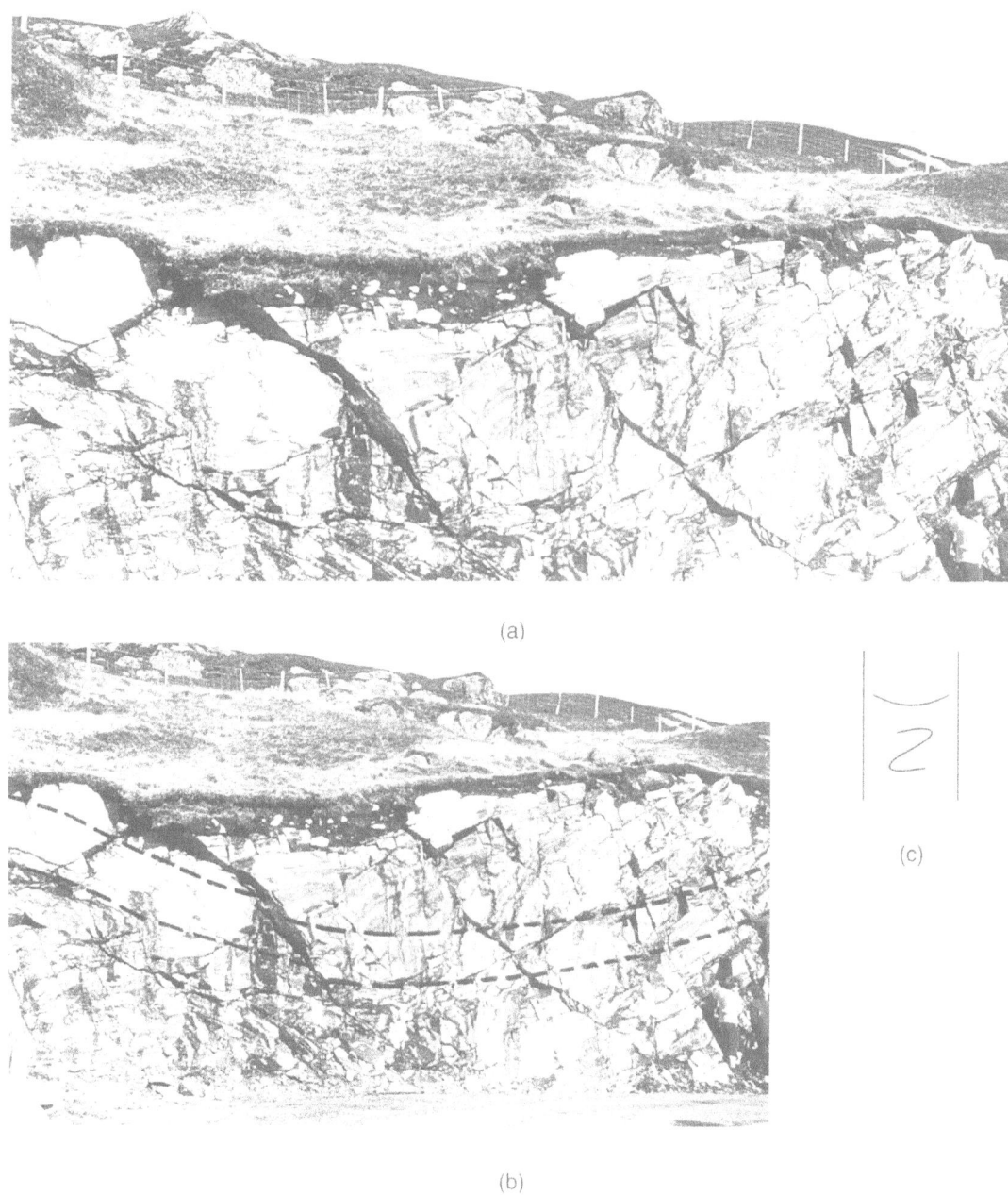

(a)

(b)

(c)

Figure 11.7 (a) Curved axial planar traces of recumbent folds in gneiss of the Lewisian complex, Bernera Bridge, Harris, Outer Hebrides, Scotland. From Figure 15, Hopgood 1980. Reproduced with the permission of the Royal Society of Edinburgh. (b) In spite of the absence of a visible structural surface related to the recumbent fold axial plane, it is nevertheless possible to demonstrate the fact that the recumbent structures have been gently warped by folding on steep axial planes as shown by the curvature of the dashed lines linking successive F_r hinges. These represent the traces of F_r axial planes. (c) Succession of structures known to contribute to (a).

(a)

(b)

Figure 11.8 (**a**) Refolded interfoliated gneiss and amphibolite. Lewisian complex, Rona, Inner Hebrides, Scotland. (**b**) Sequence of events leading to (**a**) and column showing the succession of structures known to be involved.

present, before, between and/or after the fold sets described.

On the basis of section 2.1.1 the sequence of events, following the development of the foliation and early folding, from which the structure in Figure 11.7 was derived would have been, (i) recumbent folding (F_r), (ii) warping on upright axial planes (F_w).

(e) The photograph in Figure 11.8a shows a slightly more complex structure, the interference pattern generated by the overprinted relationships between four sets of folds. These folds were previously discussed in relation to Figure 10.32. The traces (or traces extrapolated) of axial planes (S_i) of isoclinal folds (F_i) curve around tight (almost isoclinal) folds (F_t) with axial plane S_t. F_i and its axial planar trace curves in turn around an open, recumbent fold (F_r). The trace of the F_r axial plane (S_r) bisects the photograph horizontally and shows as a weak (sub-horizontal) axial planar fracture cleavage (S_r) in the dark rock (espe-

cially at the lower right) where it is affected by upright crenulations (F_u) with axial plane (S_u) dipping steeply to the left. The succession, in column form, and the deformational sequence are shown in Figure 11.8b. In this example, as in the previous cases, the successive deformation (folding) of the axial traces of earlier folds shows the sequential relationships between each fold set clearly and simply. However, the relationships of the final set (F_u) to the other fold sets may not be so obvious at first sight, although on closer inspection it can be seen that these folds have developed in a spaced cleavage parallel to the axial plane of the recumbent open fold (F_r), best shown in the lower right of the photograph. For this reason it becomes clear that F_u cannot be merely a different expression of either of the two earlier tight folds F_i and F_t which happen to have axial traces approximately parallel to that of S_u, the axial plane of F_u. This spaced cleavage, a planar fabric introduced at an intermediate stage in the deformational sequence, is an example of the 'complications' added to the succession by structures interposed at intervals in the succession. These are discussed in Chapter 4. In this case, the cleavage which developed during deformation D_r (between D_t and D_u) has constrained the possible time of formation of F_u (to post-F_r).

The fold succession is thus: $... + F_i + ... + F_t + ... + F_r + ... + F_u + ...$, with the possibility that other fold sets, not developed (clearly) on this exposure, exist prior to and/or after the four sets described (viz. F_i, F_t, F_r and F_u) and/or between any or all of these four.

The sequence of events leading to the structure of Figure 11.8a (after the formation of the dominant foliation) would have been something like, (i) isoclinal folding (F_i), (ii) tight folding (F_t), (iii) recumbent folding (F_r), (iv) upright crenulation folding (F_u). Refer again to section 2.1.1.

(f) The ease with which structural relationships can be determined on the exposure is governed to some extent by the lithology, in particular the contrast in appearance (colour or tone – dark/light) between the layers affected. In those cases where the contrast is low the fold relationships are less clear and may, in some cases, be overlooked, particularly in the early stage of the investigation and during reconnaissance. The exposure of Figure 11.9 shows such a case (discussed earlier in relation to Figure 10.22). In this example the original bedding (S_s) of metasedimentary micaceous schists is in the form of tight folds with angular, dentate hinges and axial planar traces (S_a) trending from left to right sub-parallel to S_s. These are curved into upright open fold forms (F_o) with axial planes (S_o) and these axial planes curve to form open warps (F_w) with axial planar trace almost perpendicular to S_o (Figure 11.9b). The fold succession (see column, Figure 11.9c) is thus: $... + F_a + ... + F_o + ... + F_w + ...$, with the possibility that other folds exist before and/or after these three and/or between all or any of them.

Thus, the sequence of events, after the development of the foliation, which caused the structure in Figure 11.9a was, (i) tight folding (F_a), (ii) open folding (F_o), (iii) very open folding or warping (F_w).

(g) In contrast to the previous example, the fold pattern resulting from the effects of two sets of isoclinal folds (F_i and F_{ii}), one set of tight folds (F_t) and one set of very open folds (F_o) is shown clearly in the different lithologies of the photograph in Figure 11.10. Here, the clarity of the structures is accentuated by the contrast between the thin leucocratic layers and the dark grey tonalitic gneiss. The axial planar traces (S_{ii}, S_i, S_t and S_o) are labelled in Figure 11.10b and their positions can easily be identified in the photograph (Figure 11.10a) where the refold relationships between each set can be seen clearly. The curvature of the axial planar trace (S_t) of F_t stems from folding by F_o which has an axial planar trace trending approximately ENE. S_i of F_i was folded by both F_t and F_o and S_{ii} of F_{ii} folded by all three of the later folds F_i, F_t and F_o. The fold succession (shown also in column form) is thus: $... + F_{ii} + ... + F_i + ... + F_t + ... + F_o + ...$, with the possibility that the succession is both followed by, and preceded by, other folds and that yet other folds may be present between all or any of the four folds described from the figure.

The sequence of events responsible for the structure in Figure 11.10 would have been, (i) folding, now intrafolial isoclinal (F_{ii}), (ii) folding, now isoclinal (F_i), (iii) folding, now tight (F_t), (iv) open folding (F_o). See again section 2.1.1.

(h) The structural relationships explained in Figure 11.11 have previously been shown and discussed in Figures 4.19 and 4.20. They require structural interpretation that goes beyond simply establishing cross-folding relationships. Here folded quartzofeldspathic gneiss has been intruded by a thin, more or less horizontal tabular basic body (now amphibolite). This cuts across, rather than follows, the folded surfaces, and where the traces of S_w

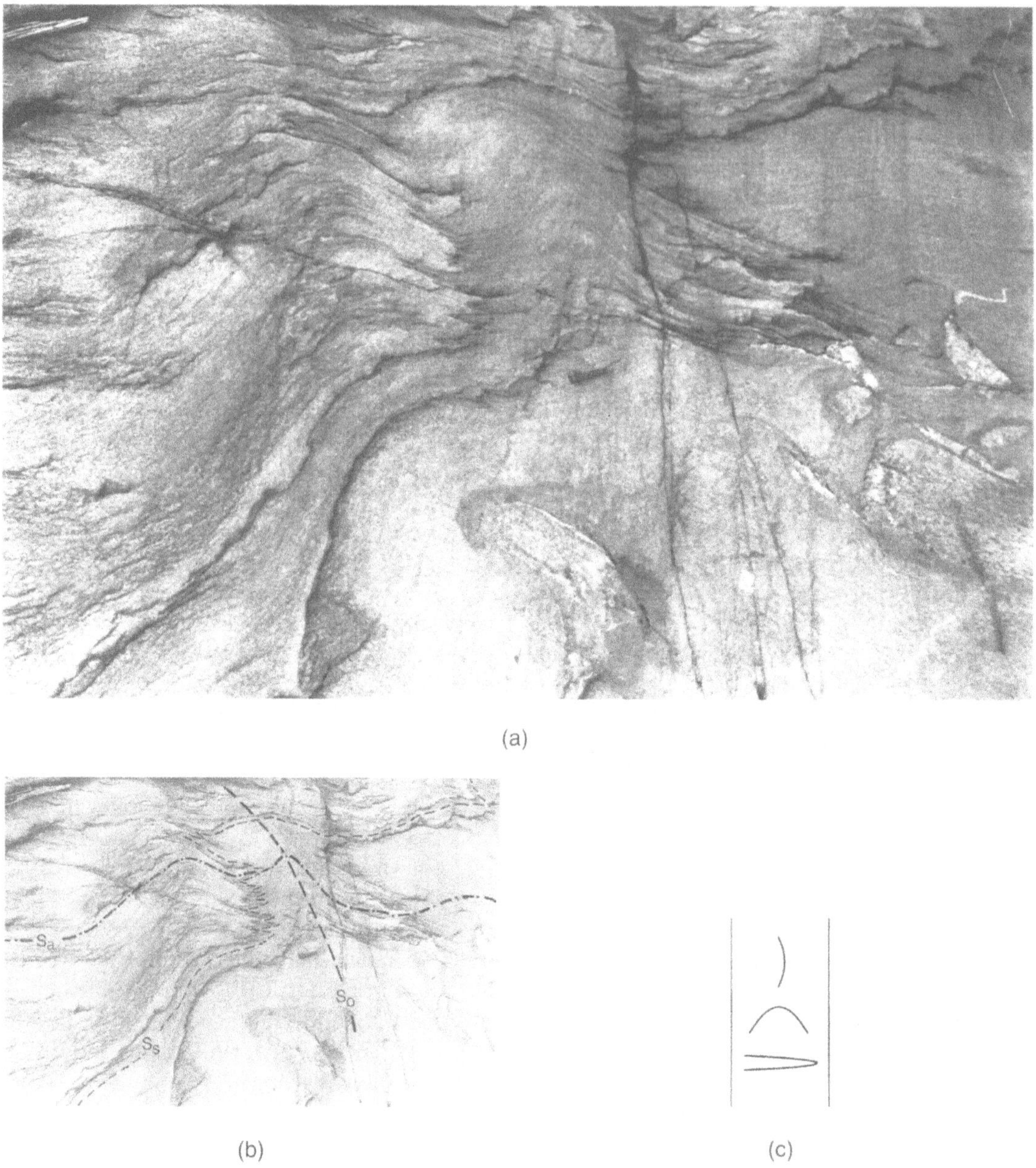

(a)

(b)

(c)

Figure 11.9   (a) Folded Svecofennian micaceous schists. (b) Traces of the original sedimentary layering (S_s), the axial planes of tight angular folds (S_a) and of upright open folds (S_o). From Figure 16, Hopgood 1980. Reproduced with the permission of the Royal Society of Edinburgh. (c) Column showing the known succession of structures. Karelia, Finland.

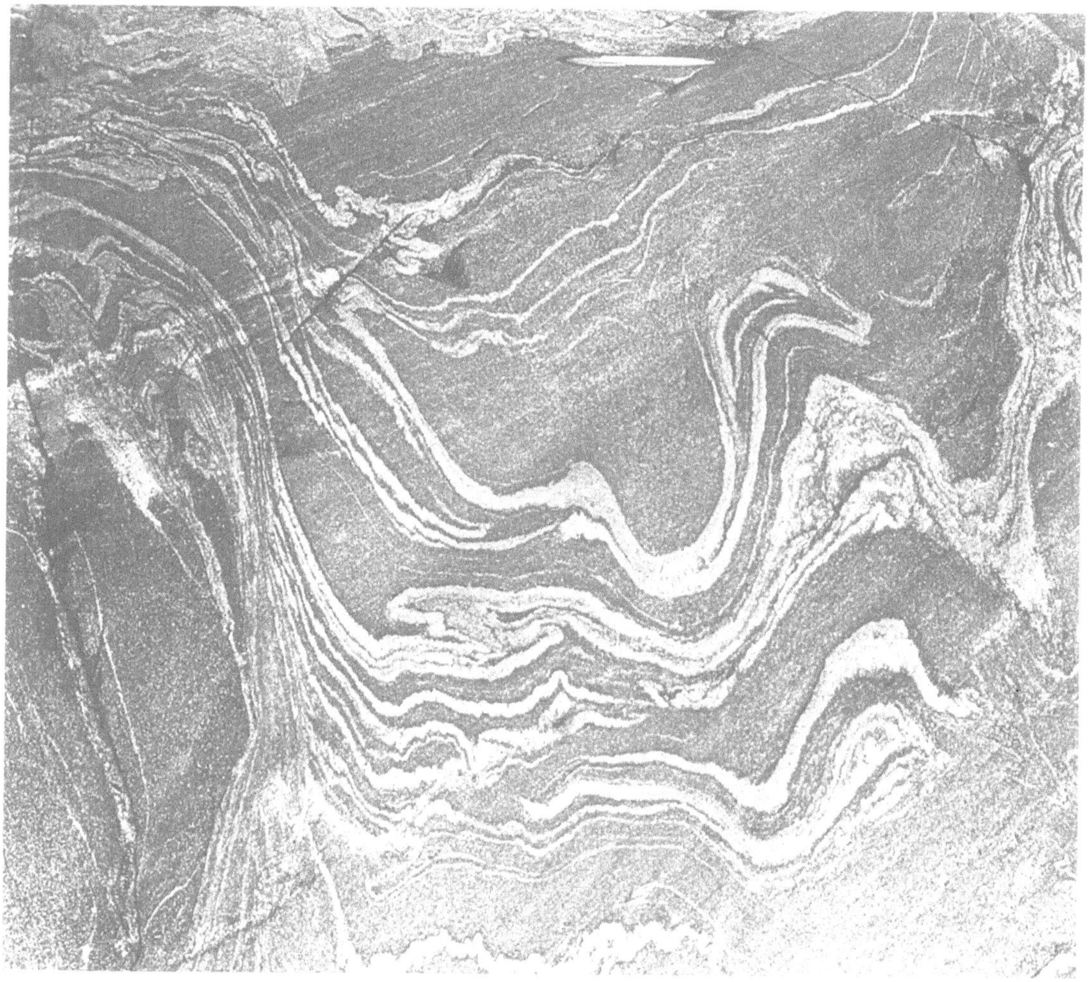

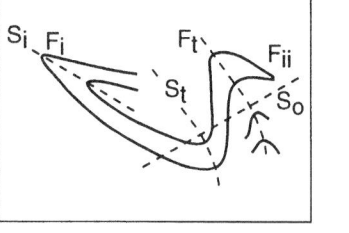

Figure 11.10 (a) Interference pattern resulting from multiple refolding in Svecofennian migmatite. (b) Sketches showing the relationships between the fold sets (F_{ii}, F_i, F_t and F_o) and the position of these in the known succession, shown in column form. See also Figure 13.6e. Gråklobben, Jussarö region, Finnish Archipelago.

(b)

(a)   (b)

Figure 11.11   (a) Relationships between folding and intrusion. Quartzofeldspathic gneiss folded by tight and open folds (traces of axial planes shown as S_t and S_o) cut by a thin horizontal basic sheet intrusion and then gently refolded by open wave-folds (axial planar trace, S_w). From Figure 11d, Hopgood 1980. Reproduced with the permission of the Royal Society of Edinburgh. (b) Succession of structures known to have contributed to (a) arranged in column form. Lewisian complex, east coast, Rona, Inner Hebrides, Scotland.

and S_t intersect the lower side of the basic intrusion the discordance shows clearly. The fold sets cut by the basic intrusion are, on the left, broad, upright structures (F_o) with an axial surface (S_o) trending upwards away to the left, tight upright structures (F_t) on the right with curved axial surfaces (S_t) trending upwards slightly away to the right. At the centre right of the picture (but not labelled on the figure) F_t is curved by an asymmetrical fold structure (F_a) with axial surface (S_a) inclined upwards to the left.

The basic rock must have been intruded as a more or less planar body, apparently after F_t, F_o and F_a and subsequently gently folded (warped) with an upright axial plane to produce folds (F_w) distinct from F_o which is cut by the basic 'dyke'. This probability and the likelihood that F_w is later than F_t and F_o are both confirmed by structural relationships in nearby exposures.

The structural succession (shown in the column of Figure 11.11b) is thus: ... + F_t + ... + F_a + ... + F_o + ... + F_w + ..., with the total events (E) recognizable (including the 'dyke' emplacement, E_d) being: ... + E_{Ft} + ... + E_{Fa} + ... + E_{Fo} + ... + E_d + ... + E_{Fw} + ..., with the possibility that other events which produced structures not observed on this exposure could extend the sequence at either or both ends or at intermediate stages within it. This is a further example of the 'complicating

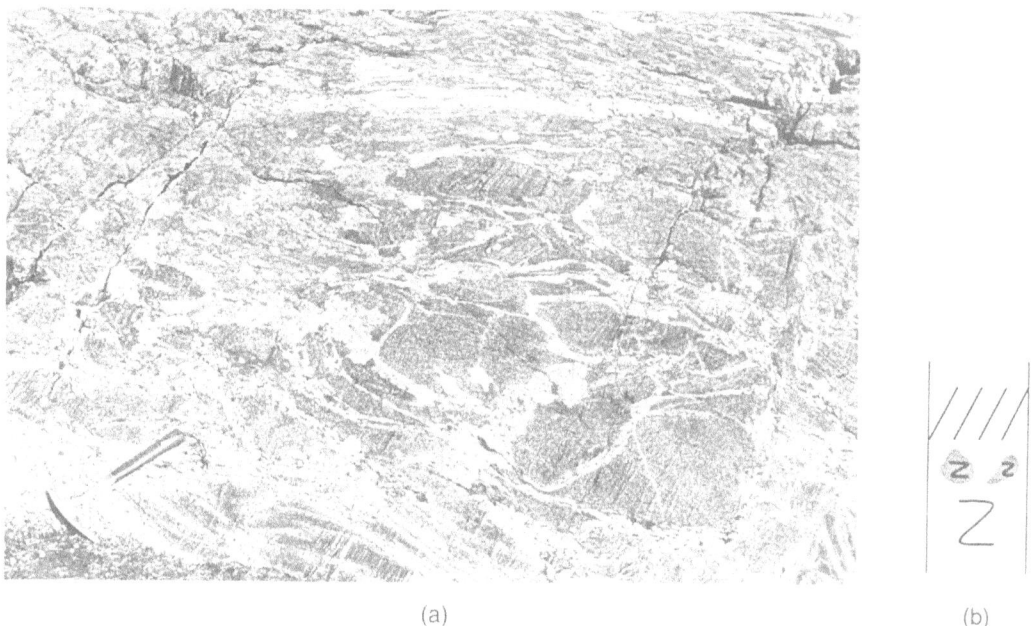

(a)   (b)

Figure 11.12 (a) Agmatite comprising palaeosome blocks of folded gneiss (folds are clearly visible in the block at the centre rear) in a leucocratic neosome, all crossed by a later overprinted cleavage whose trace trends from upper right to lower left. (b) Succession of structures known to have contributed to structure shown in (a). Lewisian complex, Isle of Barra, Outer Hebrides, Scotland. Compare Figure 1.3.

factors' referred to in the discussion of structures interposed in the succession in Chapter 4.

In line with section 2.1.1, the sequence of events (E), following the development of the foliation, that led to the interference pattern in Figure 11.11 would have been something like, (i) upright tight folding (E_{Ft}), (ii) asymmetrical folding (E_{Fa}), (iii) upright open folding (E_{Fo}), (iv) emplacement of the amphibolite body (E_d) and (v) upright warping (E_{Fw}), using notation based on events (E).

(i) Figure 11.12 shows an example of another type of structural relationship which is less obvious than some of the preceding illustrations. Here an agmatite comprising dark grey palaeosome blocks of folded (F) grey gneiss (e.g. block at centre background) in pale neosome is transected by a planar structure, a cleavage (C), and associated parallel leuco-cratic veining (both cleavage trace and veining trend away to the right of the photograph). The fact that the cleavage cuts both palaeosome and neosome and is parallel across the entire exposure demonstrates that it is the latest visible structure.

The succession of structures (also shown in the column of Figure 11.12b) is: ... + folds (F) + ... + agmatite blocks + ... + cleavage and veins + ... resulting from a sequence of events (E) such as this: ... + E_F + ... + E_A + ... + E_C + ...

Again, as the previous examples, it is possible that there are structures resulting from other events contributing to the overall sequence that are not obvious at this exposure.

Following the discussion in section 2.1.1, the sequence of events (after the development of the gneissic foliation) responsible for the

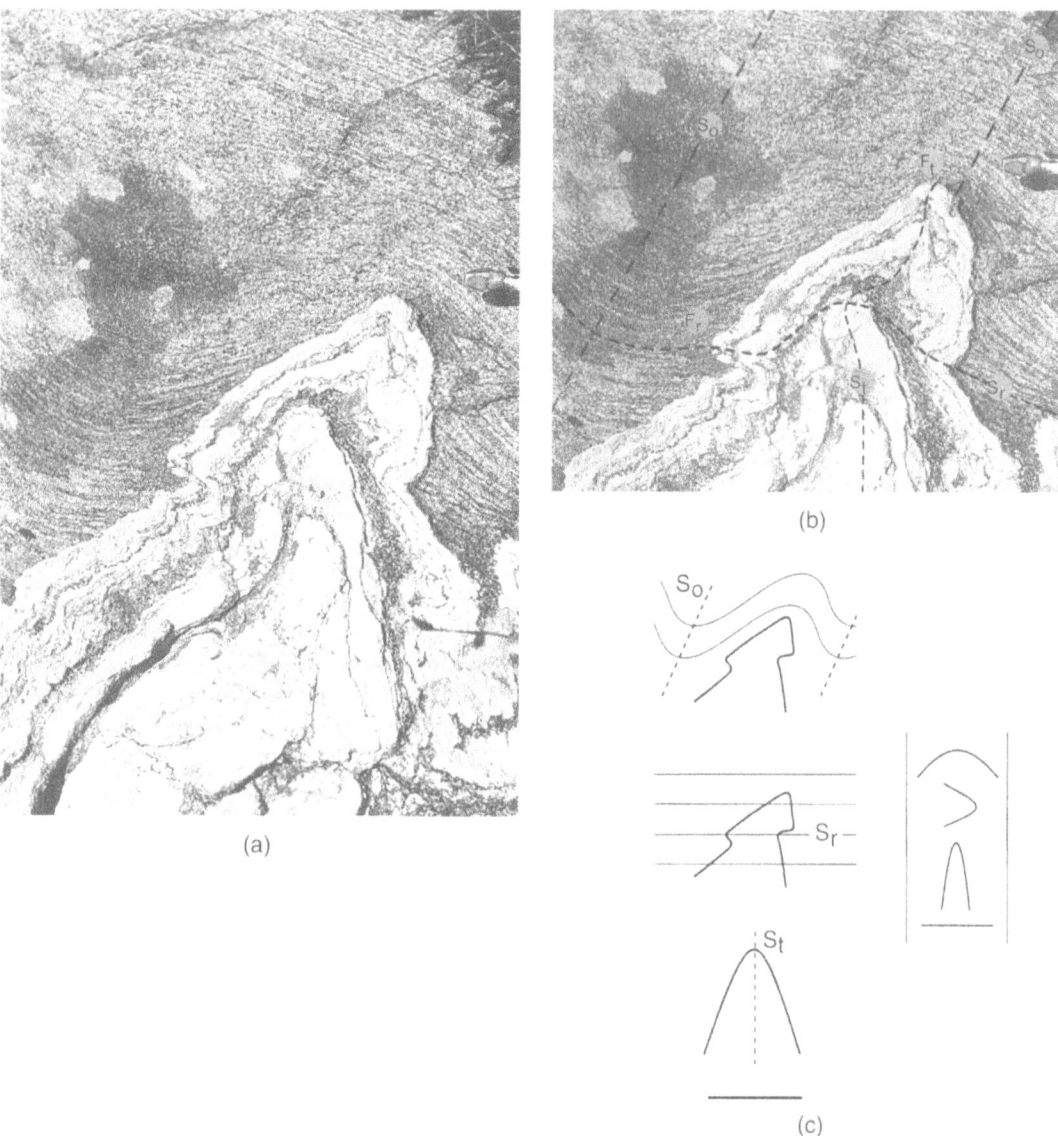

(a)

(b)

(c)

Figure 11.13   (a) Refold association between tight folds, upright folds, and recumbent folds. From Figure 12b, Hopgood 1980. Reproduced with the permission of the Royal Society of Edinburgh. (b) The folds and fold axial traces labelled. (c) Sketches showing the development of the structure in (a) from planar foliation through early tight folding (F_t), recumbent folding (F_r) to open upright folding (F_o), with a column showing the known structure succession. S_t, S_r and S_o represent the traces of fold axial planes. Lewisian complex, Rona, Inner Hebrides, Scotland.

structure shown in Figure 11.12 would have been: (i) folding (F), (ii) agmatization (A) and then (iii) imposition of cleavage (C).

(j) The interference pattern shown in the photograph in Figure 11.13a results from the overprinted relationships between several folds. The relationships between the structures are shown in Figure 11.13b. The folds range from tight to open structures, associated in one case with an axial planar cleavage. Tight upright folds (F_t) with steep axial planes (S_t) curve across the axial trace of recumbent folds (F_r) with near-horizontal axial planar traces (S_r) and associated crenulation cleavage in the amphibolite. Open folds (F_o) on steeply dipping axial planes (S_o) deform the crenulation cleavage.

Although both F_o and F_t are upright structures in this setting, F_o (and S_o) can be distinguished from F_t (and S_t) because F_o affect a structure (S_r, the cleavage in the amphibolite) which did not exist at the time F_t was formed (during deformation D_t). The cleavage was imposed during deformation D_r, **after** F_t was formed. Each of the structures comprising the succession is represented in the column of Figure 11.13b which also shows the sequential development of the observed interference pattern.

The structural succession is thus at least: ... + F_t + ... + F_r + ... + F_o + ..., with the possibility that other sets of structures also exist.

It should be noted that D_o must have affected F_t also, so that strictly, F_t is a composite upright structure, comprising $F_t + F_o$. The fact that it is composite could not have been appreciated without the intervention of S_r and the evidence provided by its subsequent deformation during D_o. Furthermore, the existence of F_o could not have been recognized had it not been for the presence of the intervening structure (S_r) which, during D_o recorded only F_o (see section 4.1).

The sequence of events, following the development of the foliation, which produced the structure of Figure 11.13a would have been:

(i) tight folding (F_t), (ii) recumbent folding and axial planar crenulation cleavage (F_r/S_r), (iii) open upright folding (F_o). See again section 2.1.1.

(k) Figure 11.14 shows field photographs (a, A–E) and sketches (b, A–E) of the relationships between folds, foliation and lineation, fractures and intrusive rocks. In the sketches, amphibolite (S) is dark grey, amphibolite (L) is light grey, aplite (P) is white and hornblendite (B) is black.

Photograph A shows that the irregularly-jointed dark country rock (amphibolite L) is strongly foliated with discontinuous thin leucocratic veins, and also contains some thicker continuous white veins, parallel to the foliation. The rock (amphibolite S) exposed on the dark vertical face nearest the camera below the notebook lacks the strong foliation and irregular discontinuous veining of L but is cut by the thicker veins (aplite P) on the right, so confirming that the foliation and veining is vertical. This can also be seen from the disposition of small, foliation-parallel faces elsewhere. A faint vertical foliation trace can, however, be seen in S on the long vertical face, particularly near the notebook.

The partial succession at A is: Amphibolite (L), foliation (? + concordant veins), amphibolite (S), foliation in S parallel to the earlier foliation (? + concordant veins).

Photographs B and C show comparable features with at least one, much wider, tabular leucocratic body (aplite P) parallel to the foliation and cutting S, and one in the form of a tight to isoclinal fold at the centre right of photograph B. Because S is unfolded the isoclinal fold form in the aplite is probably mimetic after earlier (pre-S) folding, presumably associated with the earlier foliation-forming event.

The partial succession at B and C is: Amphibolite (L), isoclinal folding + foliation (? + concordant veins), amphibolite (S), foliation in S parallel to the earlier foliation (? + concordant veins), aplite (P) concordant to the foliation.

Figures 11.14 (a) Photographs (A–E) of fold and intrusion relationships involving four rock types.

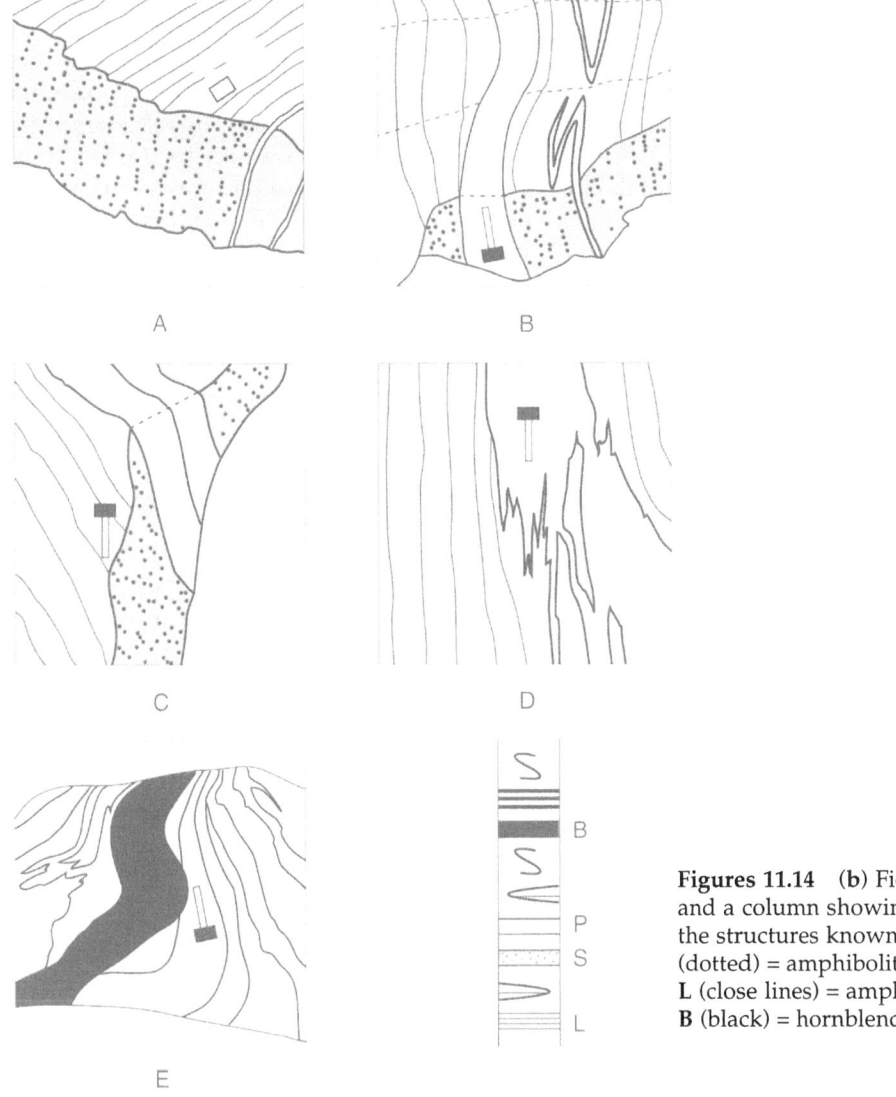

A

B

C

D

E

(b)

Figures 11.14 (**b**) Field sketches (A–E) and a column showing the succession of the structures known to comprise (**a**). **S** (dotted) = amphibolite, **P** (white) = aplite, **L** (close lines) = amphibolite and **B** (black) = hornblendite.

Photograph D shows leucocratic rock (aplite P) in tabular form parallel to the foliation on the right and interdigitating with the foliation at the centre. Between the two, and linking them, is a thin tabular vein that is distinctly discordant to the foliation.

The partial succession at D is: Amphibolite (L), isoclinal folding + foliation (? + concordant veins), aplite (P) concordant (and discordant) to the foliation.

Photograph E shows relationships between the foliation and the leucocratic bodies similar to those in photographs A–D but also shows (in the foreground) that there are strings of small, very dark lensoid bodies parallel to the foliation. In this photograph also, the foliation and leucocratic bodies are deformed (in front of the hammer) into a tight fold. Cutting across this fold is a steeply-dipping, curved, black tabular body (hornblendite B) containing a white enclave (near the end of the hammer shaft). The back body appears to be slightly folded on an axial plane that is more or less parallel to that of the tight fold in the host rock. Furthermore the black intrusion contains a foliation that is more or less parallel to the foliation in the host rock and is discordant to the margins of the intrusion. These relationships, combined with the partial successions from A–D, lead to the succession (1–9) tabulated below.

Succession
9. Asymmetrical folds affecting L, S, P and B, Figure E.
8. Foliation in B (and host rock L), Figure E.
7. Hornblendite body (B), Figure E.
6. Asymmetrical folds affecting L, S and P, Figure E.
5. Aplite (P), some discordant and some concordant parallel to (mimetic) isoclinal folds, Figure B.
4. Foliation parallel to existing foliation and lineation in S, Figure A.
3. Amphibolite (S), Figures A, B, C.
2. Isoclinal folds and foliation in amphibolite (L), Figure B.
1. Amphibolite (L), Figures A–E.

While no single entirely unambiguous interpretation of these relationships is possible from examination of the figures alone (i.e. without field work and laboratory examination) the following is one possible interpretation of the events leading to the observed succession (there are alternatives to this). The events responsible for the structure shown in Figure 11.14a could have been as follows (again see the 'The structural succession', section 2.1.1).

Sequence of events
1. Emplacement of the amphibolite (L).
2. Isoclinal folding and foliation of the amphibolite, Figures B, D, E.
3. Amphibolite (L) cut by later amphibolite (S), Figure A.
4. Metamorphism and deformation causing asymmetrical, intrafolial (shear) folds and foliation and lineation in S (parallel to the existing foliation), Figure D.
5. Amphibolite (S) cut by aplite (P), (Figure B) with some aplite intruded concordantly parallel to isoclinal folds (mimetic), Figure B.
6. L, S and P folded asymmetrically, Figure E.
7. L, S and P intruded by hornblendite body (B), Figure E.
8. Foliation-forming event affected B and country rock, Figure E.
9. L, S, P and B refolded asymmetrically, Figure E.
10. (or 8 or 9) Amphibolite facies metamorphism (B became hornblendite), Figure E.

(l) The photograph in Figure 10.15 shows an exposure in Svecofennian migmatites in the southern Finnish Archipelago. The structure of this was previously discussed with Figure 10.38 in section 10.1.10(j).

The foliation (S_o) is strongly folded and disrupted and intruded by irregular basic masses (A) and veins (on the right) and irregular masses and veinlets of leucocratic neosome (N) as well as narrow rectilinear quartzofeldspathic veins (Q).

The initial chaotic impression created by the complex appearance of the structural relationships can be resolved by considering the different features separately step-by-step, not necessarily in any particular order.

Firstly, the irregular basic veinlet (A) at the upper right, with margins approximately perpendicular to the foliation (S_o), is clearly later than the foliation which it cuts and penetrates and also apparently later than (and parallel to the axial plane of) the upright folding which affects the foliation. Furthermore the basic material (A) is affected by the leucocratic neosome (N) and together these form an agmatite (N:A).

Secondly, the neosome (N) can be seen to affect the folded foliation as well as the basic masses. It also penetrates the foliation concordantly, although there is a tendency to develop irregular discordant masses (veins) more or less perpendicular to the foliation, i.e. approximately vertical.

Thirdly, consider the upright folds (F_u). These range between tight and open with sharply rounded to angular chevron hinges. They are upright but note that their axial planar trace (S_U)is far from rectilinear (Figure 11.15b), being strongly curved about a sub-horizontal axial plane (S_r) by folding (F_r). Note too that this curvature is **not** mirrored by the more or less vertical attitudes of the basic masses and the thin leucocratic veins (Q). Therefore the curvature must precede the emplacement of these. Note also that the folded foliation is discontinuous, having been disrupted not only by the basic and acid veins but also by slip and associated foliation drag on more or less vertical surfaces (Figure 11.15b). These slip surfaces (S_s) are accompanied by incipient vein formation (grey) and by the narrow leucocratic veins (Q), forming white rectilinear traces on the outcrop (below the pencil). The slip surfaces have caused angular kinks and dislocations in the foliation and have modified the geometry of the upright folds. More significant is the fact that the slip has not only modified the existing

folds but has also separated and offset segments of folds and foliation to produce another set of upright structures (F_s). F_s are quite distinct from F_u and are expressed as tabular segments whose margins ('axial planes') are vertical (cf. the curvature of the S_u trace at the centre of photograph) and cut across the axial traces of F_u (Figure 11.15b).

Consider now the various structures. These are:

$$S_o \; N \; A \; Q \; F_u \; S_u \; F_r \; S_r \; F_s \; S_s$$

The structural succession observed at this exposure is: ... S_o ... + ... F_u:S_u ... + ... F_r:S_r ... A ... + ... N:A ... + ... F_s:S_s:Q ... + ... with the possibility that other structures not observed at this site may exist prior to, after, and at intermediate stages in this succession.

The sequence of events was as follows: S_o was folded by F_u whose axial plane (S_u) was curved by F_r on a sub-horizontal axial plane (S_r). The structure was then intruded by basic material (A) and then with the development of neosome (N) formed an agmatite (N:A). Fold (F_s) developed in response to slip (S_s) which dissected the structure into upright tabular segments separated by narrow leucocratic veins (Q).

(m) The next example (Figure 11.16) shows intricate relationships between the fine structural detail in an agmatite comprising a dioritic neosome (shown with fine dotted ornament in Figure 11.16b) with a palaeosome of blocks of foliated amphibolite (coarse dotted ornament) that have been tightly folded. This has been cut by thin, irregular grey amphibolite veins (shown black) and affected by coarse quartzofeldspathic pods (white in Figure 11.16b). The angular relationships between the palaeosome blocks appear to have remained essentially unchanged by the emplacement of the neosome.

The 'structures' exposed are named here but not labelled on the sketch. These include dark grey amphibolite (A_d, shown as coarse dots)

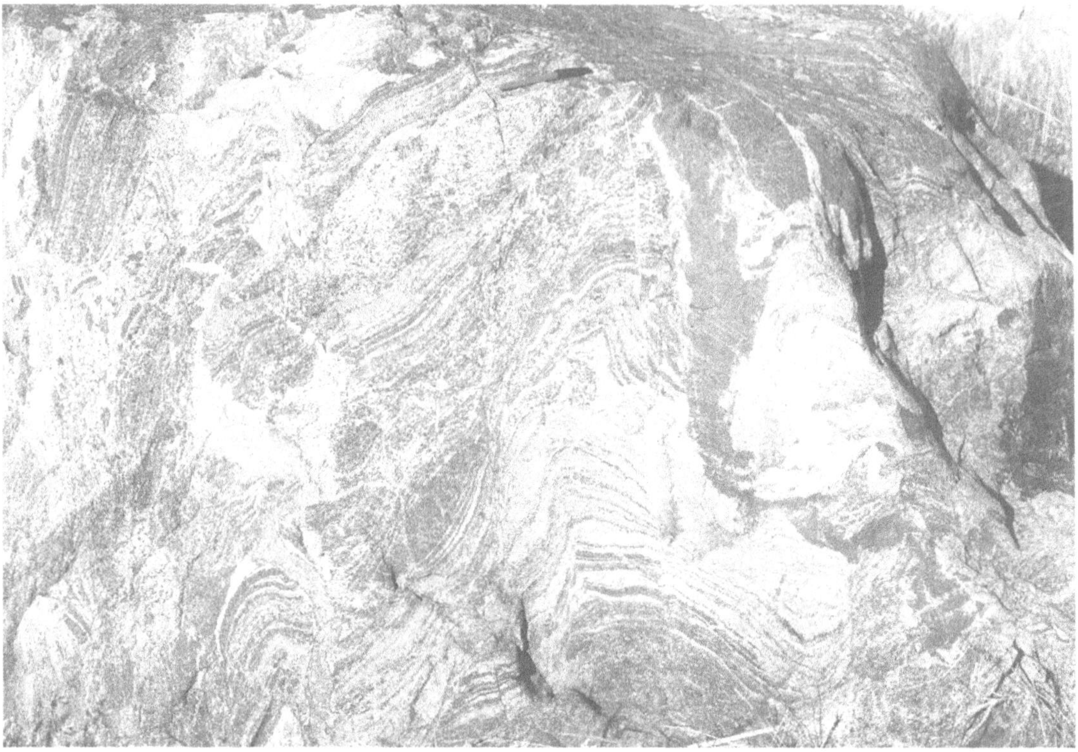

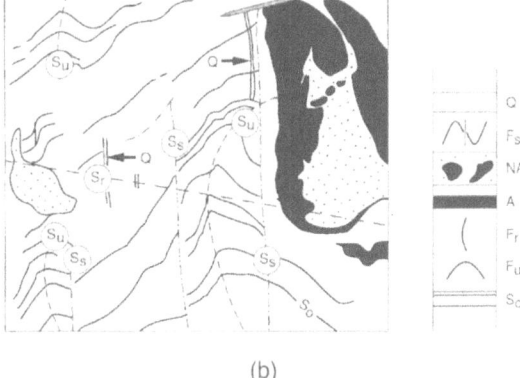

(a)

Figure 11.15 (a) Migmatite comprising disrupted grey tonalitic layers alternating with leucocratic layers and parallel amphibolite pods, that are folded, cut by small, irregular sub-vertical· basic masses (on the right), affected by anatexis and injected by thin tabular and irregular leucocratic veins (white). Svecofennides, Klobban, southern Finnish Archipelago. (b) Sketch identifying the structural features of (a) and column showing the known structural succession. From Figure 10a, Hopgood 1984. Reproduced with the permission of the Royal Society of Edinburgh. See also Figure 10.38.

(b)

foliation in the amphibolite (S$_A$), isoclinal folds in this foliation (labelled F$_{ab}$ at top in Figure 11.16b) including a relict hinge in the neosome (lower left), thin quartzofeldspathic veins (V$_{fq}$) cutting the folds with almost rectilinear traces (not shown in the sketch), irregular grey veins (V$_a$) of amphibolite (shown black) cutting through the neosome and across the foliation in the folded amphibolite (at the arrow in Figure 11.16b) and containing enclaves of dark amphibolite (circled), light grey, dioritic neosome (N$_d$, shown as fine dots) with coarse foliation (S$_N$) enclosing the palaeosome blocks of folded amphibolite and quartzofeldspathic

veins and grey amphibolite), coarse quartzofeldspathic veins (V$_{cq}$) and pods (shown white, upper left to centre).

These structures are:

$$A_d \; S_A \; F_{ab} \; V_{fq} \; V_a \; N_d \; S_N \; V_{cq}$$

The earliest structure is the dark grey amphibolite and this was foliated before being folded isoclinally, intruded by fine, 'rectilinear' quartzofeldspathic veins and disrupted to form an agmatite. Thin irregular grey amphibolite veins cut the previous structures, enclosing enclaves of the earlier amphibolite (circled), and can be seen to truncate the recti-

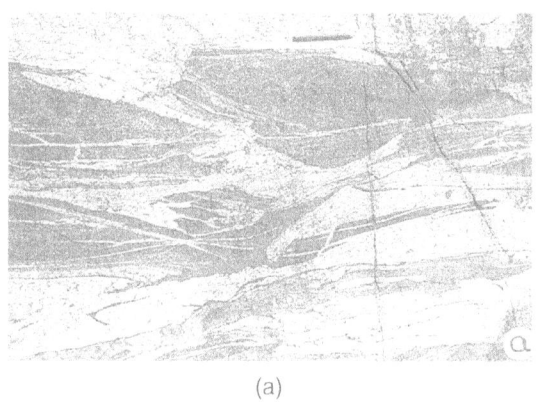

(a)

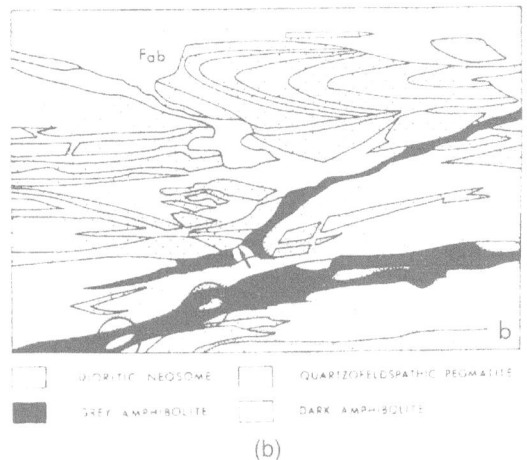

(b)

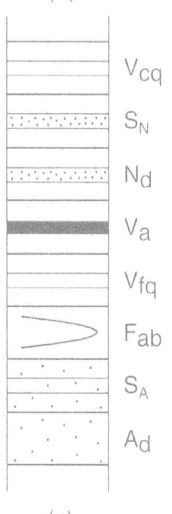

(c)

Figure 11.16 (**a**) Agmatite in Svecofennian migmatites. Baggöhamn, Skåldö, southern Finland. From Figure 2a, b, Hopgood 1984. Reproduced with the permission of the Royal Society of Edinburgh, (**b**) Explanatory sketch of (**a**). (**c**) Column showing the known succession of structures. (From Hopgood, 1984, Figure 2 (a, b).

linear quartzofeldspathic veins at the arrow. Agmatization accompanied the emplacement of dioritic neosome (the irregular shape of the grey amphibolite veins (V_a) suggests that this agmatization and vein emplacement were contemporaneous). The latest structures are the coarse quartzofeldspathic blind veins and pods. The structural succession here (Figure 11.16c) is thus: ... A_d ... + ... S_A ... + ... F_{ab} ... + ... V_{fq} ... + ... V_a ... + ... N_d ... + ... S_N ... + ... V_{cq} ... with the usual proviso that there could be other, unobserved structures before, after, or at intermediate positions within this succession.

(n) In the final example of this section, the different types of structural information that can be obtained regarding the same structure from two-dimensional and three-dimensional exposures is discussed. Considered together, the two different aspects can often make the resolution of the structure easier than it would be if the exposure was either only two-dimensional, or solely three-dimensional. Compare the two-dimensional exposure (at the left and in the foreground) and three-dimensional exposure (on the right) of the refold relationships shown in Figure 11.17. The sketch (Figure 11.17b) highlights the principal structural features (numbered) discussed below. The structures in the succession are given names in the text, but are not named in the sketch.

The picture is dominated by isoclinal, essentially recumbent folds (F_r, say, at 1a and 1b) with limbs showing a strong lineation (L, say) perpendicular to the fold hinges (2). In the centre foreground there is an oblique profile (3) which shows a distorted cross-section of F_r in two dimensions (compare the true, isoclinal profile on a vertical plane shown on the right at 1a). It also shows a 'mushroom'-shaped interference pattern (4) on the upper limb of F_r indicating overprinting between F_r and an earlier intrafolial tight fold (F_t). Like F_r, F_t also appears to be recumbent. From the symmetrical shape of the interference pattern it can be seen that the hinge of F_t trends left to right,

i.e. parallel to the lineation on the limbs of F_r. This suggests that F_t and L (L_t) are related. Note also that the lineation is reorientated with the F_r hinges where the latter change trend (6b). Exposed in three dimensions in the centre background (5) there is a group of tight recumbent folds with a trend comparable to F_t and presumably belonging to the same set.

Further, note that in the three-dimensional exposure on the right, F_r hinges curve (between 8a and 7), diverging from the predominant strong N–S trend (at 8a), through SE (at 6a) with almost the left-to-right (E–W) trend of F_t, i.e. parallel to the pen on the left where similar hinge curvature can be seen (between 2 and 6b) resulting in an almost E–W trend (6b). The (dashed) axial planar trend (7) of the folds that caused this (F_o, say) is at a high angle to the trend of F_r. As can be seen on the right, the axial plane of F_o must be steep, with the fold profile sub-horizontal in this case, in contrast to the profiles of F_t and F_r which are seen only on vertical surfaces. In circumstances where the quality of the exposure is not as good as it is here, differentiation between the two tight-isoclinal folds (F_r and F_t) would be difficult where they become subparallel with an E–W trend (at 5 and 6b).

Furthermore, there is a suggestion that both the two-dimensionally, and three-dimensionally exposed F_r hinges show slight curvature (at 8a and 8b) by very open folds (F_g, which are much less tight than F_o). These folds might be confused with F_o. However, at the top right, successive F_o-folded three-dimensional F_r fold hinges show that the trend of S_o is broadly NE (7) whereas at the centre left (between 8b and 6a), a succession of very open curved traces, representing the intersections of F_g closures with the outcrop, show that S_g trends in a broadly left-to-right direction and not NE parallel to F_o. Returning now to consideration of the top right of the photograph (7), the trace of S_o defined by a series of open F_o hinges is gently curved (concave to the upper left) apparently by a

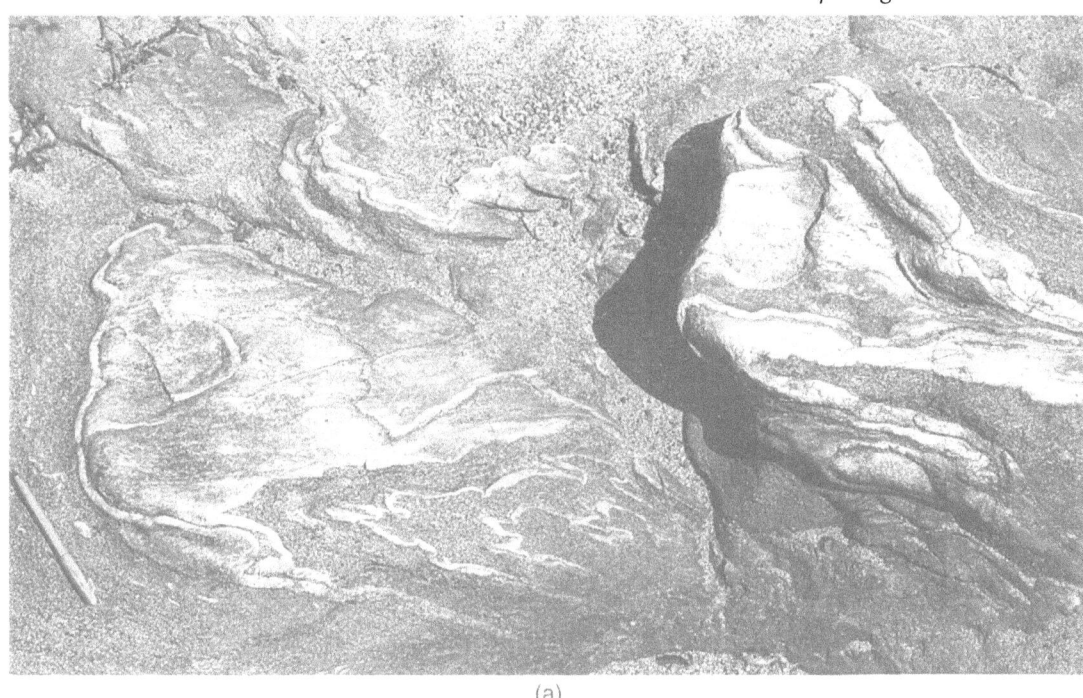

Figure 11.17 (**a**) Refold relationships between sets of tight and isoclinal recumbent folds, lineations and two sets of open upright folds. (**b**) Sketch identifying the structures in (**a**) and a column showing the succession of structures known to have contributed to (**a**). Migmatites, Mazoe River, Zimbabwe.

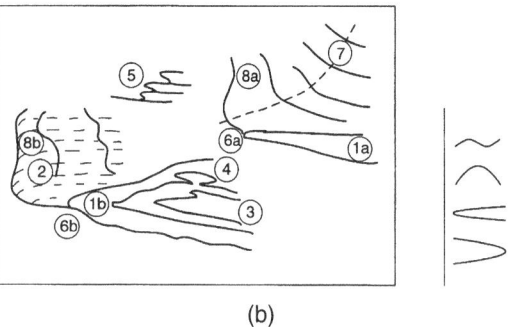

(b)

fold with a SE-trending axial plane. This suggests curvature by F_g, implying that S_g trends SE, and 'confirmation' of the existence of both F_o and F_g as separate fold sets, with F_o earlier than F_g.

The structural succession here, and shown in the column of Figure 11.17b also, is thus at least:

$$\ldots F_t \ldots + \ldots F_r \ldots + \ldots F_o \ldots + \ldots F_g \ldots$$

There could of course be other, unobserved structures before, after, or at intermediate positions within this succession.

PRINCIPLES AND PRACTICE OF DETERMINING STRUCTURAL SUCCESSIONS SUMMARIZED 12

12.1 INTRODUCTION

This chapter, and Chapter 13, summarize and integrate the content of the preceding chapters. They discuss the key points relating to the practice of establishing local successions from overprinting relationships identified at individual exposures and combining them to produce the overall succession for the structural complex under investigation.

As summaries of the concepts and principles of the methodology, it is likely that these two chapters alone will be sufficient to explain the methodology adequately for the geologist who already has some experience in the structural analysis of complex rocks. They might also be all that is required by the geologist who is unfamiliar with the approach, but who has some experience of coping with complex structural relationships and establishing structural successions, and wishes to learn the techniques. Chapter 12 treats the principles concerning the determination of structural successions while Chapter 13 formalizes the analytical procedure and presents some examples. While there is some repetition, with brief reference to aspects covered earlier, the discussion here deals not only with the theory and principles of the field methodology, but also with the practice of this approach, illustrated by examples of analysed photographs of structural relationships from a range of gneiss and migmatite complexes.

12.2 OBSERVATION

The determination of the overall structural succession leading ultimately to the proposal of a deformational sequence begins in the field with the recognition of the effects of multiple deformation, namely the identification of interference patterns (usually on more or less two-dimensional exposures) showing the effects of overprinting in three dimensions, such as those of Figure 12.1.

There are thus three stages or steps involved, viz. (i) the recognition of overprinted relationships, (ii) the establishment of the order of succession and (iii) the postulation of the sequence of events responsible for the structure. Of the three steps, the third is clearly of a lower order of certainty than the other two. However, while the second step might appear at first sight to be unambiguous we have already seen that there can always be some uncertainty regarding the order of succession, even at a single exposure.

For example, consider the three mutually interfering structures of Figure 12.2 b–d, previously discussed with Figures 8.12, 10.32 and 11.8.

The structure comprises at least three sets of folds, an **isoclinal** fold (F_i), a **tight** fold (F_t) and a **recumbent** fold (F_r). The order of succession is F_i, F_t and F_r and since it is not known whether there are other structures in the succession (that have not been recognized), the

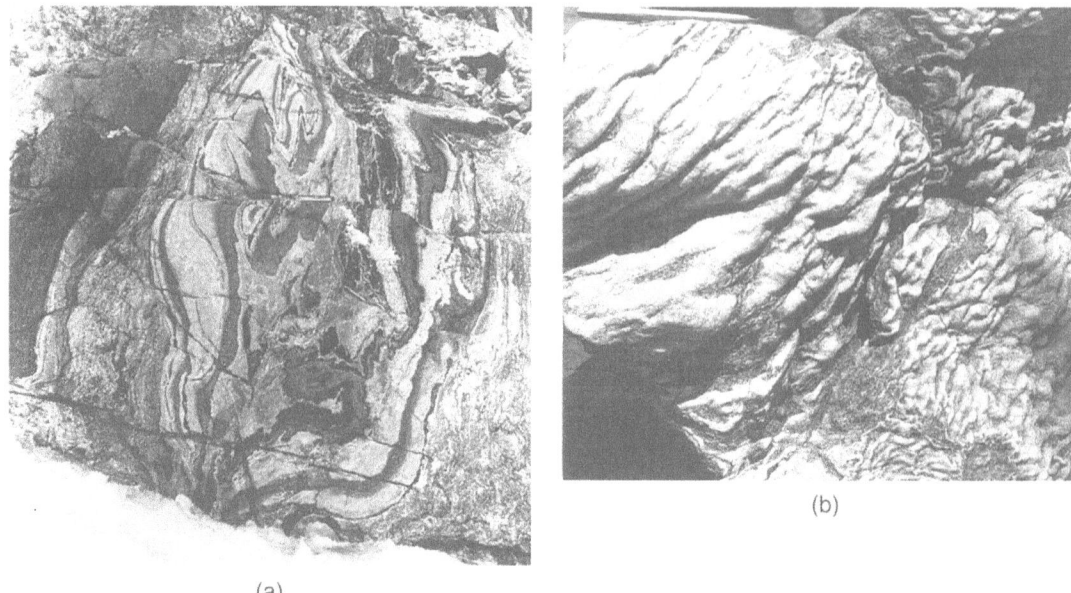

(b)

(a)

Figure 12.1 Comparison between the appearance, in two and three dimensions, of structure formed by multiple folding. (**a**) Two-dimensional interference pattern in polyphase-deformed anorthosite. Pre-Ketilidian complex, southern West Greenland. (**b**) Appearance of structure comparable to that of (**a**) exposed in three dimensions. Refolded gneiss, Zimbabwe.

succession might be interpreted in the following ways:

$$F_i + F_t + F_r + F_4 + F_5 + \ldots + F_n,$$

where the three sets are the **earliest** in the succession, or

$$F_1 + F_2 + \ldots + (F_{n-3}) + F_i + F_t + F_r,$$

where the three sets are the **latest** in the succession, or

$$F_1 + F_i + F_t + F_r + \ldots + F_n,$$

where the three sets are at positions **intermediate** in the succession.

However, these interpretations are themselves uncertain in that they presuppose that the three sets recognized are **consecutive**, without structures (not recognized here) intermediate between all or any of them.

More rigorously then, the case above could be depicted as something like the following where an unknown number of (unrecognized)

structures (represented here by $F_?$ and $f_?$) might exist at some sites among the known structures:

$$\ldots + F_i + F_? + f_? + \ldots + F_t + F_r + \ldots \text{ say, or}$$

$$\ldots + F_i + F_t + F_? + f? + \ldots + F_r + \ldots \text{ say, or}$$

$$\ldots + F_i + F_? + f_? + \ldots + F_t + F'_? + f_? + \ldots + F_r$$

say, or simply

$$\ldots + F_i + \ldots + F_t + \ldots + F_r + \ldots \text{ in general.}$$

In column form these could be depicted as shown as in Figure 12.3.

Before continuing with the discussion, recall that in previously describing this structure, which is also shown in Figure 11.8, it was seen that closer examination of the lower right of the figure reveals that besides F_i, F_t and F_r there is another group of small, **upright** folds – F_u say. The position of these structures in the succession is shown by the fact that they affect a coarse foliation in the dark rock which

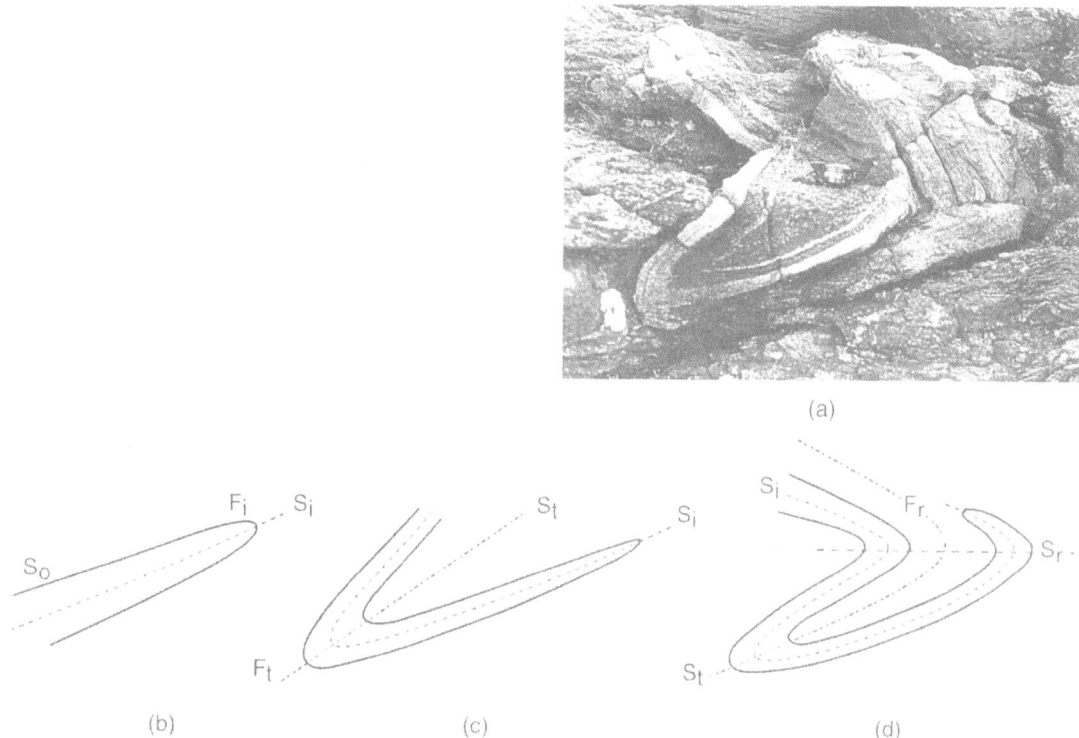

(a)

(b)   (c)   (d)

Figure 12.2 (**a**) Multiple deformation of inter-layered quartzofeldspathic gneiss and amphibolite. Lewisian complex, Rona, Inner Hebrides, Scotland. (**b–d**) Sketches showing the development of the structure of (**a**) from early isoclinal folds (**b**), through tight folds (**c**), to recumbent folds (**d**).

further inspection shows to exist in the hinge of F_r also where it is parallel to the F_r axial plane (S_r). That being so, F_u must be post-F_r and is therefore is the latest of the structures in this exposure, so the succession in general terms extends to:

$$\ldots + \ldots F_i \ldots + \ldots F_t \ldots + \ldots F_r \ldots + F_u \ldots + \ldots$$

A similar case is illustrated in Figure 12.4 (seen previously in Figure 11.13) where intrafolial folds (F_{ii}) are refolded by an upright tight fold (F_t). These are also overprinted by recumbent folds (F_r). There is yet another set of open upright folds (F_o) however, which are readily distinguishable from F_t because they are developed in a foliation that is evidently axial planar to F_r which also affects the axial plane S_t of F_t (Figure 12.4b). F_o therefore are quite distinct from, and later than, F_t because they affect a structure S_r which is later than F_t and which was imposed at an intermediate stage in the succession. Thus F_o in this example holds a position in the succession comparable to that of F_u in the previous example (Figure 12.3).

Note that because F_t at this site has an axial planar attitude (fortuitously) parallel to that of F_o it will have been modified during D_o folding to an extent that would be difficult to measure, although its influence has been recognized from its effect on F_r. The fold denoted F_t (as seen at this site) is therefore a composite structure and is strictly $[F_t + F_o]$. See also Figure 6.29.

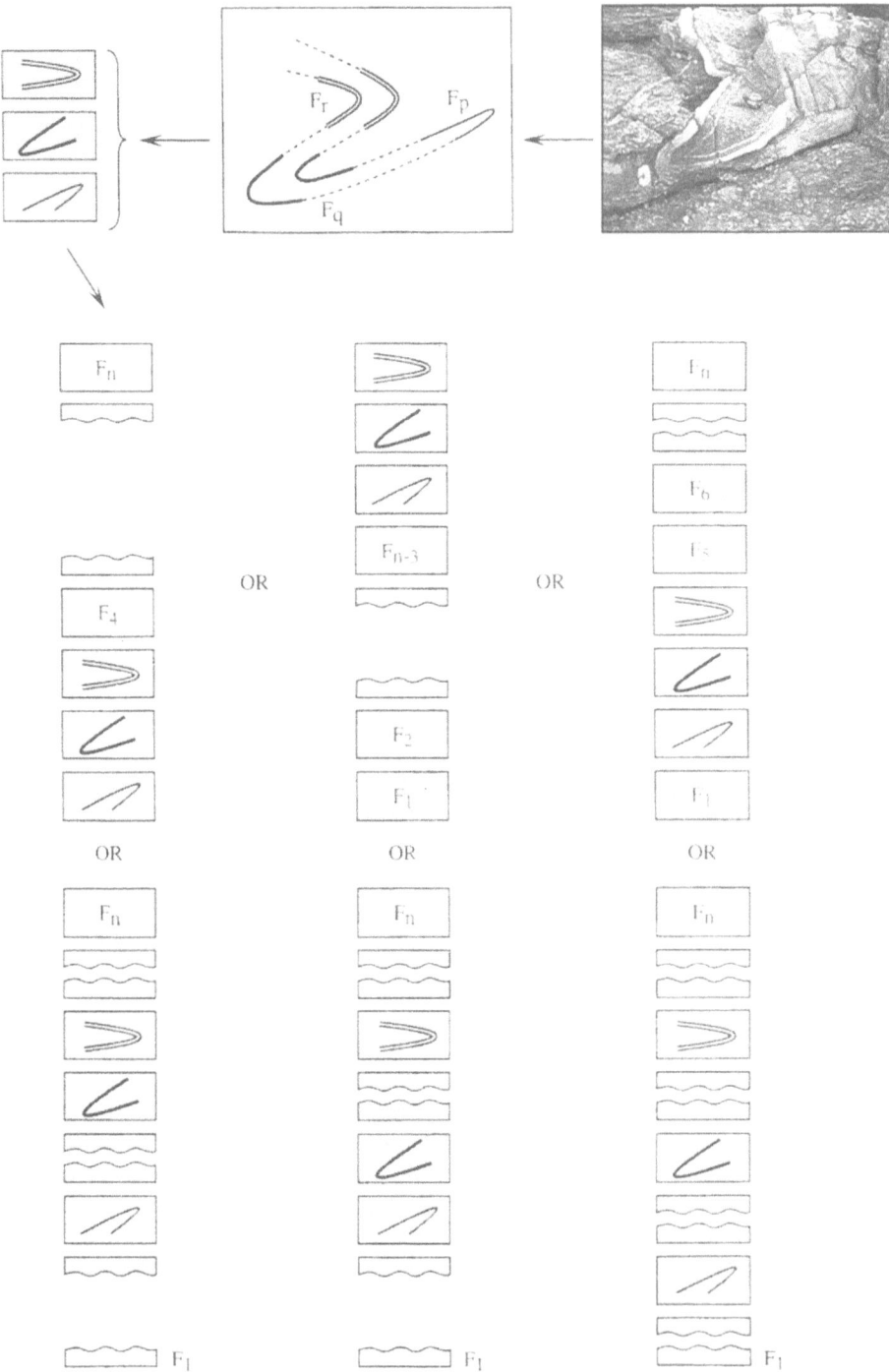

Figure 12.3 Sketches showing six possible structure successions in column form that could be derived from the known overprinted relationships of the type shown in Figure 12.2 and shown again here at the top centre and the photograph at top right.

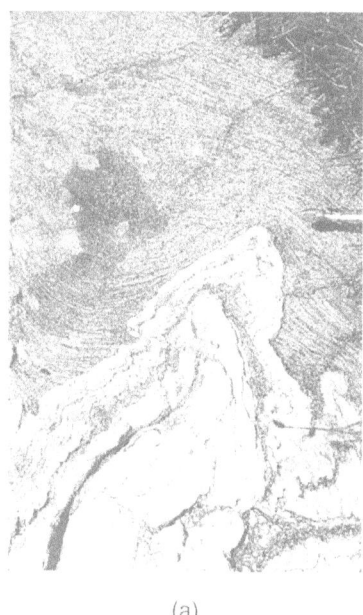

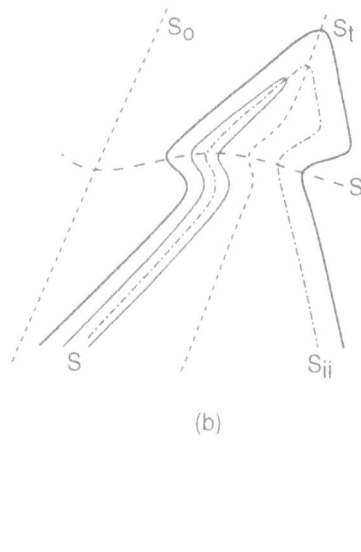

(b)

(a)

Figure 12.4   (**a**) Folded foliations (S and S$_r$) in gneiss and amphibolite. Lewisian complex, Rona, Inner Hebrides, Scotland. (**b**) Explanatory sketch of the refold relationships shown in (**a**). S = original foliation, S$_{ii}$ = axial plane of intrafolial isoclinal fold, S$_t$ = axial plane of tight fold, S$_r$ = axial plane of recumbent fold and S$_o$ = axial plane of open fold.

12.3   IDENTIFICATION OF STRUCTURES

Up to and including this stage, establishing the structural succession depends solely on observation. However, in order to determine the local succession either on a single exposure or on several adjacent exposures at a single locality something more is required. **It is necessary to be able to relate individual structures from one exposure to another**, i.e. we are dependent on being able to identify structures unambiguously.

The identity of a structure is comparable in some ways to 'identity' as used in biological classification (refer again to section 5.2, and in particular to 5.2.2) and as a reminder of what this entails consider again the group of heads shown previously in Figure 5.7, their attributes summarized in Table 5.2 and shown again here in Figure 12.5. Each of these, in the broadest sense, is comparable in that it possesses a mouth, a nose, two eyes and two

ears. On the other hand, all would be considered to be readily distinguishable from one another in respect of their head shape (short, long, narrow, broad, round, angular etc.), nose shape, lip shape, eye shape and colour, skin type, texture (wrinkled or otherwise), skin colour, ear shape and size, hair length, colour and type (straight, curly, coarse, fine etc.) and other factors such as the separation between the eyes, the nose, and the mouth. Attributes such as shape and spacing between features are more closely analogous to those used in the identity of geological structures. Distinction could be made between individuals on several counts, for example since it is likely that numbers 6, 7 and 8 would have the power of speech, while numbers 4, 5, 6, 7 and 8 might be classified as a group that are likely to have the potential to hear, and numbers 6 to 8 could be distinguished from numbers 1 to 5. Numbers 4 to 8 could be classified in terms of having the ability to see and so are

Figure 12.5 Sketches of a group of heads of different types illustrating the range of differences and similarities between individuals who, while possessing attributes in common, are nevertheless distinguishable. Previously shown as Figure 5.7.

could be found where the distinctions are so subtle as to be barely recognizable by most observers. Ultimately, such as in the case of identical twins, individuals would be quite indistinguishable in terms of appearance.

To a lesser extent the same could be said to be true of geological structures and, as has already been discussed (section 5.2), the identity of geological structures is dependent firstly on characteristic features but ultimately on the overprinted relationships of the structure to similarly identified structures that (a) precede it and that (b) follow it. The concept of identity has already been discussed in section 5.2.6, and in structural geology has as its basis the concept of style (section 6.2). The style of a structure embodies several parameters which together constitute characteristics that are distinct, often distinctive or even unique.

While the appearance and the geometry of a single set of structures can vary, this variation might not necessarily be significant and it rests with the observer to decide which, if any, features are significant. For example, on one limb of a large structure the sense of asymmetry of related small-scale structures, i.e. structures belonging to the same set as the large fold, will be the same, whereas the sense of asymmetry of similarly related small-scale folds on the opposite limb of the large fold will be opposed (i.e. a mirror image). Here the different asymmetry sense of structures of the same set on opposite limbs of the large fold is not a significant factor in discriminating between fold sets but in other structural settings it could be.

Figure 12.6 shows a number of folds showing considerable differences in general appearance but all having in common a dextral sense of asymmetry. If these folds all happened to belong to the same structural complex (in fact the folds shown here do not), then in some circumstances this sense of asymmetry could be a significant factor in deciding whether or not the folds are related, or at least a reason for deciding that further evidence should be sought in support or otherwise of such a relationship.

very clearly distinguishable from numbers 1 to 3 say.

On the other hand, however, whereas it can be said that each of numbers 6, 7 and 8 have in common the seven features, ears, hair, nose, mouth, eyes, eyebrows and eyelashes (and on that basis might be said to be identical), each is in fact readily and unambiguously distinguishable from the others by virtue of other features which they possess, such as eye colour, hair colour, hair length, nose shape etc. The distinction in this case, while clear, is more subtle and examples of individuals

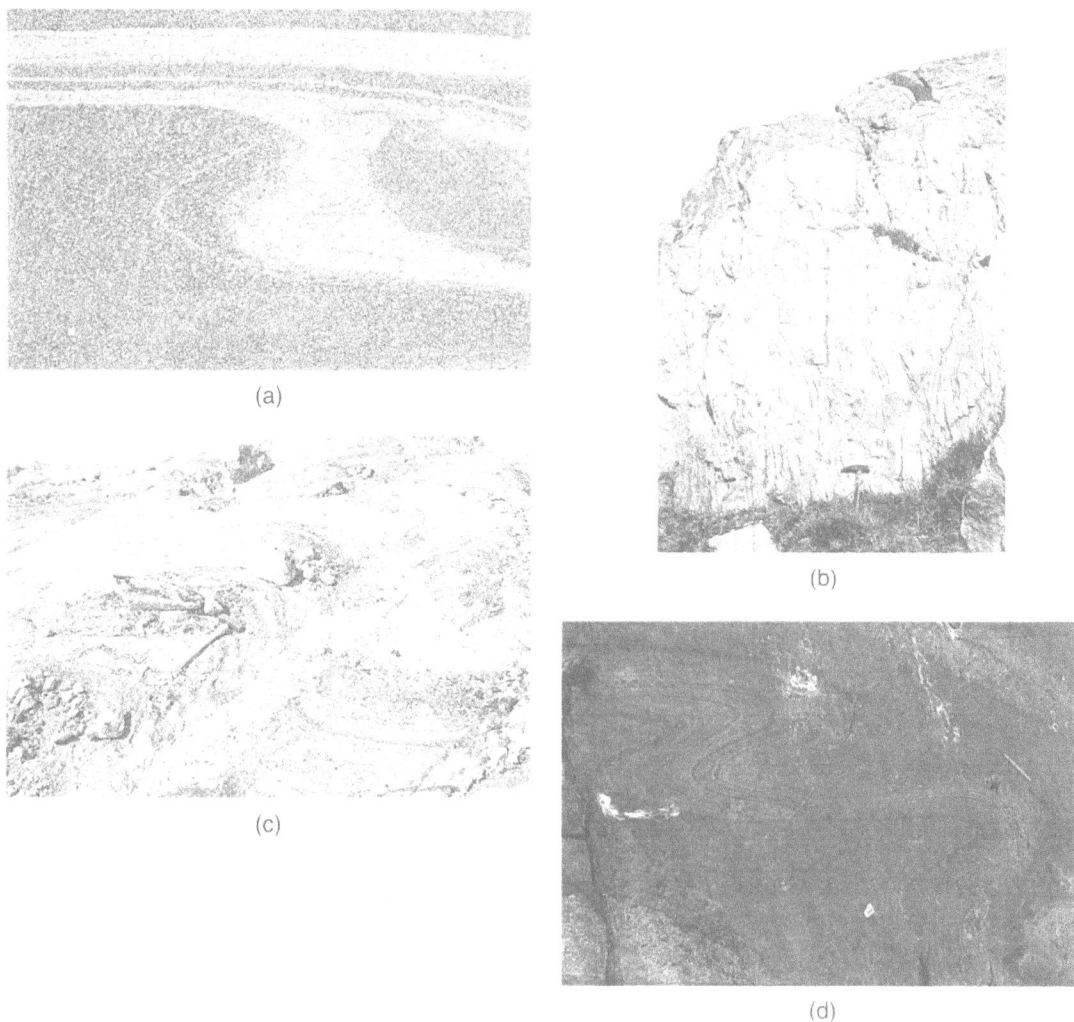

Figure 12.6 Folds whose styles differ markedly but which have in common the attribute of dextral asymmetry. On the basis of their common dextral asymmetry they might (in theory) belong to the same set. Evidence must be sought to prove or disprove a common origin. (**a**) Belemorides, USSR. (**b**) Lewisian complex, Rona, Scotland. (**c**) Pre-Ketilidian migmatites, West Greenland. (**d**) Svecofennian migmatites, Rönskär, Helsinki, Finland.

The asymmetry alone of the folds of Figure 12.6 does not of course constitute a reliable criterion for membership of a particular set but it could serve as a first step in superficially distinguishing similar-looking structures at say, the same locality.

The observer must be able to appreciate the significance of structural relationships such as these. In this case the grouping of structures has been on the basis of their dextral asymmetry, which distinguishes them from other small-scale structures at the same locality which lack this feature. These other structures, provisionally regarded as being unrelated, may for example be symmetrical, or possess a sinistral asymmetry sense.

Here it is appropriate to refer again to the discussion (summarized here) in section 6.2.2 on the significance of dip (e.g. of fold axial plane) as a distinguishing feature. It should be borne in mind that the opposite dip sense of the axial planes, while a useful characteristic in some instances, is not a **fundamental** distinguishing characteristic of the two fold sets. It is accidental, arising because of the use of the horizontal as a datum. For example, supposing the total structure had been tilted such that the axial planes of both fold sets – while still retaining their angular separation – dipped in the same sense, they would still be distinguishable because of this angular difference.

Thus it is the consistent difference in angular relationship between the attitudes of the two sets of axial planes that is significant.

12.4   KEY STRUCTURES

The identification of a structure that is both distinctive and widely developed so that it is recognizable throughout the complex provides a **datum** structure, or what may be termed a **key** structure (see also sections 6.3 and 12.5). Such a structure set is invaluable as a first step in correlating structures from one locality to another, i.e. on the basis of their overprinting relationships to the key structure (Figure 12.7).

The identification of a set of key structures during preliminary reconnaissance of the area means that the total succession is effectively divided into two, an important consideration where the succession is extensive, especially if similar-looking structures are present at more than one position in the succession. The two subdivisions comprise (i) folds (or other structures) formed prior to the development of the key structures and hence deformed by them, and (ii) structures formed afterwards and which therefore deform the key structures. Establishing these general relationships will normally be very much easier than having to determine the relationship between every set of structures (which themselves must be identified), especially if the history of deformation is a complex one.

An important attribute of a key structure is that the distinguishing features are sufficiently striking and clear, or are sufficient in number, to allow the structure to be recognized **unequivocally**, i.e. ideally the criterion for a key structure is that, besides being widespread and fairly easily recognizable, it is a multi-parameter structure. For a fold this would mean in essence that it should be distinctive in terms of factors such as geometry, attitude, style, distinctive syntectonic mineral association (e.g. quartzofeldspathic veins parallel to one limb) and clear-cut relationships to particular igneous or metamorphic events.

12.5   CORRELATION – DISCUSSION OF PRINCIPLES

The role of correlation in stratigraphic geology as discussed by Whittaker *et al.* (1991) in the following terms, was referred to on the first page of Chapter 6 and is repeated below. As

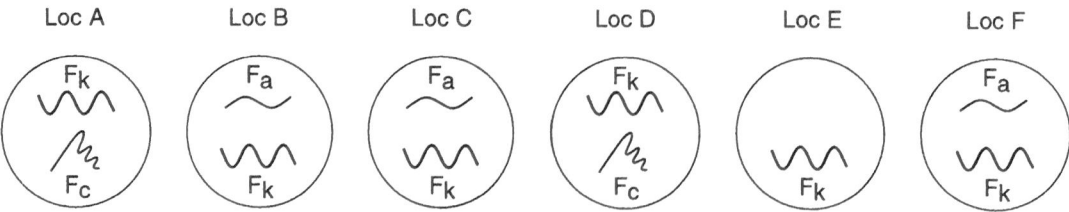

Figure 12.7   Sketches representing structural correlation between localities A to F using the relationship of structures F_a, F_c to the key structure (F_k) which is common to all localities.

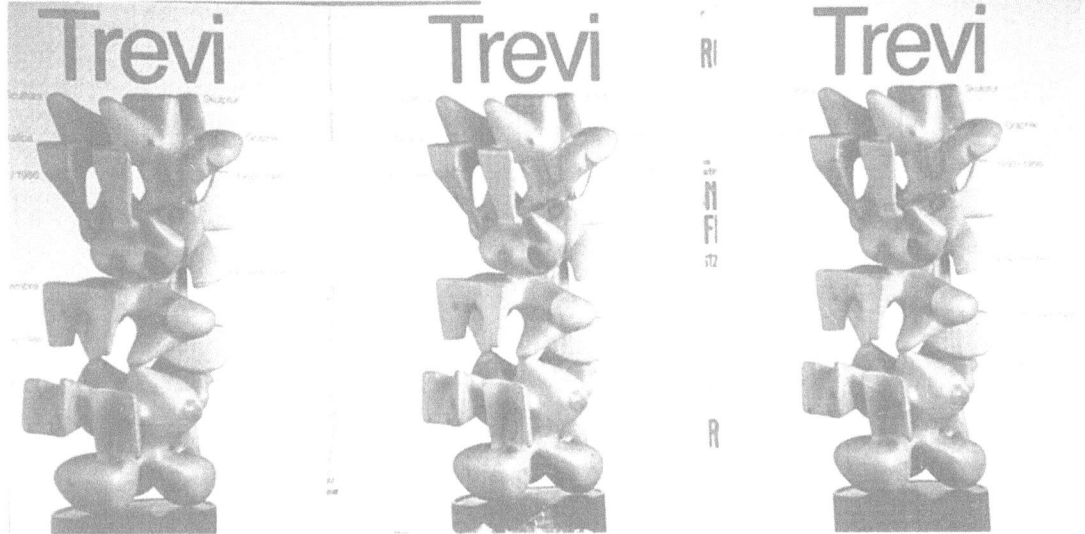

Figure 12.8 Sculptures whose shapes illustrate the basis for three-dimensional correlation. Correct matching of these three identical sculptures in terms of their components is the only possibility because of the unique form of individual components. No other form of matching is possible. Similar (more rigorous) constraints apply to the matching of formerly contiguous segments of the same tectonostratigraphic terrane, evolved during the same tectonic history, the basis for correlation being tectonic structural association in this case, rather than three-dimensional shape. Poster, Merano, northern Italy. See also Figure 12.16.

before, emphasis has been added here to certain words and phrases.

'The fundamental procedure in stratigraphy is to **observe**, **describe** and **correlate** rock successions, correlation being the demonstration of **correspondence between geological units** in some **defined property** and in **relative stratigraphical position**. When investigating an area, the first procedure is to describe the local stratigraphical succession objectively in terms of mappable or traceable lithostratigraphical units present. Currently or subsequently the local succession should be correlated, as far as possible, with the international standard stratigraphical scale or with the regional stratigraphical divisions. With most Phanerozoic rocks this correlation is most commonly achieved using biostratigraphical methods.'

The principle of correlation and its use in establishing a structure succession in structural geology is essentially the same as it is in stratigraphical geology and involves matching 'like with like' in very much the same way as the sculptures ('structures') discussed in section 8.1.3 and shown again in Figure 12.8 can be matched.

The stages of building up a succession are:

(i) Local integration between all the exposures at a single locality (of which there may be several) based on the identity of more than one structure set and mutual overprinting of these. This entails (a) the recognition of **overprinting relationships** – from interference patterns of refolded structures, and (b) the establishment of these overprinted relationships (Figure 12.9).

This is followed by (c) the **identification** of structures (see section 5.2.6) using their

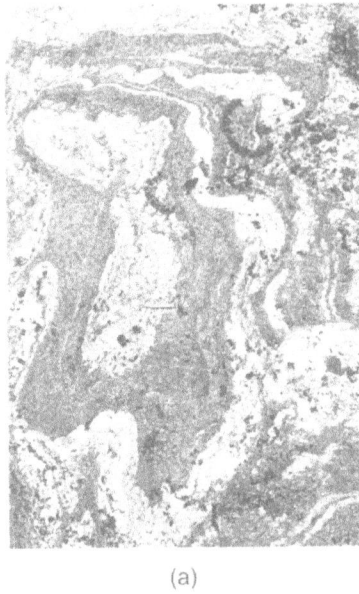

(a)

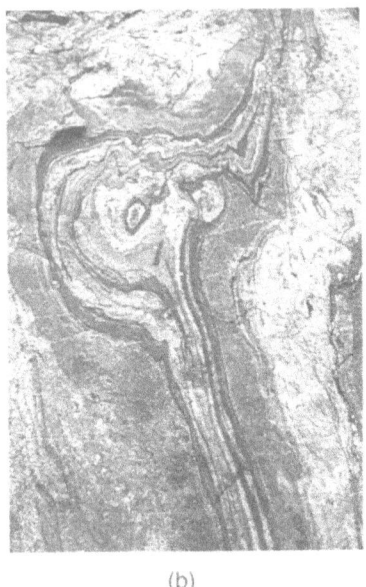

(b)

Figure 12.9 Examples of simple interference patterns resulting from overprinted folds in migmatites. (a) Structure resulting from overprinting between two fold sets with axial planar traces more or less parallel and perpendicular to the pencil and a third trending at about 45°. Anorthosite, Pre-Ketilidian complex, West Greenland. (b) Structure in migmatite probably resulting from two, mutually perpendicular fold sets affecting a third (earlier) intrafolial isoclinal set. Sand River, Republic of South Africa.

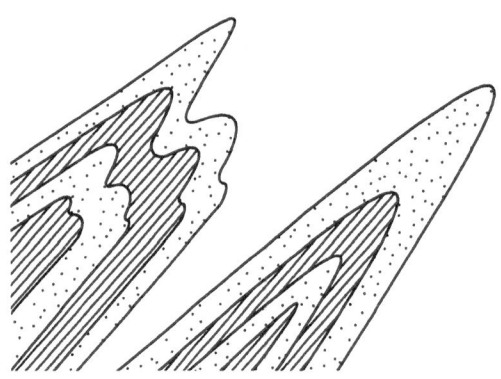

Figure 12.10 Sketches representing characteristic stylistic features which allow the identification of the two folds as members of the same set. Although the two fold profiles differ, both are associated with a distinctive axial planar structure stemming from say, a particular form of syntectonic mineral assemblage and habit characteristic of a particular deformational event.

combined characteristic features, criteria or attributes, i.e. their style (Figure 12.10).

This leads initially to the establishment of relationships by:

(ii) Correlation (section 6.1) between structures on the same exposure or nearby at the same locality, i.e. (a) nearby correlation over short distances of a few metres or tens of metres (Figure 12.11), and (b) comparable, proximal correlation of structures between different, but adjacent localities. This can be assisted greatly by the presence of key structures. Correlation in this case might entail relating structures from neighbouring localities (viz. proximal correlation) say a few hundreds of metres or a few kilometres apart (Figure 12.12), in this case relative to the key structure (F_k), or (c) comparison between structures observed at widely spaced localities, say on islands (Figure 12.13), separated for example by distances of several (tens of) kilometres (viz. distal correlation).

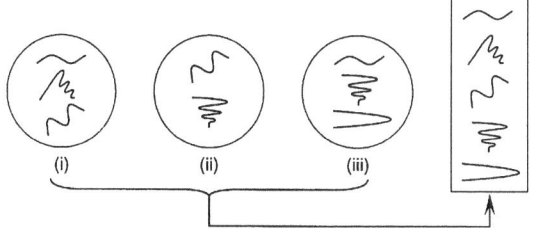

Figure 12.11 Sketches to represent close correlation (e.g. where there is no more than tens of metres between exposures) between structures whose relationships are shown at (i), (ii) and (iii), followed by integration to form the local sub-succession on the right.

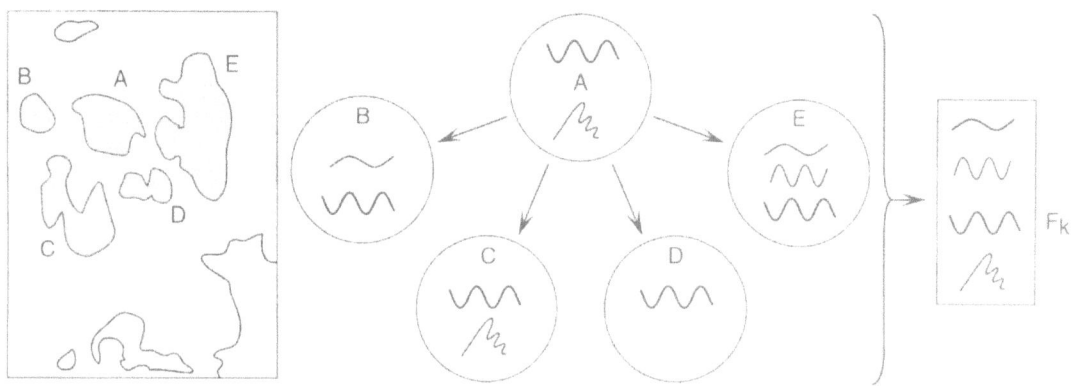

Figure 12.12 Sketches representing proximal correlation of structure F_k (between distances of hundreds of metres to, say, a few kilometres), followed by integration to form the structure succession.

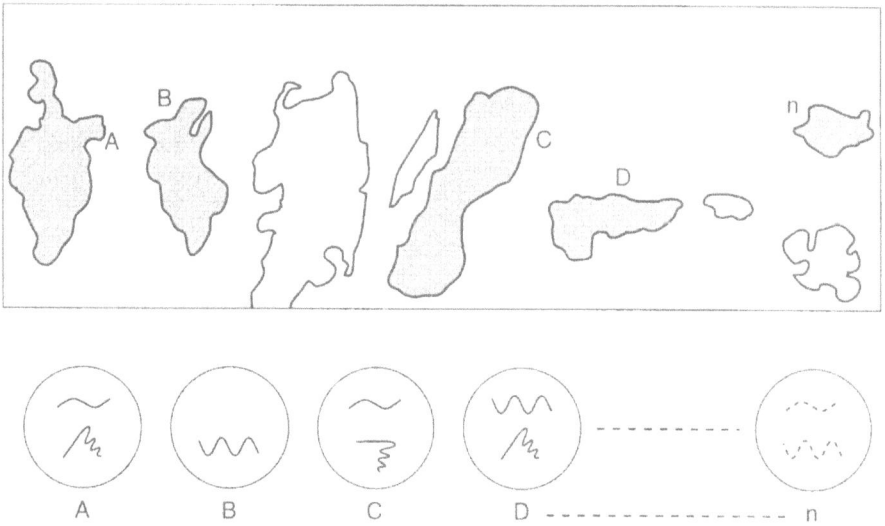

Figure 12.13 Sketches to represent distal correlation between exposures several (tens of) kilometres apart, leading to the establishment of a structure succession based on integration of the structure from n localities.

The concept of identity assumes greater importance when the establishment of structural successions depends on correlation, as is more often the case. Recall that where the establishment of relationships between structures depends on the recognition of interference patterns, this is achieved solely and unambiguously on the basis of direct observation of overprinted relationships (see again section 12.2 above). However, in all other instances relationships are determined on the basis of **correlation** and this depends on the recognition and identification of structures. Consider the relationships shown in Figure 12.14.

At locality A the refold relationship between F_a and F_b is certain and clearly demonstrable as: F_a followed by F_b. At locality B, on the other hand, whereas F_a might be identifiable the identity of F_x remains to be ascertained. While the relationship suggests that F_x could be F_b (because it is later than F_a) this is by no means certain and must remain uncertain until such

time as comparison can be made of its relationship to F_a as well as to other structures (e.g. F_z) here and elsewhere. However, by showing that F_x always bears the same relationship to F_a, F_z and to other structures as does F_b, extrapolation to other localities would confirm correlation between F_x and F_b.

As implied above, the certainty of such correlation is greatly strengthened when it involves comparison between two or more overprinted sets of structures, the identity of each of which has been ascertained (Figure 12.14). Here the number of attributes, or criteria, is considerably greater than that for a single structure, so strengthening the basis for correlation (Figure 12.15), i.e. the '**tectonic signature**' or '**fingerprint**' becomes highly distinctive.

There is a higher degree of certainty in correlating **successions** of structures than there is in correlating **single** structures because of the greater number of criteria available for comparison.

This aspect of correlation is particularly important where correlation is attempted between widely separated (distant) crustal segments (Figure 12.13) such as formerly contiguous parts of the same terrane (see section 6.1.3). Here the analogy between correlation of separate crustal segments with complex tectonic histories and the matching of a unique segment of a three-dimensional jigsaw puzzle, such as the sculptures of Figure 12.8, is apposite (Figure 12.16).

The degree of confidence in any such correlation is likely to be greater when the separation between exposures is small (metres say) than if it is measurable in terms of say tens of kilometres. Therefore it is advisable, given the choice, to extend the correlation by comparing structural relationships firstly on exposures within metres of one another, and then at exposures increasingly distant from the first one examined while monitoring the similarities and differences between them (cf. Hopgood and Bowes, 1972). Alternatively, if a reasonably confident correlation has been made

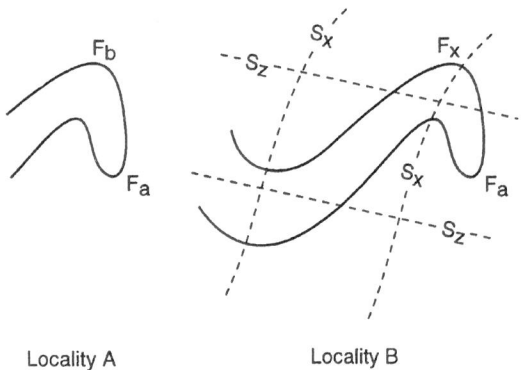

Locality A   Locality B

Figure 12.14 Recognition of the identity of overprinted structures: sketches of structural associations at two localities, A and B. At locality A, F_a is clearly earlier than F_b. At locality B the structure F_x is also later than F_a. However, the identity of F_x cannot be assumed (to be that of F_b) solely because it is later than F_a. It must first be found to have the same relationship not only to F_a but also to other structures such as F_z say, as F_b does (both here as elsewhere) before the correlation between F_b and F_x is confirmed with certainty.

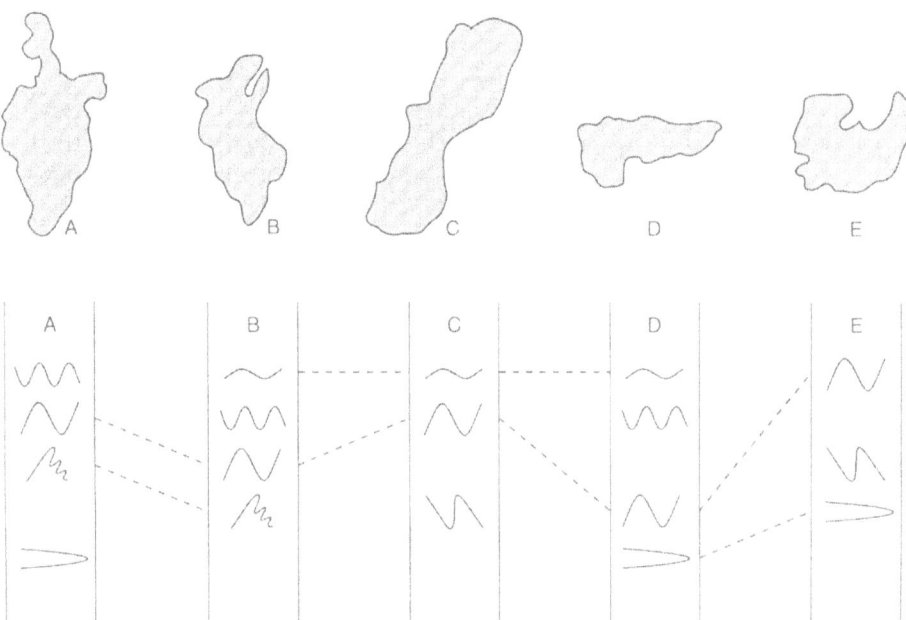

Figure 12.15 Sketches of columns representing structural successions recorded from islands A, B, C, D and E. In each case there is a high degree of certainty of correlation because it is based on overprinting relationships between a number of structures.

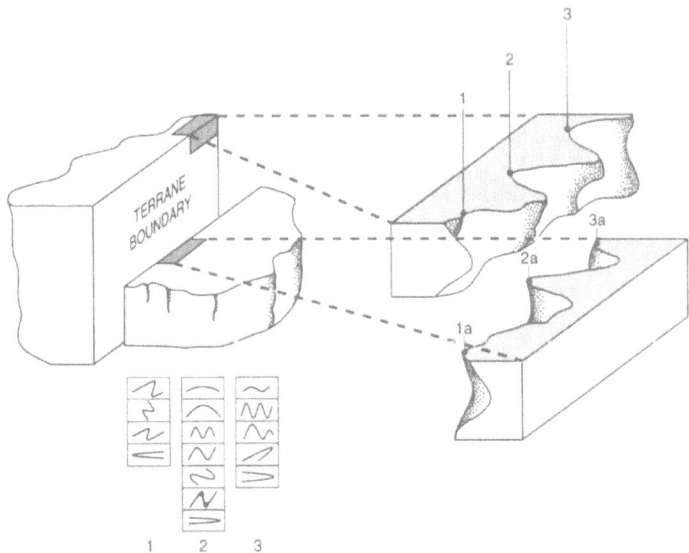

Figure 12.16 Sketch showing the matching of three-dimensional surfaces. A match between formerly contiguous terrane boundaries by correlation between complex structural successions (shown here by three pairs of structure successions from originally contiguous locations) is analogous to the matching of a three-dimensional jigsaw puzzle (cf. Figure 12.8) although the structural successions are likely to provide a match with a higher degree of certainty because their complexity renders them unique, or very nearly so.

between two localities at a distance, test (the validity/certainty of) this by examining structural relationships at other localities intermediate between the first two (Figure 12.17).

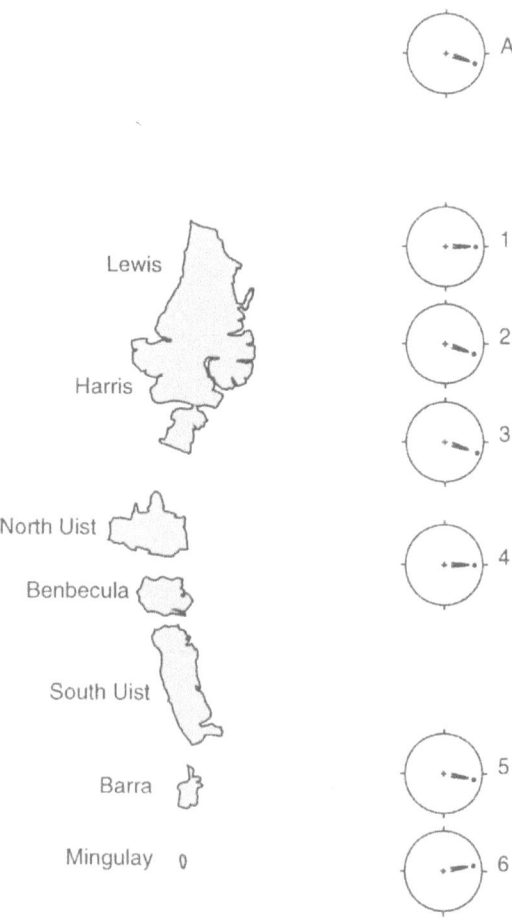

Figure 12.17 Sketches showing the gradual change in orientation of structures over a distance of 200 km across a crustal segment. Variation in attitude of the fifth fold set (F_5) identified in the Lewisian complex between North Lewis and Mingulay in the south, Outer Hebrides, Scotland. Changes in plunge and trend of the F_5 axis are shown on the right; 1, North Lewis. 2, West Lewis. 3, southwestern Lewis and Harris. 4, North Uist and Benbecula. 5, Barra. 6, Mingulay. A is the average of 1–6. Derived from Figure 2, Hopgood and Bowes, 1972.

The stages involved in correlating structures and structure successions are summarized in the flow sheet of Figure 12.18.

The synthesis of a structural succession begins in the first instance with the recognition of the existence of overprinted relationships (1 in Figure 12.18), followed by the recognition of structures (2) and their refold/overprinted relationships (3) within a single structure (3a), on a single exposure (3b) and/or on several exposures at a single locality (3c). These steps are all based on **observation**. The next stage is dependent on knowing the **identity** of structures and is the process of building up the succession on the basis of **correlation** of structures, at a single outcrop (4) or between nearby outcrops (5) or between widely separated outcrops (6). At locality A (7) the relationship between F_a and F_b is certain and their identities can be established, whereas at B the identity of F_a (and of F_b) must be established on the basis of correlation between F_x and F_b somewhere else say, and hence in terms of their relationships to other, earlier and later structures such as F_z. By this stage of the overall synthesis, correlation will have become possible between successions of structures (8). There is a much higher degree of certainty associated with correlating successions of structures than there is in correlating single structures (structure sets).

12.6   INTEGRATION OF LOCAL SUCCESSIONS – DISCUSSION OF PRINCIPLES

12.6.1   INTEGRATION OF LOCAL SUCCESSIONS

The principles of building up an integrated local succession from (say) four exposures A, B, C and D are illustrated in Figure 12.19 which shows representations of a hypothetical partial fold succession at each exposure. The total observed local succession is integrated in Figure 12.19b by correlating fold sets in the

1. Recognition of overprinting

2. Identification of structures

3. Refold (overprinted) relationships

   a. On a single structure

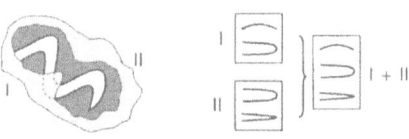

   b. On a single exposure

   c. On several exposures at a single location

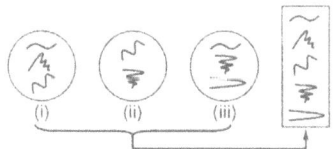

4. Close correlation

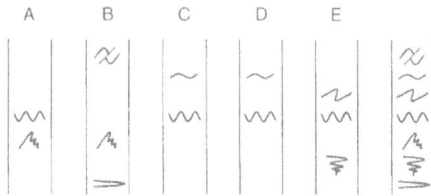

5. Nearby correlation (in relation to F_b)

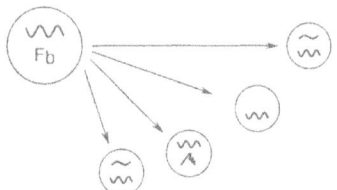

6. Distant correlation

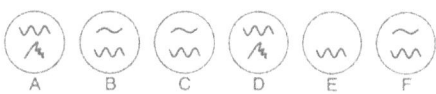

7. Correlation after identification of F_x

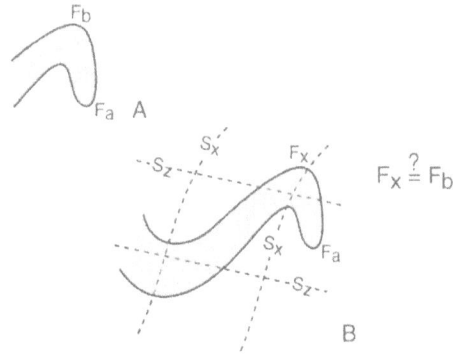

8. Distant correlation of successions

Figure 12.18   Chart summarizing the stages involved in correlating structures and structure successions beginning with the initial observation and recognition of structural overprinting at 1.

partial successions from one exposure to another.

In Figure 12.19a five different hypothetical fold sets are represented diagrammatically in order of overprinting at each of the four exposures (1–4), two at exposure 1, three at exposure 2 and two at exposures 3 and 4. The order of each exposure is based on observed refolding relationships and the fold sets are the only ones seen. They therefore comprise partial successions. At exposure 1 recumbent folds (F_r) are folded by inclined tight folds (F_i). At exposure 2, inclined tight folds (F_i) are folded by more open asymmetrical folds (F_a) and both are folded by upright warps (F_w). At exposure 3 upright open folds (F_o) are folded by upright warps (F_w). At exposure 4 asymmetrical folds (F_a) are folded by upright open folds (F_o).

Comparison of the fold relationships at all four exposures shows that F_r (see only at exposure 1) is clearly the earliest structure; F_i (seen at exposures 1 and 2) is certainly the second fold set; F_a, from exposures 2 and 4, is the third set; F_o, at exposures 3 and 4, is the fourth set; and the open warps (F_w), seen at exposures 2 and 3, is the latest fold set.

The result of correlating fold sets from one exposure to another (in columns 1–4) and the integration of the total observed succession (in the right hand column) is shown in Figure 12.19b. Although only four exposures are used here to illustrate the method, in practice, observations at several exposures are usually necessary to confirm the order of folding.

12.6.2 STRUCTURAL SUCCESSION FROM OVERPRINTED RELATIONSHIPS

The following procedure is the one normally adopted in establishing the structure succession from overprinted relationships observed on the outcrop at a series of localities. The objective is to:

A. Determine (i) the number of structures at each of the localities and (ii) their overprinted relationships. This can be done by

sketching each structure in succession, for example as in the boxes of Figure 12.19a. The boxed structures in Figure 12.20 show the relationships between just two structures that have been observed.

B. Place each structure in the correct position in the succession as was shown in Figure 12.19.

(a) Procedure

(1) Assume that the folds are more or less symmetrical, i.e. that the axial planes are equidistant between the fold limbs, and draw in the traces of the fold axial planes (Figure 12.21).

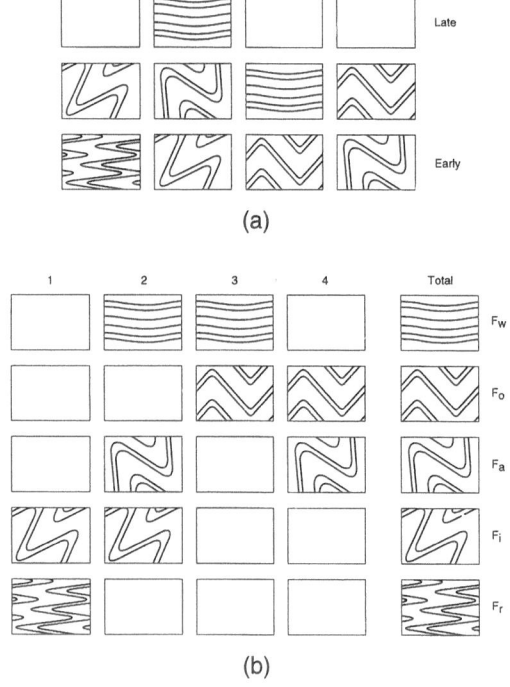

(a)

(b)

Figure 12.19 (a) Sketches representing local partial successions of a a hypothetical series of fold sets (F_r, F_i, F_a, F_o and F_w) observed at four different exposures (1–4). In (b) these are first correlated and placed in their correct positions in columns 1–4, and then integrated in the column on the right which represents the total observed local succession.

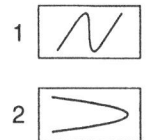

Figure 12.20 Sketch of two structures in order of succession, youngest at top.

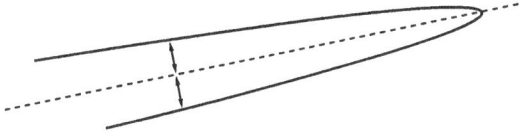

Figure 12.21 Sketch of a single (isoclinal) fold showing the trace of the axial plane approximately equidistant from the limb traces.

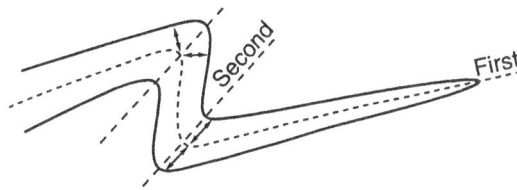

Figure 12.22 Sketch showing the overprinting relationships between two fold axial planar traces. The first is curved while the later is rectilinear, because it is undeformed.

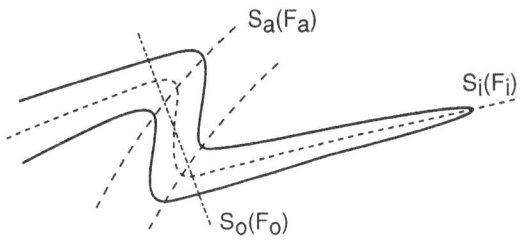

Figure 12.23 Sketch showing the refold relationships between three fold sets (F_i, F_a and F_o) showing decreased curvature of axial planar traces from S_i to S_a, then to S_o, which, being undeformed, is rectilinear.

(2) Thence determine which traces are deformed by which, i.e. which traces are later and which earlier, as in Figure 12.22.

(3) Where the structures are more complex than these, use the curvature of the simplest (nearest to rectilinear) axial traces rather than the folded 'foliation' of the limbs to determine the latest axial planar trace (Figure 12.23).

(b) Example

The photographs of Figure 12.24a show remnants of fold hinges in structures resulting from the relationships between five fold sets (F_i, F_t, F_a, F_r and F_o) at a locality in southern Karelia, Finland. There are two types of isoclinal folds, one with sharp, attenuated hinges and the other with round hinges. These have been affected by open folds with almost planar limbs and curved axial planes, and elsewhere at this locality, by small, angular asymmetrical folds, also with curved axial planes. Figure 12.24b shows sketches of folds from four exposures at the locality, together with four columns of empty boxes into which the folds identified at each exposure can be placed in order, as shown in Figure 12.19. Figure 12.24c shows that there are three sets of folds at exposures I, II and III and at least four sets at exposure IV.

Figure 12.24c(i) shows individual folds placed in order of overprinting for each locality. At exposure I it can be seen that F_r overprints F_a which overprints F_i. At exposure II F_r overprints F_t which overprints F_i. At exposure III F_o overprints F_r which overprints F_a, whereas at exposure IV overprinting relationships confirm what would already have been determined by integrating the relationships at exposures I, II and III, viz. that F_o overprints F_r which overprints F_a which in turn overprints F_t which overprints F_i.

Finally Figure 12.24c(ii) sets out individual structures not only in their correct **order** in the succession but also in their correct **positions** in the succession and proposes a tentative numerical order of succession. Figure

12.24c(ii) shows that the total observed partial succession (shown at locality IV) could in fact have been determined with certainty only from observations at exposures I–III. This is because the relationship between F_a and F_o is not absolutely clear at exposure IV. The succession determined corresponds to the partial succession of Figure 1.31.

12.6.3   INITIALLY UNCERTAIN RELATIONSHIPS

Often (in fact usually) it is the case that the relationships observed between structure sets are uncertain to begin with, as in the following stylized hypothetical example which shows the stages in resolving this uncertainty.

Sketches representing structure sets at four exposures A, B, C and D, recorded on four islands are shown in Figure 12.25. This case differs from the preceding ones in that, while the overprinting relationships between many of the sets are unambiguous, relationships between others are not, so that over-

all the relationships between the sets are less clear than they were in the previous examples.

A preliminary examination shows that at exposure A the succession includes four sets of structures whose order is clear (shown in square boxes labelled 1–4) and also two sets (a and b in circles) whose position is uncertain. At exposure B three sets (1, 2 and 3 in square boxes) show unequivocal overprinted relationships while the position of a fourth set (c) in the succession is uncertain. At exposure C two sets (1 and 2) show overprinted relationships clearly while the position of a third set (d) is unknown. At exposure D it is possible to establish the order of succession of three fold sets (1, 2 and 3) and this leaves a fourth set (e) whose position is in doubt. The relative positions of A_{1-4} with respect to B_{1-3}, to C_{1-2}, and to D_{1-2}, which can be determined on the basis of known relationships, are arranged accordingly with dashed 'tie lines' in Figure 12.25. This leaves five structures a, b, c, d and e whose positions remain to be resolved.

(a)

Figure 12.24   (a) Overprinting relationships in folded Svecofennian carbonates. **Left**, Tight fold showing gentle curvature (concave downwards) of its axial plane. **Right**, intrafolial isoclinal fold hinges folded by a sinistral asymmetrical fold (top arrow), angular folds (left and below pencil) and a sinistral asymmetrical fold (lower arrow). South side of Highway 186 *ca* 10 km NW of Highway 104 junction, west of Helsinki, Finland. (**b, c**) Procedure for establishing the structural succession based on overprinting relationships recorded from these rocks at four adjacent exposures I, II, III and IV.

(**b**) Empty boxes in columnar arrangement for each exposure. In the boxes of (**bi**) the structures would be placed in order of overprinting (as in (**a**) of Figure 5.18) and then they would be placed in their true relative position in the overall succession in the boxes of (**bii**), as in (**b**) of Figure 12.19.

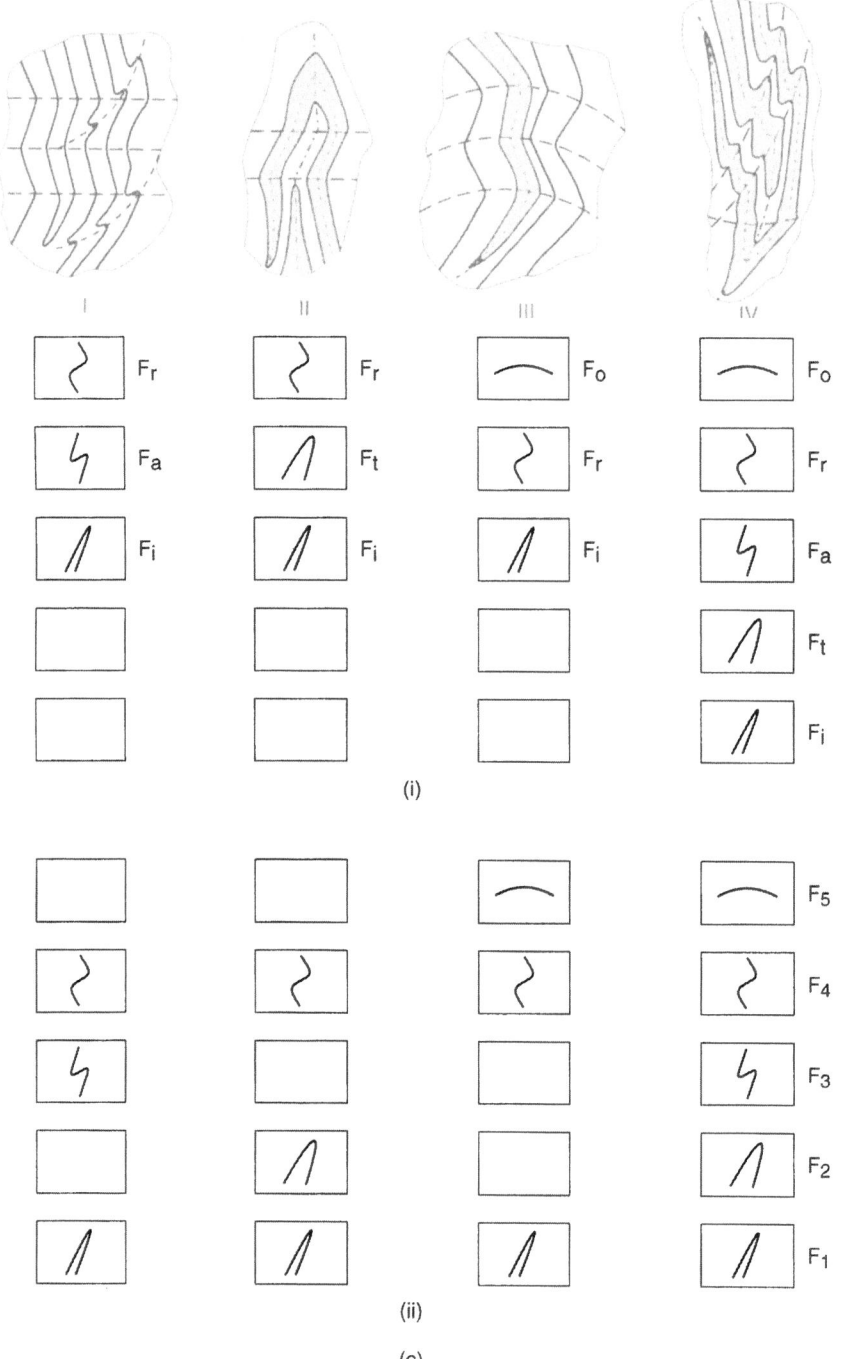

(i)

(ii)

(c)

(c) The steps (**b**i, **ii** above) carried out for the structures of (**a**). F_i = isoclinal, F_t = tight, F_a = asymmetrical, F_r = recumbent, F_o = open. (See pages 298 and 299 for Figures 24(b) and 24(c).)

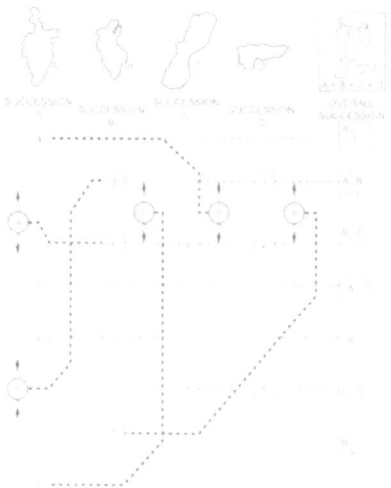

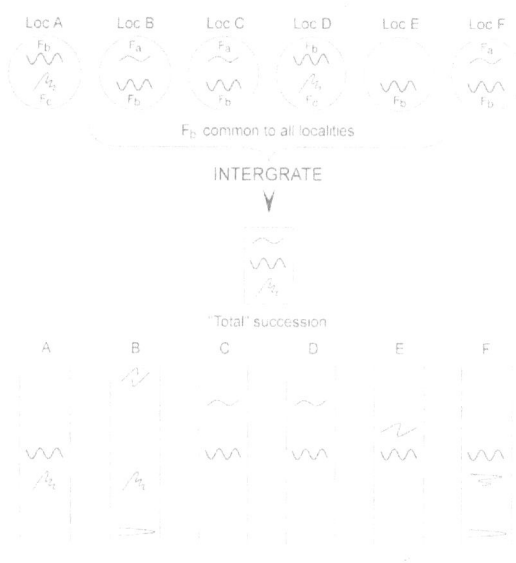

Figure 12.25 Sketch representing the stages in establishing the structural succession from known overprinted relationships between structures A_{1-4} with respect to B_{1-3}, to C_{1-2}, and to D_{1-2} (in square boxes) and unknown relationships between structures a, b, c, d and e (circled) on four islands A, B, C and D. The lighter dashed 'tie lines' link structures to the overall succession whose relationships are known and the heavier lines join those structures whose relationships need to be resolved.

Figure 12.26 Summary chart of the procedures involved in correlating single structures and partial successions. Correlation between **single** structures (F_b) among a number of localities (A–F) above leading to the integration of these to the 'total' succession compared with correlation between local **partial successions** (lower) from localities A–F.

Re-examination and close comparison of all structure sets reveals that the five whose positions could not be established at the exposures where they were first observed can in fact be identified with sets whose positions in the succession have been determined elsewhere, at one or more of the other exposures (Figure 12.25).

As a result it is found that:

$$a = B_2 = C_2 \text{ (also} = D_2),$$

i.e. it is equivalent to the second set observed on each of the islands (localities) B, C and D.

$$b = B_3 = D_3,$$

i.e. it is equivalent to the third set observed on each of the islands (localities) B and D.

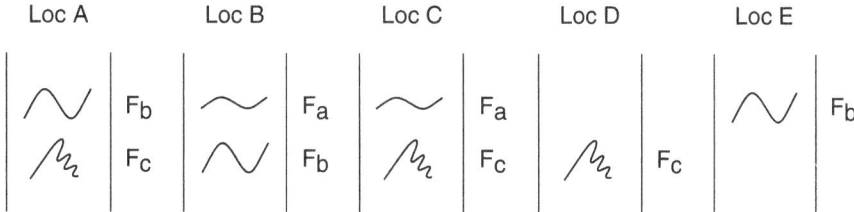

Figure 12.27 Representation of fold successions comprising relationships between fold sets F_a, F_b and F_c recorded at five localities, A–E.

$$c = A_1,$$

i.e. it is equivalent to the first set observed on island (locality) A.

$$d = A_4,$$

i.e. it is equivalent to set four on island A and

$$e = B_1,$$

i.e. it is equivalent to the first set observed on island (locality) B.

This allows the integration of the total local succession of eight sets of structures as shown in the 'overall succession' for the islands (A + B + C + D) in the right hand column, a fact confirmed by the consistent overprinted relationships between all the sets observed at other localities visited subsequently. **Again it must be stressed that the relative positions of the 'unknown' structures in the succession must ultimately be confirmed by their overprinting relationships**.

The correlation procedure discussed so far is summarized in Figure 12.26.

12.6.4   GAPS IN THE SUCCESSION

The foregoing examples illustrate the principle of assembling in the correct order structural information collected from several localities. However, at this stage one should be reminded that there is an important aspect of this procedure whose significance ought to be re-emphasized.

Consider again structures recorded at a number of localities, A–E say. The local structural successions determined for each of these localities on the basis of overprinting is shown in Figure 12.27.

These are simple successions with no more than two sets in each, and in the cases of D and E, only one set is shown. There is sufficient evidence to show that none of the successions is complete, and all are different. But the important factor is that, although none of the successions is complete, **none of the successions is inconsistent**. As can be seen,

F_a never lies below F_b and F_c, and F_b never lies below F_c in the successions. This factor is of fundamental importance when it comes to integrating the overall successions because **no matter whether or not one or more sets is missing from local successions the relationship between the sets is always consistently the same** and this provides the basis for their integration.

Take, for example, the two partial successions from localities I and II shown in Figure 12.28.

Set F_x lying above F_y can, together with F_y, be correlated between localities I and II and because the partial successions show relationships that are consistent, the two can be integrated to a composite (I + II) succession as shown in Figure 12.29.

This principle applies to the integration of any number of partial successions observed in a structurally homogenous terrane, whatever the number of sets recorded (from 2 to n) **provided their relationships are consistent** (Figure 12.30). This means that, no matter how complete or incomplete they are, local successions can be correlated so as to build up the overall known succession.

The lack of appreciation of the principles just discussed by those unfamiliar with this type of correlation is often expressed by their doubt, or even disbelief, that correlation can be effective or reliable, particularly in those cases where the effects of strain partitioning are likely to be significant. Because of this widespread uncertainty it is worth referring back to section 8.1.9 where the relevance to structural correlation of the effects of strain partitioning were considered in greater detail.

12.6.5   SUMMARY OF STAGES IN THE INTEGRATION OF STRUCTURAL SUCCESSIONS

An overall succession is built up through a series of steps beginning with observations of overprinted relationships at say single

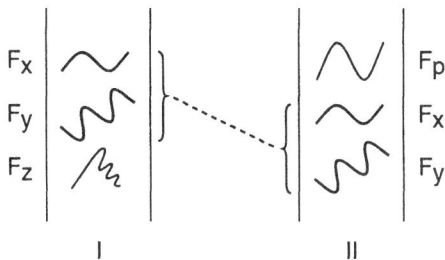

Figure 12.28 Comparison between partial successions comprising the relationships between fold sets F_p, F_x, F_y and F_z at localities I and II.

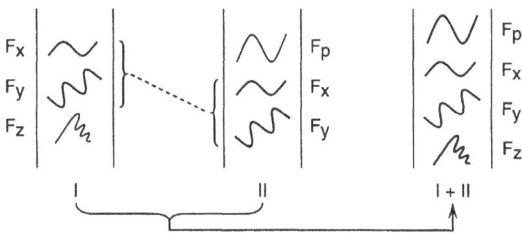

Figure 12.29 Integration of the partial successions comprising relationships between F_p, F_x, F_y and F_z from localities I and II shown in Figure 12.28.

exposures to give local **partial** successions. Two or more local partial successions are integrated to provide a **composite** succession, and all the composite successions together comprise the **'overall'** succession, the complete **known** succession (Figure 12.31).

For an illustration of the principle of integration of local partial successions to form the 'total' known succession see section 13.4.3, 'Structural succession from Lake Baikal, Siberia'. Four local sub-successions recorded from localities at the southern end of the lake are represented by four columns in the table of Figure 13.17 and the integration of these is shown in a fifth column, the overall **known** succession for the study area.

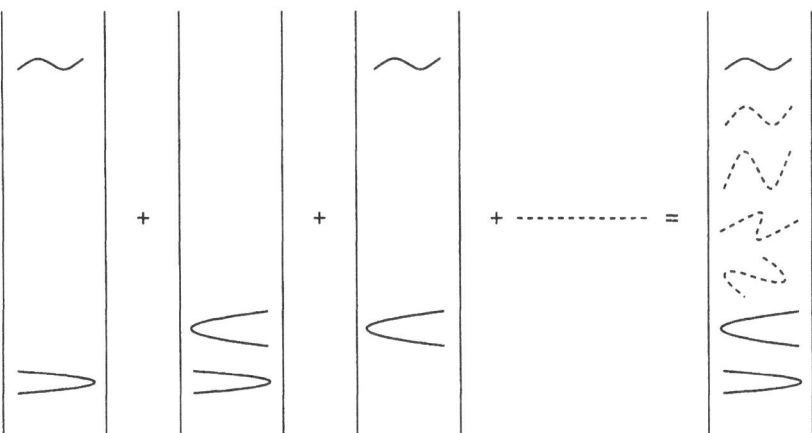

Figure 12.30 Integration of related partial successions from several localities. Although none is complete (implied missing structures are shown as dashed folds) the order in each is consistent, not only with that of the other localities, but also with that of the known overall succession shown on the right.

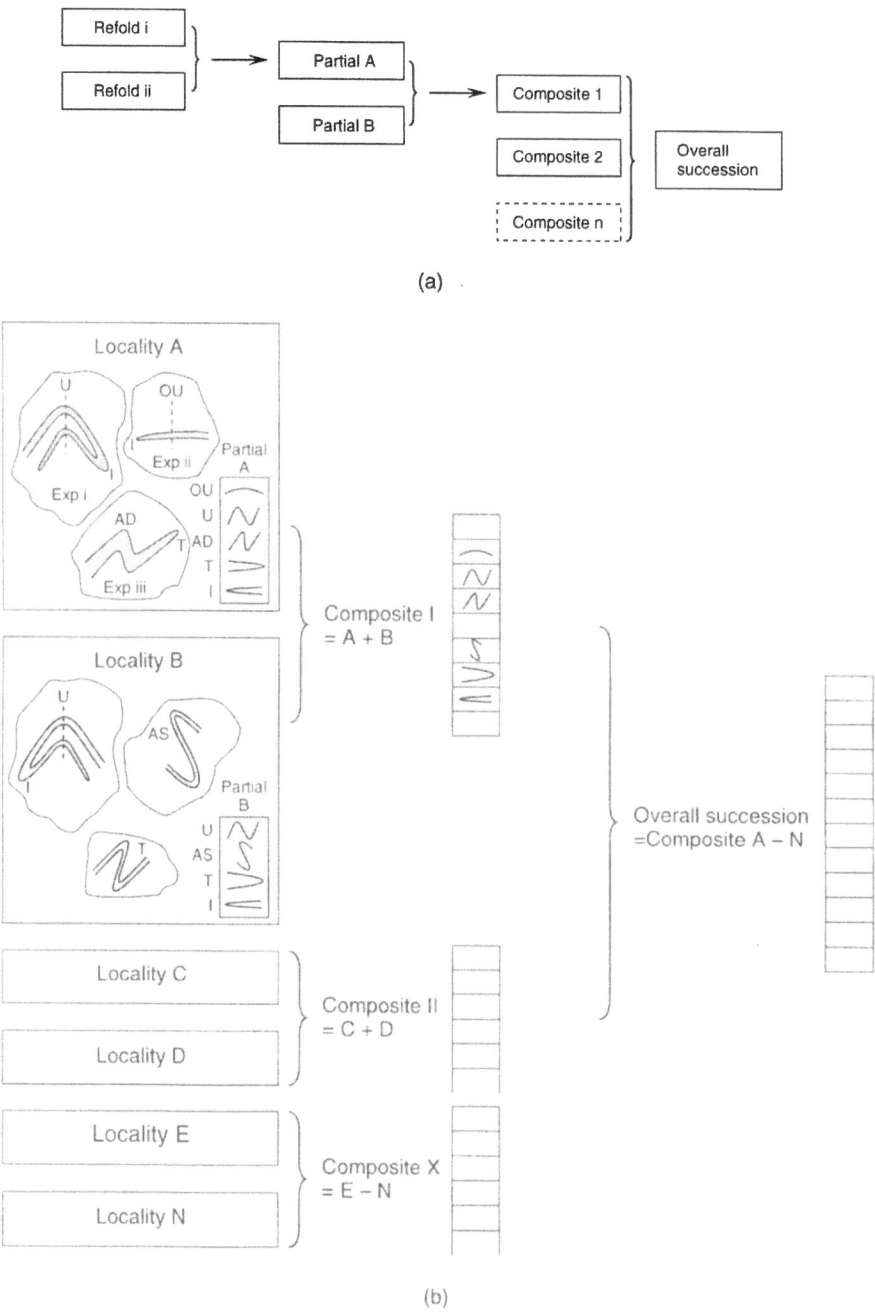

(a)

(b)

Figure 12.31 Summary chart of the steps involved in establishing the structural succession for a study area. (**a**) Outline of the stages shown schematically in (**b**). At the left, the refold relationships recognized on different exposures at a particular locality can be integrated to form a **partial succession** for that locality. In the centre, the combination of any number of partial successions (A, B, C, ... N) gives a **composite succession**. On the right, the integration of all composite (or partial) successions (I, II III, ... X) provides the **overall** (known) **succession** (1–N).

ANALYTICAL PROCEDURE SUMMARIZED

13

13.1 INTRODUCTION

The procedures for determining structural successions that were discussed in earlier chapters are formalized and summarized in this chapter, and three structural successions determined from complexly deformed migmatites in different terranes are presented as illustrations of what can be achieved by applying these procedures.

13.2 CORRELATING LOCAL SUB-SUCCESSIONS: FORMAL PROCEDURE

As we have seen, so long as the overprinted relationships at each exposure are reasonably clear, partial successions can be integrated to produce a total observed succession. Correlation, the prerequisite to integration, is illustrated by Figure 13.1 which shows correlation of structures between the partial structural successions observed at six localities, I to VI. These are the only structural relationships seen at each of the localities and each represents only a small part of the whole of the structural succession in the complex.

However, as discussed previously (section 12.6.3), not every structure observed shows clear overprinting relationships and the integration of the total succession is not always straightforward (see also section 8.1.9). Nevertheless, provided the approach is systematic and the procedure is followed in a step-by-step fashion, integration of even a long and complex structural succession should not present an insurmountable problem.

Consider now the overprinting relationships shown between six sets of structures at a number of exposures at six localities A–F. These are represented in outline in Figure 13.2a. At localities A, B and D overprinted structural relationships are shown at three exposures (i–iii). At localities C and E observations have been recorded at each of two exposures (i–ii) and at locality F overprinted relationships have been observed at only one exposure (i).

The results of preliminary attempts to integrate the partial structural successions at each locality, largely on the basis of geometry and style (assuming the latest folds to be more open and more upright etc.), are shown in the right hand side of the figure (Figure 13.2b) but in all cases there is ambiguity between the positions of some of the sets. Part of this ambiguity, that at localities A and B, can be resolved by referring to the succession at locality F where overprinted relationships, not seen at A and B, between the two structures ($\vee\wedge\wedge$ and $\sqrt{}$) have been observed (Figure 13.2). This increases the precision of the correlation and, while maintaining broad consistency between the relationships of all structures, it allows the succession at A to be rationalized to that shown in Figure 13.3a, with the exact positions of two structures ($\wedge\!\!\!\wedge$ and $\rightleftharpoons$) remaining uncertain, and the succession at B to be restricted to that shown in Figure 13.3b, with the exact site of one structure ($\rightleftharpoons$), which must lie somewhere

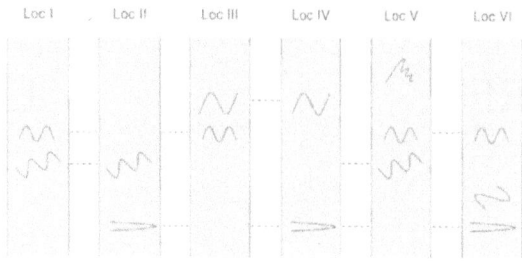

Figure 13.1   Correlation of the six partial successions recorded from localities **I–VI**.

between the lowest and highest structure, still uncertain. Comparison between Figure 13.3a and 13.3b shows that the succession for A and B can be restricted to that shown in Figure 13.3c leaving the position of only one fold set (viz. ⧩) uncertain but nevertheless constrained between ⋀ and ⟹.

Overprinting relationships at locality C do not provide any further information but exposure (i) at locality D shows that ⧩ lies below ⋁⋀ further limiting its position to between ⋁⋀ and ⟹ (Figure 13.3d). At locality E this ambiguity is finally resolved by the structural relationship shown at exposure (ii). Here it can be seen that ⧩ precedes ⋀ and the order of succession of all of the six structures observed has now been determined (Figure 13.3e) with the provisional approximate integration of Figure 13.2 now modified to a succession (Figure 13.3e) whose order is **consistent in terms of all observed overprinting**.

Although the basic procedure remains the same, the approach to integrating the total observed succession can vary. In the example just discussed the integration of the 'whole' succession was carried out **after** the integration of all the local partial successions but it can also be done concurrently as a kind of 'running total'. With increasing experience the second approach is likely to be adopted as a

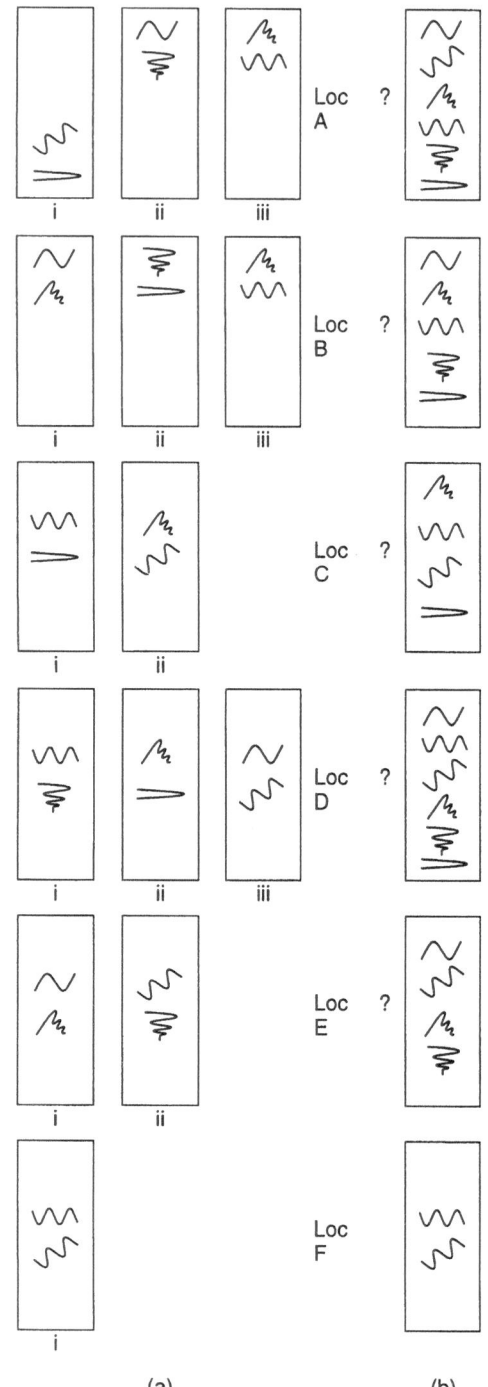

(a)   (b)

Figure 13.2   (a) Structural relationships observed at exposures (**i–iii**) at locations A to F integrated to form the sub-successions for each locality (**b**).

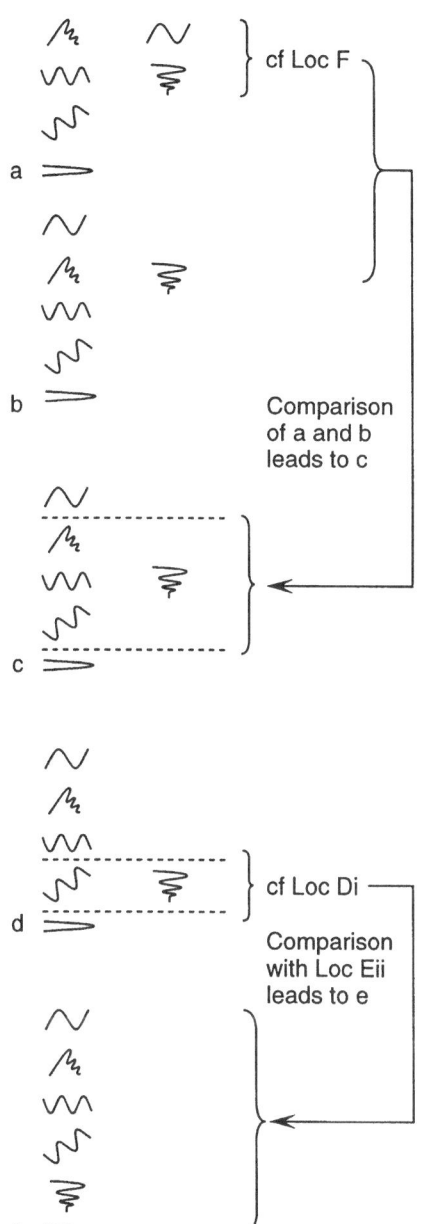

Figure 13.3 Integration of local sub-successions **a–e** to form the 'overall' succession.

matter of course, at least at some stages of most integration exercises.

A variation of the procedure is shown in Figure 13.4 where structural relationships at three localities A, B and C, with overprinting from two exposures at locality B, are set out. The later structures are placed above those they overprint.

At locality B the refold relationships from exposures (i) and (ii) can be rationalized to the succession shown because one structure (⌒𝓛) is common to both exposures. Comparison of this succession with the relationships at locality A, which has the same structure in common with locality B, allows the arrangement shown for localities A and B. The positions of ⇒ and ⌒𝓛 are unambiguous but at this stage the relative positions of ⋀ and ⌒ cannot be resolved without additional information. The necessary information is provided by the overprinting relationships at locality C where ⌒ is seen in relation to ⋀ which is common to A and B, and this allows the succession for A, B and C to be determined unambiguously (lower right).

An extension of the method used above is illustrated in Figures 13.5, 1–5, where the overprinting relationships of six fold sets from six localities A–F are represented in stylized form. The structures and structural relationships have been observed at one or more exposures (i–iii) at each location and are shown in boxes in their correct order (earliest at the base), having been determined on the basis of overprinting. At localities A and D, overprinting relationships are shown for three separate exposures, at localities B, C and E, observations recorded in each case from two exposures are shown, and at locality F overprinting relationships between two sets of folds are recorded from only one exposure.

The structures and their relationships on the five exposures at localities A and B are com-

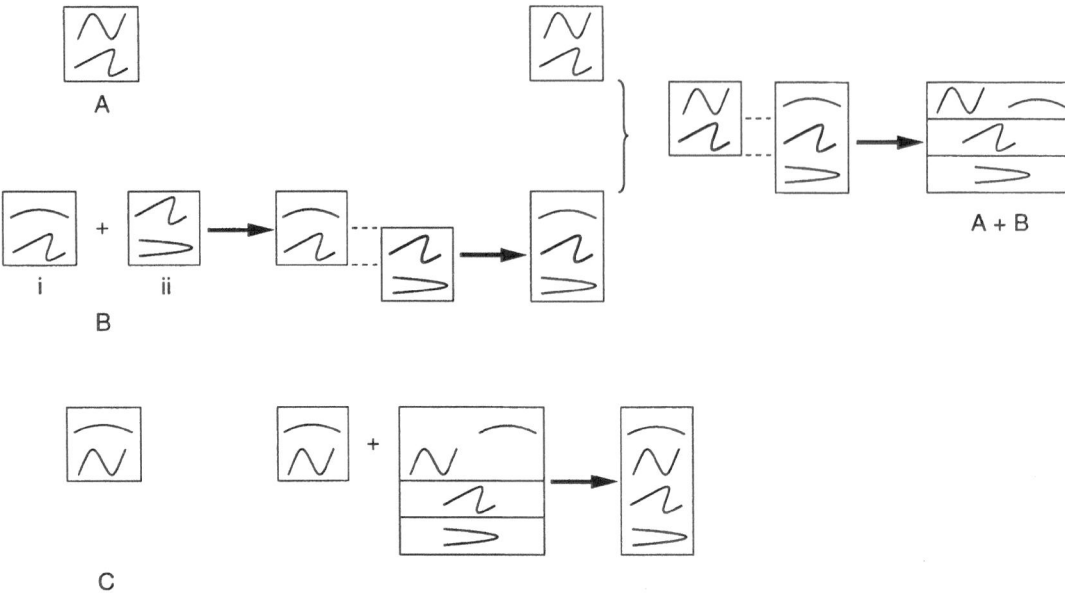

Figure 13.4 Integration of structural relationships to form a succession based on overprinted relationships at three localities, A, B and C. The data at B are combined and rationalized and then added to the data from A. The data from A and B are similarly combined and rationalized to form A + B and these data are then combined with those from C and rationalized to form the succession shown in the column at lower right which represents the structural relationships from the three localities.

pared and, where possible, fold sets are correlated (Figure 13.5, 1a, 1b). The comparison demonstrates the unambiguous overprinting relationships between one pair of fold sets, and two groups of three fold sets. The correlation is shown in Figure 13.5, 1b with like sets matched; A(ii) and B(ii) (with set ⤳ common) overlap (Figure 13.5, 1c. A(iii) and B(i) have set ⤳ in common and consequently overlap. As a result the five boxes (exposures) become three (Figure 13.5, 1c, upper) and because set ⌒ is common to both the centre and right-hand boxes, and set ⟵ is common to both the centre and left-hand boxes, the three boxes can be combined as shown in the lower part of Figure 13.5, 1c, to show all that is known so far about the overprinted relationships between the six sets

using the information from localities A and B. At this stage, set ⌒ is known to be later than sets ⤳, ⤳, ⋀ and ⟵ but its relationship to set ⤳ but its relationship to set ⤳ is not yet known. Set ⟵ is known to be earlier than sets ⤳ and ⤳ but its relationship to sets ⤳ and ⋀ is not yet known.

The information acquired at localities A and B is added to observations made on the two exposures at locality C and comparison is made between the different sets of structures (Figure 13.5, 2b). With the relationship between ⟵ and ⋀ now established and now also that between ⌒ and ⟍, the succession recorded so far can be integrated as shown in Figure 13.5, 2c, showing

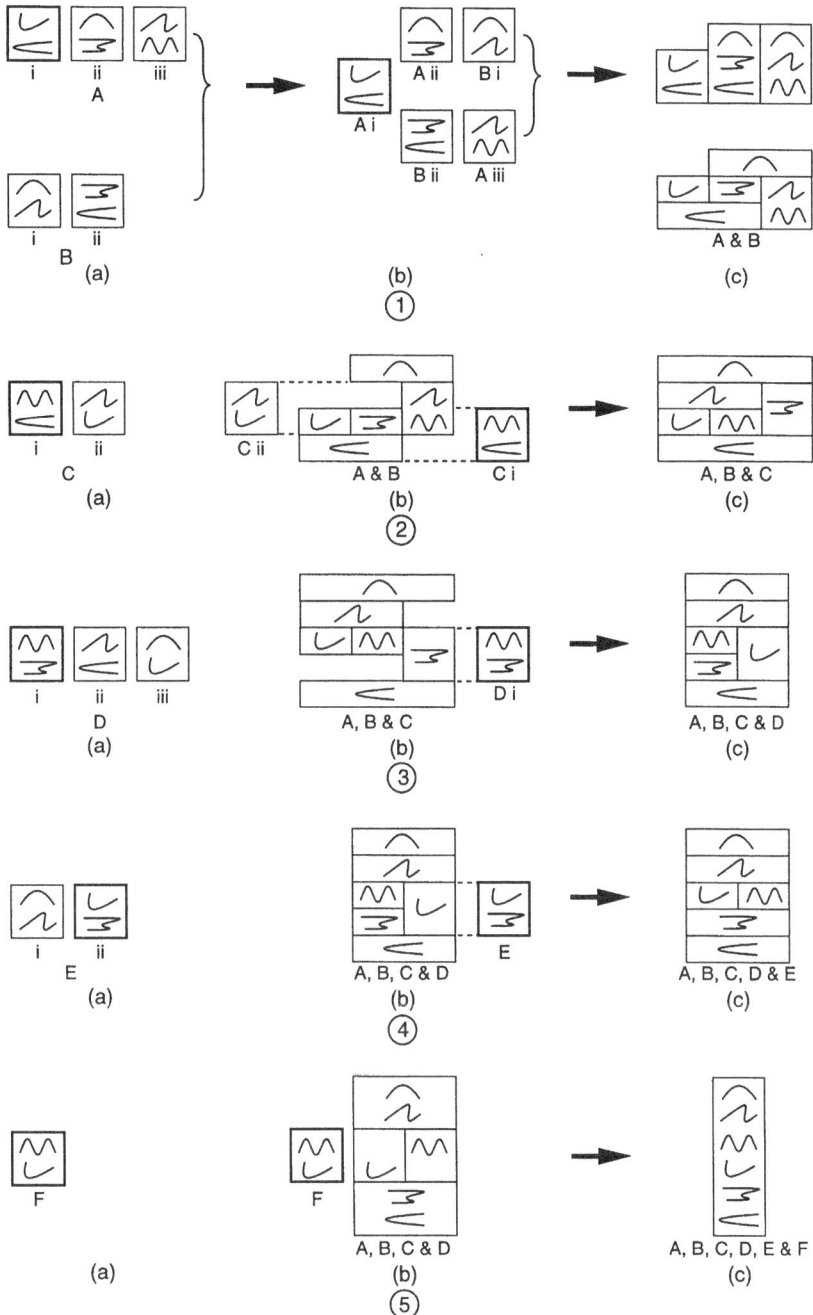

Figure 13.5 Integration of structural relationships from six localities A, B, C, D, E and F, in three steps from (a) to (c), beginning by combining the data from localities A and B. The combined data are then added to those from C and rationalized in (2c) and the data recorded from subsequently observed localities D to F are combined successively in the same way to form the overall structural succession shown in the column at lower right.

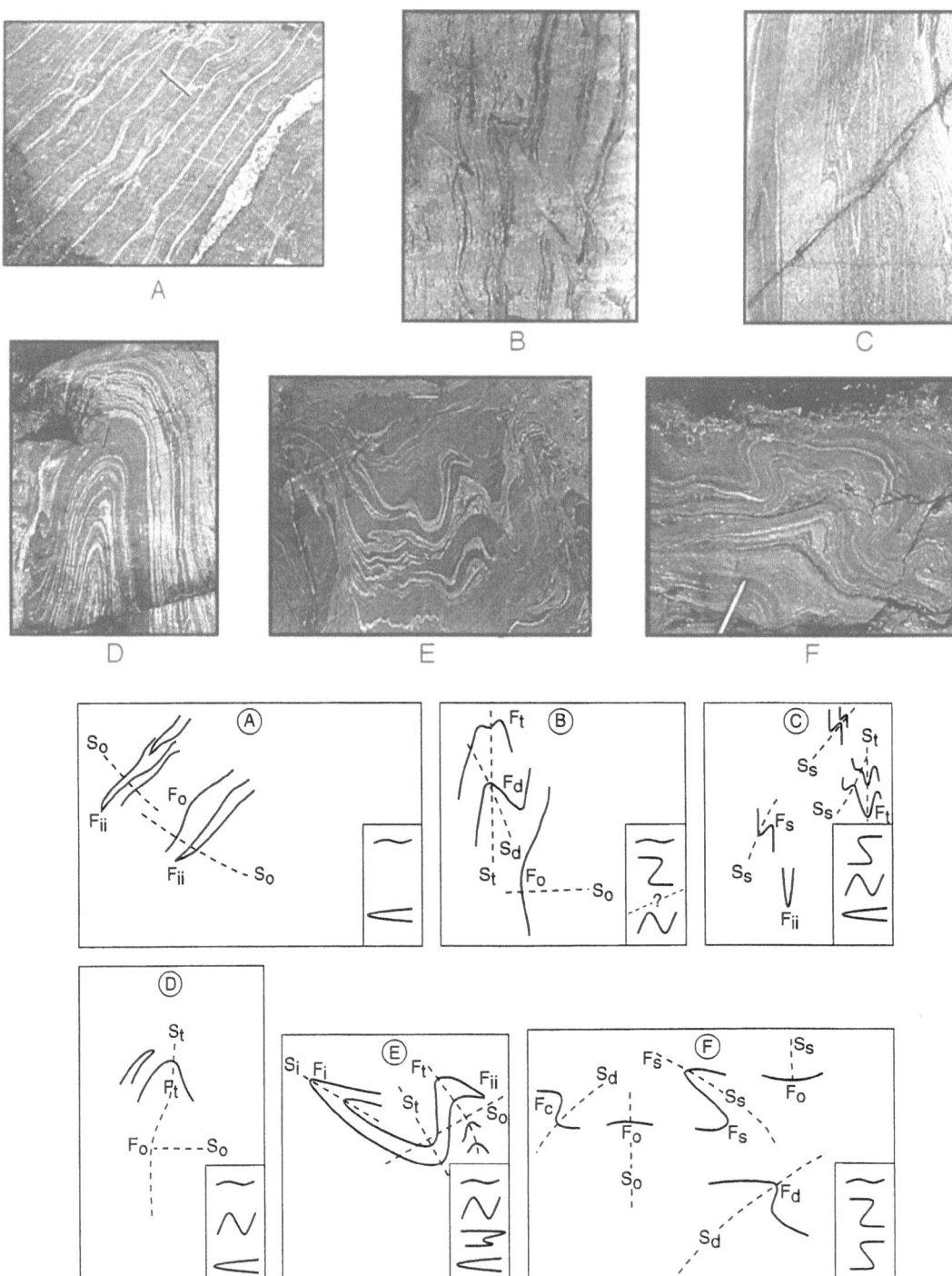

Figure 13.6   Photographs from six localities (A–F) of refolded folds, with corresponding sketches of the photographs identifying the fold relationships and showing the local sub-successions.

that ⌒ is the youngest of the sets observed and ⇐ the oldest.

The procedure is followed through successive localities (Figures 13.5, 3; 13.5, 4) until such time as the relationship between all known fold sets has been determined (Figure 13.5, 5). Note that while this succession **appears** to differ (in terms of sets three, four and five) from that established provisionally at the outset when evidence was used only from locations A and B, there is no inconsistency in the relative order of the structures in each case.

Initially, pending the discovery of further overprinting evidence, one might attempt a provisional order of succession based on the correct relationships shown at locations A and B, integrated in terms of geometry (with more open folds later in the succession) and axial planar attitude (with more upright axial planes later), and this would in fact be as shown in Figure 13.5, the succession ultimately derived in terms of overprinting relationships. See the discussion on this approach in section 13.3.1 and compare this with the procedure discussed with Figures 13.2 and 13.3.

13.3 ILLUSTRATIONS OF PROCEDURE

The application of this procedure to actual fold structures is illustrated in Figure 13.6. Here, photographs of structural relationships at half a dozen localities A–F are shown. The structures and their relationships to one another are identified in the sketches of the photographs and the local sub-successions determined on the basis of overprinting relationships are shown for each. The key to the symbols used for the structures is shown in Figure 13.7.

At locality A, F_o (⌒) clearly affects F_{ii} (⇐) which it post-dates. At locality B while F_t (⋁) clearly precedes F_o (⌒), the relationship between F_d (⟋) and the other folds sets is certainly not shown unambiguously on the exposure. At locality C the

relationships between F_{ii} (⇐), the earliest, F_t (⋀) and F_s (⟅), the latest, are clear. At locality D, F_{ii} (⇐) precedes F_t (⋀) which is followed by F_o (⌒) and at locality E the relationships between the sets are F_{ii} (⇐), the earliest, followed by F_i (⟋), F_t (⋀) and F_o (⌒), the latest, in that order. Finally at locality F the relationships between F_s (⟅), the earliest, F_d (⟍) and F_o (⌒), the latest, are shown.

The integration of the local sub-successions for localities A–F to establish the complete succession of structures recognized at the six localities is summarized in Figure 13.7 using the same procedure as that already discussed in relation to Figure 13.5.

13.3.1 USE OF GEOMETRY FOR PLACING FOLDS IN PROVISIONAL ORDER OF SUCCESSION

At the outset, in those cases where no clear grouping into sets is immediately obvious, and before the finer details of the structure are sought as an aid to identifying the folds of particular sets, the general shape (and to some extent the attitude) of folds can be used as an indication of their relative ages. Such might be the case where in a complex structure it is not possible to subdivide the succession on the basis of a 'key structure' (see below). This, it must be stressed, is only an interim working hypothesis on which to begin the study. It is subject to verification and continual review and has as its basis the fact that folds, whatever their initial shape and attitude, continue to be modified by subsequent deformation (see section 7.4). As deformation progresses, the flattening of early folds means that they tend to become tighter (with inter-limb angles tending towards zero), and sometimes intrafolial. Therefore, as a rule, early folds are unlikely to be open and upright structures. Overprinting

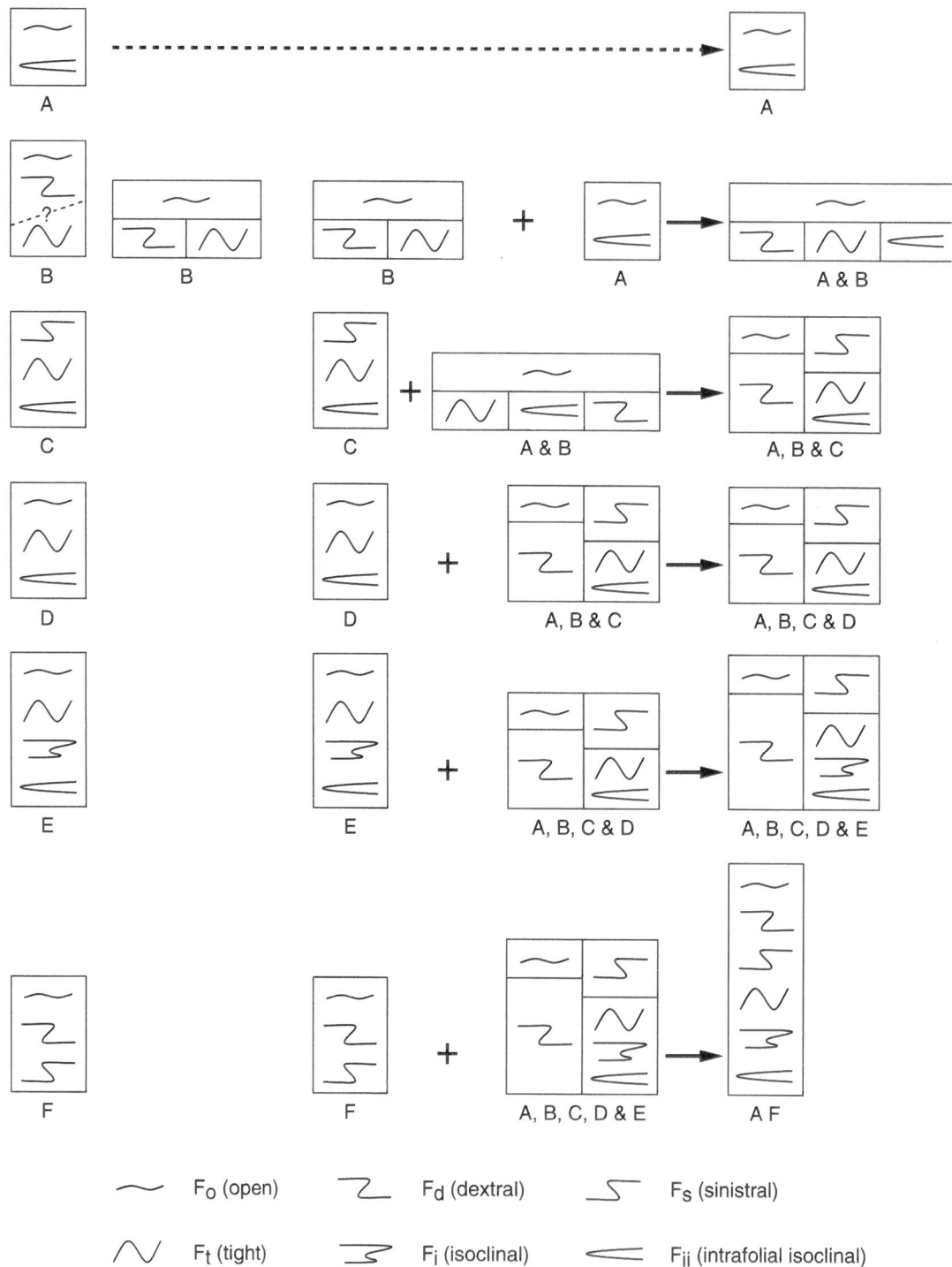

Figure 13.7   Integration of the structural relationships recorded from localities A to F (Figure 13.6). The partial successions at each locality are shown on the right. These are successively combined to form the overall observed structural succession (AF) at lower right.

(refolding) relationships between different tight and isoclinal folds (whether or not their axes are parallel) are usually more easily recognized than are overprinting relationships between more open structures (see Figures 10.1 and 10.2 in section 10.1.2), although this depends on some extent on the attitude of the exposure surface with respect to that of the folds being considered (Figure 13.8).

If in the first instance, following reconnaissance, it has not been possible to single out a set of folds of distinctive style consistently exposed over the whole area of study, i.e. a 'key structure', then it will be necessary to endeavour to establish the relationships of all fold sets on the basis of refolding throughout the whole succession. Overprinting relationships between individual folds will have to be determined, and on the basis of these relationships, the folds systematically integrated into sets (see Figures 12.19–12.24 and 13.2–13.7).

If that approach is unsuccessful, an attempt could be made to relate folds on the basis of intuition, using fold geometry as a guide in order to set up a provisional order of succession. However, the existence of these 'hypothetical' fold sets must always be confirmed or disproved in terms of the presence or absence of consistent refold relationships. In practice

this would mean that fold sets might have to be 'identified' in terms of groups with like characteristics such as axial planar mineral associations, leucocratic veining, hinge shape, limb shape and length, and any other distinctive features (orientation, except in certain circumstances, is unlikely to have any special significance).

As a very broad generalization, the geometrical variation in a succession of folds which tends to develop in highly deformed terranes subjected to polyphase folding is similar to that shown in Table 13.1 and Figure 13.9. Although the shape of a fold can be a useful indication of its approximate position in the succession, the table should be used as a guide only in the initial stages of a study until firm evidence is accumulated to confirm any provisional classification.

When the folds of a succession are combined on a single compound structure, such as those shown schematically in Figure 13.10 (Bowes and Hopgood, 1975), this gives some indication of the intricacy of the overall geometrical pattern. From this it can be seen that it is advisable to avoid attempting to work on outcrops exposing very complicated structural relationships, particularly in the initial stages of the study, because geometry is unlikely to be a

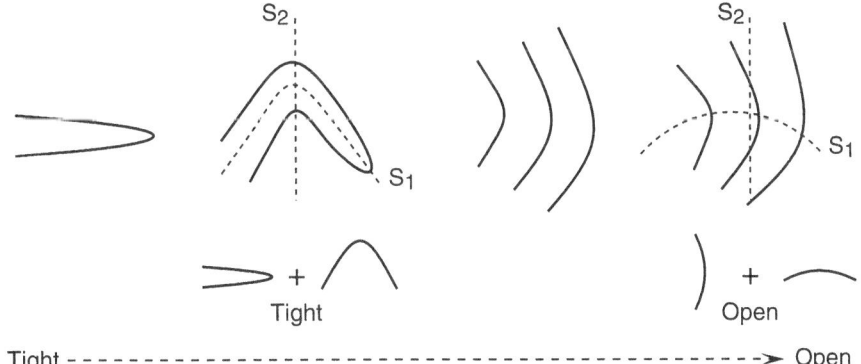

Figure 13.8 Sketches to show variation in the ease of recognition of overprinted relationships between tight structures (relatively easy) and between open structures (relatively difficult). The effect of tight folding superimposed on isoclinal folds (axial traces S_1 and S_2 on the left) is easily recognized, whereas the effect of open folding superimposed on an existing open fold (axial traces S_1 and S_2 on the right) is more difficult to identify.

Table 13.1 Potential range of fold types in a succession (cf. Fleuty, 1964).

Fold	Axial planar attitude
7. Very open warps	Upright (vertical axial plane)
6. Very open folds	Very steeply inclined to upright
5. Open inclined/upright	Steeply inclined to upright
4. Open asymmetrical	Steeply inclined
3. Tight asymmetrical	Gently inclined
2. Isoclinal	Parallel to foliation
1. Isoclinal intrafolial	Parallel to and within foliation

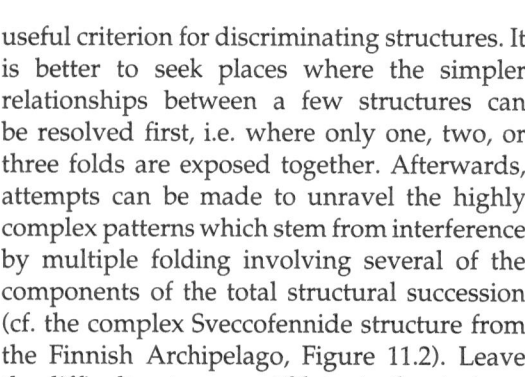

Figure 13.9 Stylized sketches representing the fold forms described in Table 13.1.

useful criterion for discriminating structures. It is better to seek places where the simpler relationships between a few structures can be resolved first, i.e. where only one, two, or three folds are exposed together. Afterwards, attempts can be made to unravel the highly complex patterns which stem from interference by multiple folding involving several of the components of the total structural succession (cf. the complex Sveccofennide structure from the Finnish Archipelago, Figure 11.2). Leave the difficult outcrops until later in the study.

Remember that ultimately the validity of grouping structures on the basis of geometrical and stylistic features would still have to be tested to see if the folds within each group consistently show the same relationships to the folds of other groups or sets – groups whose identity has also been tested or is subject to testing in a similar fashion. For example, if the existence of three fold sets F_a, F_x and F_i say, is proposed, this must be confirmed by establishing their overprinting relationships. This will entail determining whether or not folds of set F_a, say (or their axial planes or their associated lineation) are refolded by folds of set F_x, or vice versa, and whether comparable relationships can be recognized between the folds or fold elements of sets F_a and F_i or between F_x and F_i etc. The necessity for this proof cannot be avoided.

Finally, to restate what was said at the very beginning, although some of the methods discussed are exacting, and although the initial impression might be that some are even tedious, practice and experience will eventually lead to their being used with little more conscious effort than that required by the experienced commuter in following a complicated yet familiar city route to work. In time, the rules for the resolution of structural complexity should grow to be instinctive and their successful application will become close to 'second nature'.

13.3.2 USING AGMATITES TO IDENTIFY STRUCTURAL RELATIONSHIPS

Before proceeding to consider some examples of field cases where the procedure has been applied, it is important to keep in mind that the same approach is adopted in determining the succession of structures whether the overprinting relationships are amongst fold structures or between folds (or other structures) and intrusive bodies. Therefore the order of succession of sets of folds, especially in migmatites, can sometimes be established in other ways, e.g. in terms of their relationships to partial melts (neosomes). This is illustrated by the relationships shown in Figure 13.11. Here migmatites deformed by fold F_b

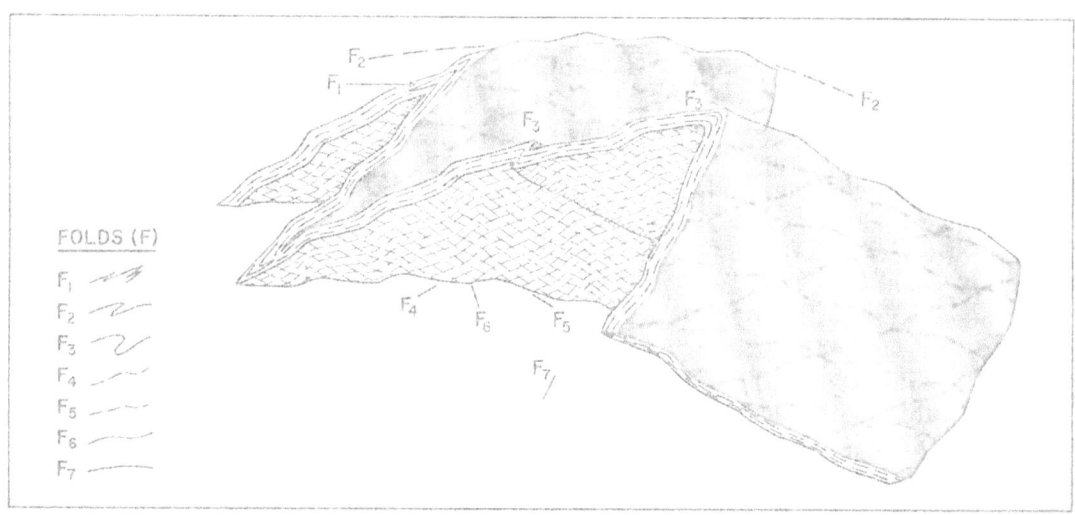

Figure 13.10 Schematic representation of seven overprinting sets of folds (F_1–F_7). Lewisian complex, Inishtrahull, Donegal, Eire. (Bowes and Hopgood, 1975, Figure 1b. Reproduced with the permission of The Royal Irish Academy).

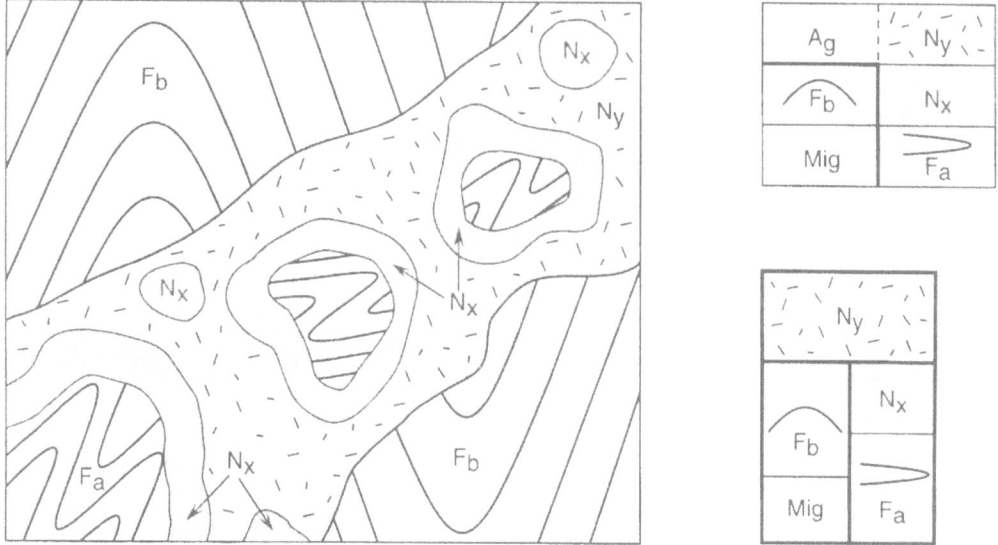

Figure 13.11 Sketch showing folds (F_b) in migmatite cut by a leucocratic neosome vein (N_y) containing palaeosome blocks comprising earlier neosome (N_x) enclosing earlier palaeosome folds (F_a). The grouped 'boxes' on the right summarize the relationships of the structures in the sketch. The top group shows the structure in the agmatite in boxes separated by heavy lines from the succession in the folded migmatite. The lower group shows the known relationships between all the structures, migmatite structures on the left and agmatite on the right. N_x is known to be the latest structure but the relationships between F_a and F_b cannot be established without further information. Compare Figures 4.38, 5.20, 5.37 and 5.38.

are cut by an agmatite vein comprising leucocratic neosome N_y containing palaeosome
blocks comprising an earlier neosome N_x
enclosing folds F_a. In this case, complete resolution of the succession is not possible because
the relationships to F_b of both fold F_a and
neosome N_x remain ambiguous, i.e. there is no
evidence to show with certainty whether, for
example, F_a and F_b are the same or not. More
information is needed to establish these relationships unequivocally. Such information
might be provided by the discovery that both
F_a and F_b are associated with a distinctive
syntectonic mineral assemblage (see section
5.2 and especially section 5.2.5) in which case
F_a and F_b belong to the same set, or it might be
found that F_a and F_b each have some distinctly
different associated feature, in which case
they belong to separate sets.

Note: It should be noted that while in some of
the examples shown, the resolution of half a
dozen fold sets has resulted from observations
at only half a dozen localities, the determination of the relationships between this number of sets in reality is normally likely to entail
the examination of exposures at a great many
more localities. It would be an unusual situation where as few as six localities were needed
to establish a structural succession of this
length.

13.4 APPLICATION OF THE PROCEDURE – EXAMPLES FROM DIFFERENT SETTINGS

To illustrate the application in the field of the
methods explained in the preceding chapters,
three examples of structural successions established in different settings are presented here,
the first from the Baltic Shield, the second from
China and the third from eastern Siberia.

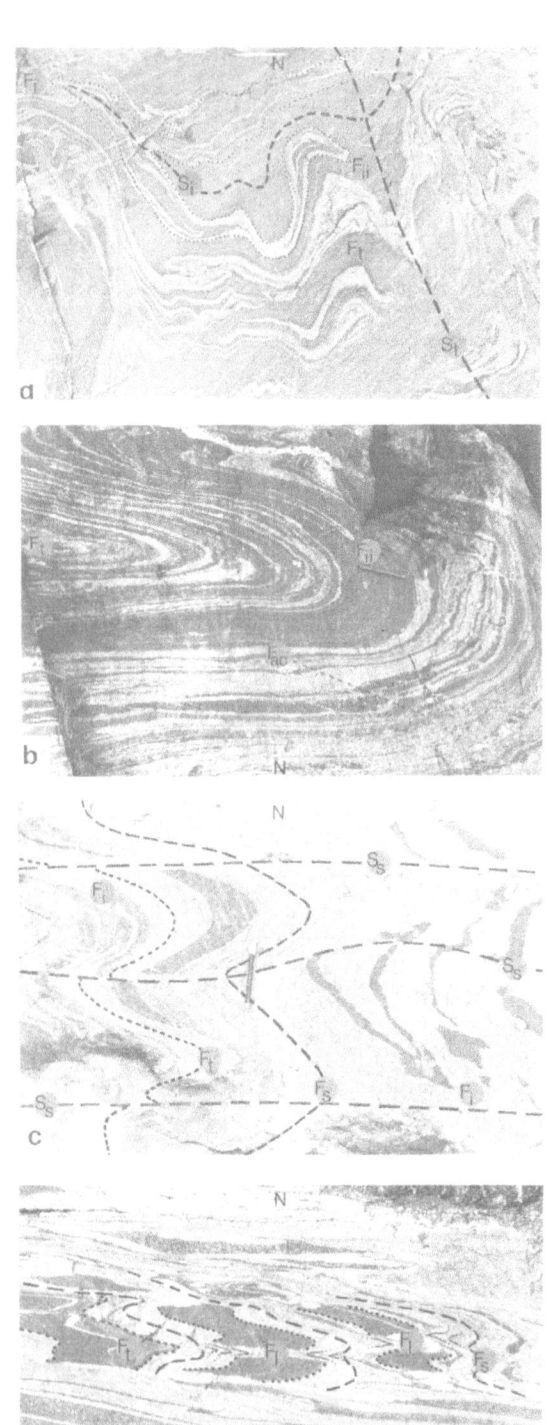

Figure 13.12 Photographs (**a–i**) of structures and
structural relationships in Svecofennian migmatites
in the Jussarö region, southern Finland. For an
explanation of these see the text. From Figure 19,
Hopgood 1980. Reproduced with the permission of
the Royal Society of Edinburgh. ▶

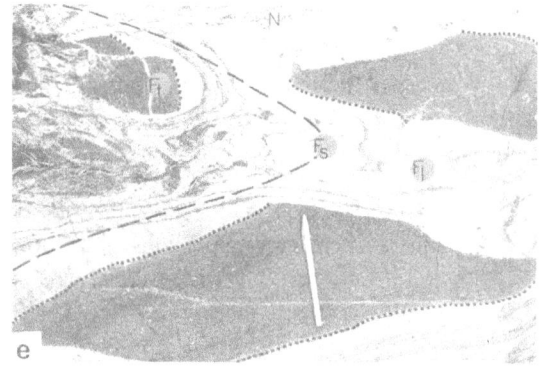

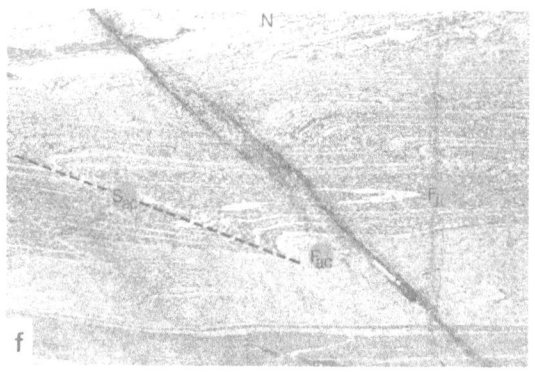

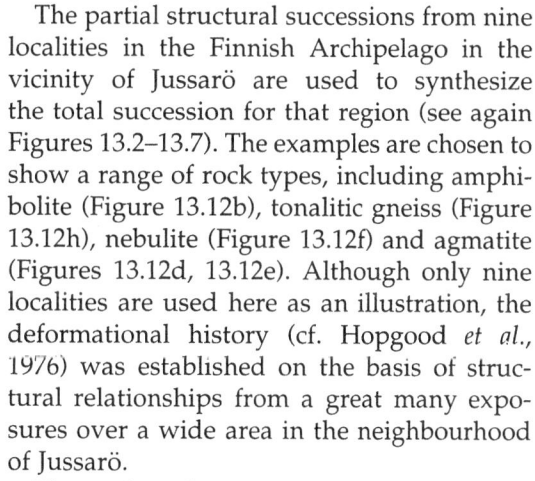

13.4.1 STRUCTURAL SUCCESSION IN THE FINNISH ARCHIPELAGO

The structural succession (Figure 13.12) described in Svecofennian migmatites (Hopgood, 1984) is from the area of Sederholm's classic studies in the early 1900s in southern Finland (Sederholm, 1967). The exposure is extensive and of extremely good quality as can be seen from a number of illustrations in the figures of earlier chapters.

The partial structural successions from nine localities in the Finnish Archipelago in the vicinity of Jussarö are used to synthesize the total succession for that region (see again Figures 13.2–13.7). The examples are chosen to show a range of rock types, including amphibolite (Figure 13.12b), tonalitic gneiss (Figure 13.12h), nebulite (Figure 13.12f) and agmatite (Figures 13.12d, 13.12e). Although only nine localities are used here as an illustration, the deformational history (cf. Hopgood *et al.*, 1976) was established on the basis of structural relationships from a great many exposures over a wide area in the neighbourhood of Jussarö.

The earliest banding, commonly accentuated by concordant quartzofeldspathic veins, is seen to have been folded at least twice, forming what are now isoclinal folds (Figure 13.12a), and these two sets of isoclinal folds (F_{ii} and F_i) were refolded (F_t) by a set of now tight structures (Figure 13.12a, 13.12b) whose axial planes are now parallel to a strongly developed E–W foliation. Following F_t folding,

Figure 13.13 Sketches of folds forming part of the Svecofennian structural succession in the Jussarö area, southern Finland.

still, the foliated matrix was squeezed between the agmatite blocks to produce folds (F_1) with opposed hinges analogous to scar folds, around boudins (Figure 13.12e); these tend to disrupt the continuity of the earlier F_s folds (Figure 13.12d, 13.12e) and accentuate an increasingly strongly developed foliation parallel to the axial planes of F_t and F_s (Figure 13.12d). This foliation was then folded by two sets of asymmetrical folds (Figures 13.12f, 13.12g), the first with an anticlockwise rotation sense (F_{ac}) and a later set with a clockwise rotational sense (F_r). Later in the deformational sequence the foliation was deformed by more open folds (F_o) exhibiting a tendency towards shear on conjugate surfaces sub-parallel to the fold limbs, with associated concentration of quartzofeldspathic material, shown dotted in Figure 13.12h. The latest deformation recognized in the sequence resulted in broad open warps (F_w) (Figure 13.12j) affecting the orientation of all the earlier structures. The total succession is summarized as simplified profiles in column form in Figure 13.13.

It was also possible to determine the positions in the structural succession of several igneous bodies emplaced within the Svecofennian complex, so providing a rigorous framework for their isotopic dating. These include early volcanogenic rocks, mafic pegmatoid masses and later quartzofeldspathic veins, aplites, coarse muscovite-bearing pegmatites and a later discordant micro granite. This allowed not only the determination of the time intervals between some of the deformational episodes which contributed to the structural succession, but also the establishment of the time span of the Svecofennian Orogenic Episode and, as a result, an indication of the rates at which Proterozoic tectonic processes took place (Hopgood *et al.*, 1983).

13.4.2 STRUCTURAL SUCCESSION IN THE DABIE COMPLEX, CHINA

The succession just described from the Jussarö region of the Finnish Archipelago was estab-

the development of a leucocratic neosome resulted in agmatization of the basic amphibolite palaeosome (Figures 13.12c, 13.12d, 13.12e), leaving isolated folds (in amphibolite) of earlier generations 'floating' in a leucocratic matrix. Subsequent deformation, the result of slip on discrete surfaces (S_s), more or less parallel to the axial planes of F_t folds, caused further disruption of the structure and produced folds composed of deformed foliated matrix and modified F_t folds (Figure 13.12c). Later

(a)

(b)

Figure 13.14 (a) View of the terrain in Hubei Province, China, showing the quality of outcrop in Archaean migmatites. (b) Exposures typical of the area close to the dam at Feng Huang Guan where the structural succession in the Dabie complex was determined. Hubei Province, China.

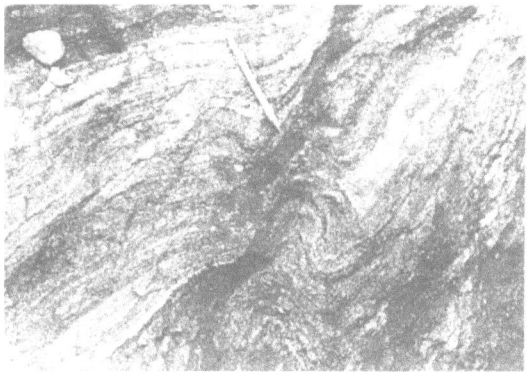

Figure 13.15 Photograph of F_{ene}, the key structure used in the establishment of the structural succession in the Dabie complex at the Feng Huang Guan Dam, Hubei Province, China (see table 13.2) showing its distinctive lobate hinge.

lished from several localities in an area of exceptionally good exposure. However, it is also possible to determine complex successions where the number of exposures is restricted to only a few in a limited area of inland outcrop, and indeed where these are not necessarily of particularly good quality. The exposure shown in Figures 13.14a and 13.14b is typical of the kind used to establish the extensive local structural succession (at least 20 structural features) in Archaean migmatites of the Dabie complex in Hubei

Province, China (Hopgood *et al.*, 1989). Here, as in other cases where the structural succession is extensive, the process of determining the succession was aided considerably by the recognition of a structure suitable for use as a key structure, in this case the set of distinctive folds designated F_{ene} (Figure 13.15).

With F_{ene} as a key set (Figure 13.15), the structural succession was built up in the field in the following manner. As each of the structures described in the succession were recognized they were classed initially as belonging to one of three groups, viz., **earlier** than F_{ene} (i.e. overprinted by F_{ene}), **equivalent** to F_{ene} (e.g. F_{ene} neosome – N_{ene}) or **later** than F_{ene} (i.e. overprinting F_{ene}). Once this was done, the interrelationships between folds of the earlier sets and those between the later sets were determined by further, more detailed examination of refolding and overprinting relationships involving fold axial planes, cleavage, lineation, veining etc. to produce a succession of more than 20 structures. The position of the key structure (F_{ene}) in the succession is shown in Table 13.2.

The structural succession established for the Dabie complex provides the potential for

Table 13.2   Structures in the Dabie complex, China

$$N_{ap}$$
$$S_i$$
$$S_m$$
$$F_w, S_w$$
$$N_{nne}$$
$$N_g$$
$$\mathbf{F_{ene}} \text{ (key structure)}, S_{ene}, N_{ene}$$
$$F_n, S_n, L_n, N_n$$
$$N_t$$
$$F_s$$
$$F_b$$
$$F_a$$
$$F_{iv}, S_{iv}, N_{iv}$$
$$S_c$$
$$F_{ib}, S_{ib}, N_{ib}$$
$$F_{ii}, S_{ii}$$
$$S_t, N_t$$
$$S_o$$

The structures are arranged in order of succession with the youngest at the top. F = fold, S = foliation, L = lineation, N = neosome.

undertaking further petrological work and the structural framework necessary for controlled isotopic study comparable to that carried out for the Svecofennian complex, referred to above.

13.4.3   STRUCTURAL SUCCESSION AT LAKE BAIKAL, SIBERIA

As a further example of a succession determined by employing the procedures discussed earlier, but this time using observations recorded over a much wider area than in the previous case, the results are presented of a study undertaken in the Precambrian Sharyzhalgay complex exposed at the southern end of Lake Baikal, Eastern Siberia. Observations were made at a series of exposures in railway cuttings along the north shore of the southwestern arm of the lake between Kultuk and Port Baikal, south of Irkutsk.

The methods employed are the same as those used in the previous example but in this case the succession was determined from the integration of sub-successions (see Figure

12.31 in section 12.6.5) from several separate sites extending over a much wider distance. Here advantage was taken of the almost continuous exposure in cuttings along a former stretch of the Trans-Siberian Railway to collect data and determine local sub-successions at intervals of several kilometres for comparison and to establish a 'total' overall succession. The localities were visited using a Soviet Academy of Sciences research vessel operating from Port Baikal (Figure 13.16).

The study showed the structure to be the result of deformation in two different tectonic regimes, so providing the basis for isotopic study and the determination of the ages of the complex (Aftalion *et al.*, 1991). This allowed the subdivision of the complex in this area into Proterozoic and Archaean units. It also made possible the establishment of the timing of charnockitization in the complex and the demonstration of the fact that it was structurally controlled. Furthermore it enabled the

Figure 13.16   Railway cutting on the branch line (formerly of the Trans-Siberian railway) along the lake shore, viewed from a Soviet Academy of Sciences research vessel, *Professor Treskov*. This shows the quality of the exposure from which the observations of structural relationships were made and which formed the basis for determining structural successions. The complex structure seen in the photograph comprises large, inclined isoclinal folds, open upright and inclined folds, asymmetrical folds and shears dipping to the left. Sharyzhalgay complex, north shore, southwestern Lake Baikal, Eastern Siberia.

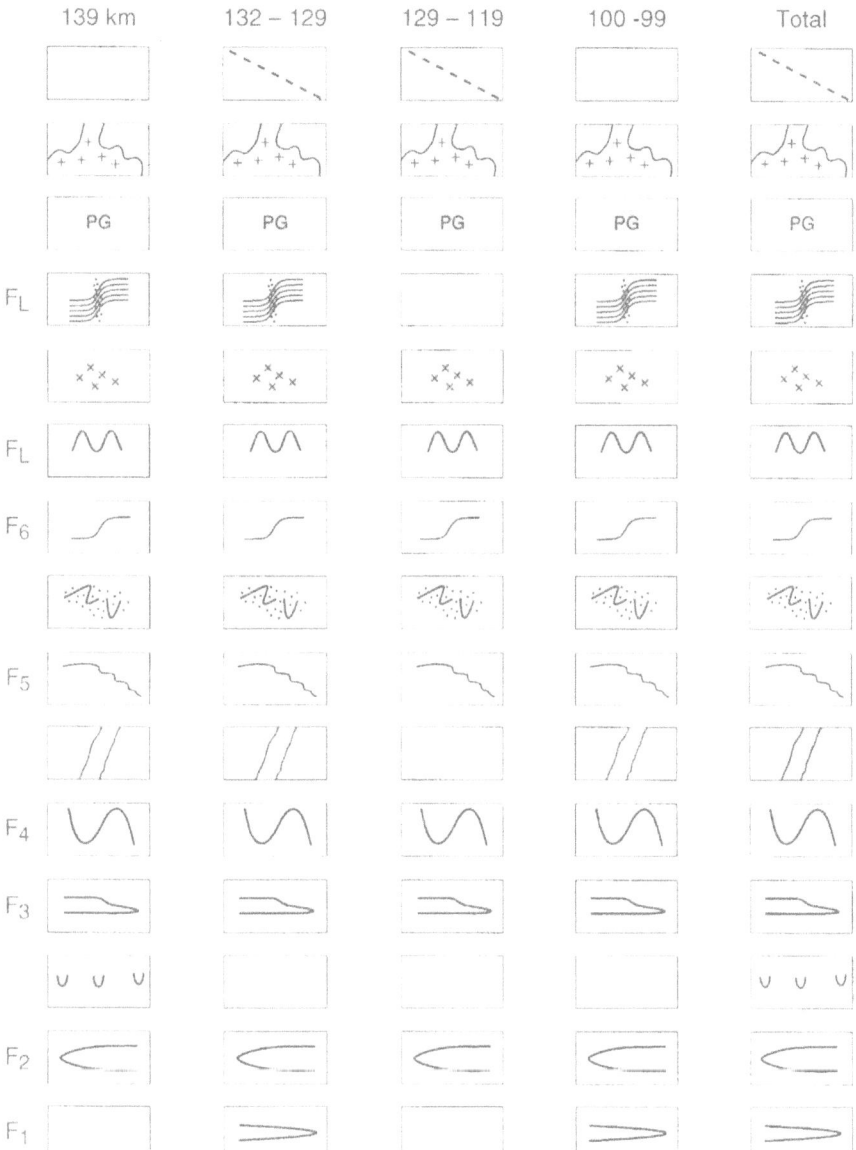

Figure 13.17   Structural successions recorded from the Sharyzhalgay complex at four localities along the west shore at the southern end of Lake Baikal, Eastern Siberia. The 'total' succession integrated from these four localities is shown in the right-hand column. Figures at the top of the columns show distances from Irkutsk along a disused spur of the Trans-Siberian Railway. Fold sets are labelled from F_1 through F_6 to F_L (= F_{Late}) and various sets of intrusions (granitoid, charnockitic and granulitic (PG = pyroxene granulite, U = ultramafic)) whose positions in the succession have been ascertained are represented in the columns by different ornament. The dashed lines at the top of some columns represent dislocations. (Based on Figure 11, Hopgood and Bowes, 1990.)

structures that controlled this to be identified (Hopgood and Bowes, 1990).

Using the techniques previously discussed here, local successions were determined and recorded from several cuttings along a 40km traverse along the shore. Those showing the fullest successions, from exposures along four stretches ranging between 1km and 10km in length, are presented in Figure 13.17, together with the 'total' succession integrated from the four sub-successions. The table summarizes the information used in the interpretation of the geological history and includes deformational structures (folds and dislocations) and shows the relative positions of emplacement of ultramafic, charnockitic and granitic material in the successions, some of which was used in the isotopic dating of the complex.

14.1 ACADEMIC APPLICATIONS

In addition to being used to determine structural successions and, by extension, deformational sequences, and thus providing a basis for establishing the evolutionary history of migmatites and related rocks, the methods discussed have a number of other broader applications and these are summarized at the end of this chapter.

Although structural relationships in rocks such as gneisses and migmatites which formed at the elevated temperatures and pressures (often close to melting conditions) associated with deformation at deep crustal levels are usually exceptionally complex, resolution of this complexity is essential to allow controlled isotopic age-dating of crustal complexes and the identification of the processes responsible for crustal development as well as their timing. As has been said earlier, it is paradoxically the very complexity of such rocks which holds the key to the realization of their significance because their structure is unique and characteristic, **and** resolvable. However, where such structural work is ignored, the results of expensive and time-consuming isotopic studies can be ambiguous or not meaningful in terms of earth history.

Understanding of the structural relationships of the rock units ensures the rigorous control essential in selecting key units for isotopic dating. In collaborative research with geochronologists the method has provided the basis for determining the time spans of Precambrian and later orogenic episodes. This is because it provides the means of determining the order of events contributing to the structural successions (or frameworks) even in complexly deformed rocks. This not only affords the opportunity for geochronologists to date specific events during the operation of deep crustal processes, but also permits the determination of time intervals during orogenesis, as well as giving the rates at which processes took place. By using the methods discussed here, it was established that the length of time of an orogenic episode relatively early in earth history was similar to that of episodes later in the geological record, and this disproved the presumption of major differences in the nature of processes operative during at least the last 2,000 Ma – so lending some support to the application of 'uniformitarian' plate tectonic theories. Furthermore, the establishment of the relationship of structures to metamorphic and igneous events provides a succession whose complexity is such as to make it characteristic (if not unique), so providing a powerful tool for correlation with other previously related, but now dismembered and separated, crustal segments. This approach represents the deep crustal aspect of terrane correlation and provides another line of evidence to be used, together with isotopic evidence, in continental reconstruction such as that involved in reassembling Gondwanaland.

There are other, more specific applications of the methods. One is where knowledge of the structural succession could be used in the investigation of 'post-basement' structural relationships, something that has implications for neotectonics studies. Such an example is that where, as a result of structural study in Uganda, the regular spatial variation of the attitude of an identifiable ('key') structure (F_4) in migmatites adjacent to the Western Rift Valley led to the discovery of a probable link between variation in structural orientation to basement warping associated with rifting. The axial plunge of the fourth fold set identified in the succession shows dispersion in a vertical great circle girdle perpendicular to the NNE trend of the Rift wall (Figure 14.1). Such regular distribution of orientation data is characteristic of structures late in the succession (compare Figures 2.11 and 5.15). But, while the parallelism of the axial direction of the dispersion with the trend of the Rift might possibly be entirely fortuitous, the fact that the elliptical spread of the F_4 maximum within the girdle is to some extent mirrored by that of the later structure, F_5 (which similarly shows

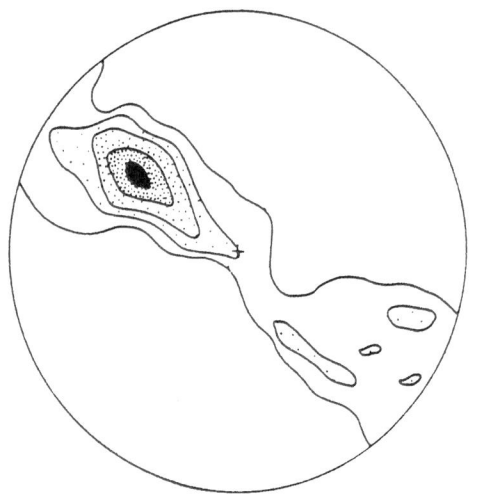

Figure 14.1 Contoured stereo plot showing the distribution of the axial plunge of a fold structure late in the succession in migmatites adjacent to the east wall of the Western Rift Valley, Ankole District, Uganda. The dispersion suggests reorientation about an axis parallel to the Rift wall and by implication, a response to crustal warping associated with the rifting. (After Figure 10, Hopgood, 1970).

dispersion of a few degrees about an axis parallel to the Rift wall) prompts the suggestion that reorientation of F_4 is not entirely the result of overprinting by folds later in the succession but could include a component related to the crustal upwarp associated with the rifting (Hopgood, 1970, Figure 10). This implies that further, more extensive study of the structural relationships, specifically aimed at determining whether or not a link can be established between crustal warping and associated regular reorientation of basement structures, would be rewarding.

Although application of the methods described in the previous chapters have been developed as a result of work on deep crustal rocks of high metamorphic grade and Precambrian age, and their application tends to have been concentrated predominantly on establishing structural relationships in these areas, it should be noted that they can be, and have been, used to study the structure of a wide range of rocks. These include middle and upper crustal rocks of lower metamorphic grade and various ages. The methods have been applied to studies in the following settings. Archaean rocks: Southeast China, Eire, West Greenland, Northwest Scotland, West Australia (Wheatbelt). Early Proterozoic rocks: Finland, Siberia. Late Proterozoic rocks: Czech Republic, southern Western Australia. Precambrian rocks (undifferentiated): Czech Republic, India, Northern Rhodesia (Zimbabwe), Uganda. Palaeozoic rocks: Czech Republic, Poland, Scotland. Mesozoic rocks: New Caledonia, New Zealand.

14.2 ECONOMIC APPLICATIONS

As well as leading to an understanding of the processes, time intervals and physical conditions that existed during the development of the deep crust, the methods can provide an insight into the development and shape of highly deformed ore bodies, including those in Precambrian terranes. The ability to unravel the complex structural relationships which constitute an integral part of the character of migmatites and some gneisses is critical to the understanding of the form of deformed ore bodies in such high-grade metamorphic terranes and is an essential precondition to the successful exploration and efficient exploitation of such bodies.

When integrated with recently-developed, very low level, airborne remote sensing techniques, the methods allow recognition of the structural configuration of concealed outcrops in regions where exposure is severely limited by lakes, forest and glacial drift deposits. This is particularly relevant in forest areas (often remote) in Scandinavia and North America.

14.3 SUMMARY

Some of the applications are listed below.

1. Determination of the order of events contributing to the structural successions (or frameworks) even in complexly deformed rocks.
2. Use of the characteristics of the structural successions so identified to correlate between separated rock units with comparable deformational histories including fragments of dismembered supercontinents. Determination of the order of deformational events in relation to metamorphic and igneous events provides a succession whose complexity is such as to make it characteristic (if not unique), so providing a powerful tool for correlation with other previously-related, but now dismembered and separated, crustal segments.
3. Reliably identifying the relative positions in the structural succession of rock units suitable for isotopic dating, so providing the means to date specific events during the operation of deep crustal processes.
4. Providing the basis for determining the time spans of Precambrian and later orogenic episodes.
5. Determining the time intervals during orogenic episodes.
6. Determining the rates at which processes took place.
7. Comparing the lengths of orogenic episodes early in earth history and those later in the geological record (see Hopgood *et al.*, 1983).
8. Understanding the processes, time intervals and physical conditions that existed during the development of the deep crust.
9. Study of the effects of deformation at deep crustal levels at elevated temperatures and pressures, often close to melting conditions.
10. Providing an insight into the development and shape of highly deformed ore bodies, including those in Precambrian terranes.
11. Unravelling the complex structural relationships which constitute an integral part of the character of migmatitic gneissic hosts for ore bodies.
12. Integration of the method with recently-developed, very low level, airborne remote sensing techniques allows recognition of the structural configuration of concealed outcrops in high latitudes where exposure is limited.

15.1 CONCLUDING REMARKS

The correlation of structures and their assignment to structure sets is an integral part of a process that embodies a series of closely interrelated steps involving continual modification and refinement of observations made in the field. These entail repeated checking of the significance of characteristic features, as well as comparison of individual structures and their overprinting relationships with structures belonging to other sets. As this nearly always involves comparison of structures between separate exposures it is important to have as much information about the structures as possible. This means that **every** aspect of the structure, attitude and style, as well as overprinting relationships, needs to be taken into consideration. The ease with which correlation between structures can be effected depends on the number of structural features they possess and how easily recognizable they are. If there are several and if they are distinctive, then there is less likelihood of ambiguity. A structure with particular distinctive characteristics which is widely developed is suitable for use as a datum or reference structure – what has been called a 'key structure' here – (see section 6.3).

Fortunately one is seldom forced to rely solely on a single structural feature as a means of identifying folds of a particular set. Even in such a case, the identification of those particular folds will be (indeed must be) supported by their consistent relationship to other structures, i.e. in terms of the ultimate criterion for distinguishing fold sets, viz. that of **overprinting**. In fact, even in such extreme cases, **overprinting** of two identified fold sets means that even though each fold of two sets is distinguishable in the field in terms of only a single feature, each always bears the same relationship to the other. In effect the correlation is based not only on this one identifying feature, but on **two** consistently interrelated factors, viz. (i) all folds of set F_A have feature 'A', while folds of the other set have not and (ii) folds with feature 'A' are always overprinted by those of set F_B.

Sometimes, in those cases where there is an element of ambiguity and structures of a particular set cannot be recognized in the field, their identity can be resolved only after thin section examination. For example, two sets of similar structures that are apparently identical on outcrop scale in the field may have distinctive micro-textural differences. Having separated the two sets on the basis of micro-textural evidence in the laboratory it can happen that consistent **mesoscopic** differences previously not recognized then become apparent. This new information can then be used with confidence to distinguish the two structures so that in future they can be recognized in the field with the need for laboratory examination. An example of such a difference might be the association with one fold set of a fine lineation related to preferred mineral elongation. Experiences of this kind tend to sharpen the perception and increase the confidence of the observer. With further experience, mesoscopic features of folds thought to be insignificant in

Figure 15.1 Similar-looking intrafolial isoclinal folds belonging to two sets. Although both have parallel axial planes, the two sets can be distinguished because of the differing attitudes of their axes and also because the hinges of the earlier set (F_1) are sharp (top centre arrow) whereas F_2 have round hinges (upper right arrow). At the outcrop shown, the hinge of one set is sub-horizontal and the other plunges steeply. Lewisian complex, Rona, Inner Hebrides, Scotland.

the initial stages of the study (merely minor structural variations in folds belonging to the same set) can be separated, if not consistently, at least in most cases. Observational perception is particularly important in cases where different fold sets have parallel or nearly parallel axial planes. Sometimes the critical distinguishing features, overlooked initially, are shown to be quite distinctive once they have been recognized. An example of such a case is the divergence in hinge attitude of the two iso-

clinal folds shown in Figure 15.1. This shows intrafolial folds in quartzites of the Lewisian complex of the Isle of Rona, Scotland. There are two sets of folds, not obviously different on cursory inspection, but seen to have consistently different axial plunges and, on closer examination, distinctly different hinges. The hinges of one set are sharp while the hinges of the other set are round.

Experience has shown repeatedly that, in any work where observation plays an important part, once a particular feature has been identified it tends to become increasingly obvious to the observer familiar with the particular group of rocks. In contrast, for the newcomer, even one experienced in fieldwork, the basis for the identification is likely to remain obscure until the structures become more familiar. Therefore it has to be borne in mind that structural investigation of repeatedly deformed rocks, especially gneisses and migmatites, entails careful, detailed observation with painstaking consideration of field relationships and frequent reappraisal of the evidence (see back to the discussion in section 12.6.3, and Figure 12.25). Accordingly, conclusions based solely on hastily-assembled data, which have not been carefully appraised, derived from reconnaissance-type fieldwork over a short period, are more often than not a waste of time. Worse, they could lead either to the early rejection of any possibility of resolving the structural complexity (see section 8.1.11), or the acceptance of an incorrect (or less seriously, an incomplete) structural succession.

The study of each new terrane should always be approached in the knowledge that a considerable amount of fieldwork may be necessary before any recognizable pattern begins to emerge from the structural complexity. Furthermore, in the early stages especially, and usually even until the end of the study, frequent revision is likely to be necessary in the light of new evidence. Perhaps one of the most difficult aspects of a study using this 'trial-and-error' approach is having to reject earlier cherished, apparently very reasonable and even elegant, hypotheses established only after lengthy and painstaking investigation, particularly if the reinterpretation follows the discovery of new evidence that has been pointed out by another observer! Neverthe-less, maintaining an open mind throughout the study is an essential prerequisite to the suc-cessful establishment of the correct structural succession. This in turn leads ultimately to the interpretation of the deformational sequence in the polyphase-deformed terrane. In this respect it is no different from any other (geo-logical) study of this type, whether it is the establishment of a stratigraphical, lithological or palaeontological succession. Of course it must always be accepted that in some cases the establishment of a final 'definitive' structural succession may prove to be an impossibility for a number of reasons including either, (i) because the exposure is insufficient to allow resolution of the structural complexity or (ii) simply because the time available is insuf-ficient to carry out the necessary observations.

The fundamental steps in the procedure (recognition of overprinting, identification of structures, correlation and integration etc.) leading to the establishment of a structural succession for rocks that have been subjected to multiple deformation are summarized pictorially in Figure 15.2.

15.2 POSTSCRIPT: SOME POINTS TO BEAR IN MIND

1. In spite of the effects of anatexis and tectonism it is possible within limits to tackle the resolution of the structure of polyphase-deformed gneisses and migmatites on the basis of overprinting relationships of small-scale (mesoscopic) structures and correlation between them, with the use of form surfaces on a larger scale.

2. It is important to recognize the structures seen, i.e. to identify them, and if that is not possible, at least recognize that as yet uniden-tified structures exist (whatever they are), and

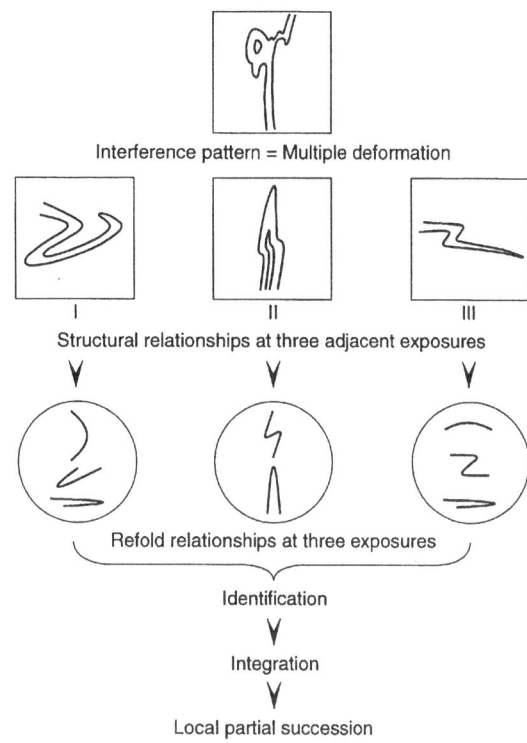

Figure 15.2 Summary chart of the essential stages in determining a structural succession, from the initial observation of overprinting relationships on the outcrop (cf. Figure 12.31b) to their integration to form a local partial succession.

that they differ from one another, i.e. they are (potentially, at least) distinguishable from one another (section 5.2.3).

3. Make a point of recording structures in separate categories if they possess even the slightest recognizable distinguishing features – at least until such time as there is evidence to combine them with other structures, or conversely, confirm their separate identities (section 5.2.3 and in particular Figure 5.8).

4. Avoid using genetic terms for labelling structures. Use purely descriptive terms and leave the interpretation of their origin until last, e.g. avoid words like 'stretched' for deformed objects. Instead use 'elongate' (section 6.2.1).

5. Avoid making assumptions about rela-tionships (angular, spatial, temporal or genetic)

between, for example, lineation and fold axes, cleavage and fold axial planes, foliations etc. but look carefully for such relationships nevertheless! (See section 10.1.5).

6. Beware of assuming parallelism between strike of foliation, axial plane and fold axial plunge direction.

7. Note the time and date, as well as the direction of view, when taking photographs. Shadows and highlights often emphasize significant structural detail which may have been overlooked in the course of recording the structure for which the photograph was taken but which can be seen clearly on the photograph, usually after leaving the field! (See sections 9.2.2(d), 9.2.2(e) and Figures 10.23 and 10.24.

8. Make a habit of checking compass readings if only roughly in terms of the sun direction in case of local magnetic anomalies, especially when readings appear to be significantly different from what might be expected.

9. Be particularly careful about speculating on the degree and sense of plunge or dips from smooth two-dimensional surfaces, especially if the structure is complex. It is possible for such inferences to differ by as much as 90° from the true attitude. Always endeavour to confirm any such inference on a three-dimensional exposure as near as possible to the surface examined but bear in mind that even this is unlikely to be satisfactory where small-scale multiple folding may have caused attitudes to change over very short distances. (See Figure 10.13 and sections 10.1.4 and 10.1.5).

10. Remember that departure from planarity of the exposure surface will affect the shapes of interference patterns seen, and in extreme cases will cause gross distortion of fold profiles. (See section 3.1.4, Figure 3.15, and section 10.1.5.).

11. Note carefully the attitude of veins, such as leucocratic pegmatites, in relation to fold limbs or axial planes. Distinctions such as these, i.e. whether or not a vein is parallel to the axial plane, or to one limb or the other of a fold, can provide criteria which may be definitive. (See Figure 5.8 in section 5.2.3, Figures 4.8a and 4.8b in section 4.3.1, and Figures 4.15 and 2.1.

12. Examine sand-abraded surfaces in particular, for indications of small-scale structures etched by differential erosion (see Figure 1.28 in section 1.1.4).

13. As a matter of course begin as soon as possible to try to build up a succession of structures. Once this is begun, attention is focused on relationships that require clarification, such as the need to distinguish between two folds, one of which refolds the other and which are otherwise apparently indistinguishable, or apparently anomalous relationships between folds, gaps in the succession etc. The need to find the evidence necessary to resolve these problems can then be borne in mind while continuing with the study of other aspects of the structure.

14. Remain mentally adaptable. During the course of the study it is likely that, from time to time, it will become evident that some aspects of the structural framework being built up require modification in the light of new discoveries.

15. Lastly, and this is every bit as important as any of the foregoing, it should be remembered that proof of the reliability of the methods discussed here is provided by the fact that reproducible results are obtained by the same and also by different observers attempting to resolve the same structural problems. This of course is the ultimate test of any hypothesis or method (See section 8.1.4).

Many of the difficulties encountered in attempting to resolve the structural complexity caused by multiple deformation stem from lack of evidence rather than flaws in the principles. This can be the result of insufficient exposure or a consequence of inadequate observation. In this respect it is important to distinguish between the decreasing order of certainty associated with the following: **observation**, followed by **interpretation** and then **speculation**, where the certainty of the last is considerably less than that of the other two.

REFERENCES

Aftalion, M., Bibikova, E. V., Bowes, D. R. Hopgood, A.M. and Perchuk L.L. (1991) Timing of early Proterozoic collisional and extensional events in the granulite-gneiss-charnockite-granite complex, Lake Baikal, USSR: A U-Pb, Rb-Sr, and Sm-Nd isotopic study. *Journ. Geol.* **99,** 851–61.

American Geological Institute (1976) *Dictionary of geological terms.* Anchor Press.

Ashworth, J. R. (1985) Introduction in *Migmatites.* Atherton (ed). Blackie, Glasgow.

Berthelsen, A., Bondersen, E. and Jensen, S. B. (1962) On the so-called wild migmatites, *Krystallinikum* **1,** 31–49.

Borradaile, G. J. (1978) Transected folds: A study illustrated with examples from Canada and Scotland. *Bull. Geol. Soc. Am.* **89,** 491–93.

Boulter, C. A. (1983) Compaction sensitive accretionary lapilli: a means for recognizing soft-sedimentary deformation. *Proc. Geol. Soc. Lond.,* **140,** 789–840.

Bowes, D. R. and Hopgood, A. M. (1975) Structure of the gneiss complex of Inishtrahull, Co. Donegal. *Proceedings Royal Irish Academy,* **75,** 369–90, vii plates.

Bowes, D. R., Hopgood, A. M. and Tonika, J. (1992) Structural succession and tectonic history of the Mariánské Láznê complex, central European Hercynides, western Czechoslovakia. In Kukal, Z. (ed). *Proceedings of the 1st International Conference on the Bohemian Massif, September 26-October 3,* 36–43. Prague; Geological Survey of Czechoslovakia.

Bowes, D. R., Hopgood, A. M. and Tonika, J. (1994) Discrimination by structural criteria of the southern Fichtelgebirge tectonic domain from the Bohemian Forest and Mariánské Láznê domains in the Bohemian Massif, northwestern Czech Republic. *Zlb. Geo. Paläont.* **1992** (7/8), 773–78.

Bowes, D. R., Hopgood, A. M. and Tonika, J. (1995) Structural succession at Vysoky kámen in the Czech part of the southern Fichtelgebirge tectonic domain of the Central European Hercynides. *Journ. Czech Geol. Soc.***40,** 103–14.

The Cambridge Encyclopedia of Earth Sciences. Cambridge University Press, Cambridge.

Clough, C. T. (1897) In Gunn, W., Clough, C. T. and Hill, J. B. *The Geology of Cowal, including the part of Argyllshire between the Clyde and Loch Fine.* Mem. Geol. Surv. Scotland. H.M. Stationery Office, Edinburgh.

Davis, G. H. and Reynolds, S. J. (1996) *Structural Geology of Rocks and Regions.* John Wiley and Sons, Inc., New York.

Duncan, A. C. (1985) Transected folds: a re-evaluation, with examples from the 'type area' at Sulphur Creek, Tasmania. *J. Struct. Geol.* **7,** 409–19.

Epard, J-L. and Escher, A. (1996) Transition from basement to cover: a geometric model. *J. Struct. Geology,* **18,** 533–48.

Fleuty, M. J. (1964) The description of folds. *Geological Association Proceedings,* **75,** 461–92.

Funder, S., Bennike, O., Mogensen, G. S. *et al.* (1984) The Kap København Formation, a late Cainozoic sedimentary sequence in North Greenland. *Rapp. Grønlands Geol. Unders.,* **120,** 9–18.

Ghosh, S. K. (1993) *Structural Geology.* Pergamon, Oxford.

Ghosh, S. K. and Ramberg, J. G. (1968) Buckling experiments on intersecting fold patterns. *Tectonophysics* **5,** 89–105.

Ghosh, S. K., Khan, D. and Sengupta, S. (1995) Interfering folds in constrictional deformation. *J. Struct. Geol.* **17,** 1361–73.

Haeckel, E., (1910). *Anthropologenie oder Entwickelungsgeschichte des Menschen,* Volume I, Keimesgeschichte des Menschen. 6th Edition. Wilhelm Engelmann, Leipzig.

Halden, N. M., Bowes, D. R. and Dash, B. (1982) Structural evolution of migmatites in granulite facies terrane: Precambrian crystalline complex of Angul, Orissa, India, *Trans. R. Soc. Edinburgh Earth Sci.* **73,** 109–18.

Hatcher, R. D. (1992) *Structural Geology: Principles, Concepts and Problems.* Merrill Publishing Company, Columbus.

Hawking, S. (1988) *A Brief History of Time*. Bantam, London.

Heim, A., (1919) See Hills, E. S. (1963), p. 433.

Hendry, H. E. and Stauffer, M. R. (1977) Penecontemporaneous folds in cross-bedding: Inversion of facing criteria and mimicry of tectonic folds. *Bull. Geol. Soc. Am.* **88**, 809–12.

Hills, E. S. (1963) *Elements of Structural Geology*. Methuen & Co. Ltd, London.

Hobbs, B. E. (1966) The structural environment of the northern part of the Broken Hill ore body. *J. Geol. Soc. Australia.* **13**, 315–38.

Hobbs, B. E., Means, W. D. and Williams, P. F. (1976) *An Outline of Structural Geology*. John Wiley & Sons, New York.

Holland, J. R. and Lambert, R. St J. (1969) Structural regimes and metamorphic facies. *Tectonophysics* **7**, 197–217.

Hopgood, A. M. (1966) Theoretical consideration of the mechanics of tectonic reorientation of dykes. *Tectonophysics*, **3** (1), 17–28.

Hopgood, A. M. (1970) Structural re-orientation as evidence of basement warping associated with rift faulting in Uganda, *Bull. Geol. Soc. Am.* **81**, 3473–80.

Hopgood, A. M. (1971a) Structure and tectonic history of Lewisian Gneiss, Isle of Barra, Scotland. *Krystalinikum*, **7**, 27–60. Figures 1–29.

Hopgood, A. M. (1971b) Correlation by tectonic sequence in Precambrian gneiss terrains. *Geol. Soc. Australia Spec. Publ.* No. 3, 367–76.

Hopgood, A. M. (1973) The significance of deformational sequence in discriminating between Precambrian terrains. *Spec. Publ. Geol. Soc. S. Africa* **3**, 45–51.

Hopgood, A. M. (1976) Structures in an area northeast of Fiskenaesset, West Greenland. *Rapp. Grönlands Geol. Unders.* **73**, 16–21.

Hopgood, A. M. (1980) Polyphase fold analysis of gneisses and migmatites. *Trans. R. Soc. Edinburgh: Earth Sci.* **71**, 55–68.

Hopgood, A. M. (1984) Structural evolution of Svecokarelian migmatites, southern Finland: a study of Proterozoic crustal development, *Trans. R. Soc. Edinburgh: Earth Sci.* **74**, 229–64.

Hopgood, A. M. and Bowes, D. R. (1972) Application of structural sequence to the correlation of Precambrian gneisses, Outer Hebrides, Scotland. *Bull. Geol. Soc. Am.* **83**, 107–28.

Hopgood, A. M. and Bowes, D. R. (1978) Neosomes of polyphase agmatites as time-markers in complexly deformed migmatites. *Geol. Rundsch.* **67**, 313–30.

Hopgood, A. M. and Bowes, D.R. (1980) Structural analysis of folded cooling joints in a near-concordant minor intrusion. *Geol. Rundsch.* **69**, 84–93.

Hopgood, A. M. and Bowes, D. R. (1987) Structural succession and tectonic history of the gneiss – amphibolite – granulite – mantle peridotite association near the eastern margin of the Moldanubian zone, central European Hercynides. *Acta Universitatis Carolinae – Geologica* No. **1**, 51–88.

Hopgood, A. M. and Bowes, D. R. (1990) Contrasting structural features in the granulite – gneiss – charnockite – granite complex, Lake Baikal, U.S.S.R.: evidence for diverse geotectonic regimes in early Proterozoic times. *Tectonophysics*, **174**, 279–99.

Hopgood, A. M. and Bowes, D. R. (1995) Matching Gondwanaland fragments: the significance of granitoid veins and tectonic features in the Cape Leeuwin-Cape naturaliste terrane, SW Australia. *Journ. Southeast Asian Earth Sciences*, **11**, 253–63.

Hopgood, A. M., Bowes, D. R. and Addison, J. (1976) Structural development of migmatites near Skåldö, southwest Finland. *Bull. Geol. Soc. Finl.* **48**, 43–62.

Hopgood, A. M., Bowes, D. R., Kouvo, O. and Halliday, A. N. (1983) U-Pb and Rb-Sr isotopic study of polyphase-deformed migmatites in the Svecokarelides, southern Finland, *in* Atherton, M. P. and Gribble (eds). *Migmatites, Melting and Metamorphism*, Shiva, Nantwich, 80–92.

Hopgood, A. M., Bowes, D. R., You Zhendong, et al. (1989) Structural features and relationships in the Precambrian Dabie complex, Feng Huang Guan area, Hubei Province, China. *Earth Science*, **14**, No. 3, 221–38 (8 plates). (In Chinese with English abstract).

Hopgood, A. M., Bowes, D. R. and Tonika, J. (1992) Application of deformational sequence to discrimination of tectonic domains and elucidation of geological history in crystalline rocks of the Central Hercynides, western Czechoslovakia, *in*: Vrana, S. (ed). *Geological model of western Bohemia in relation to the deep borehole KTB in the GR*. 12–13. Prague; Geological Survey of Czechoslovakia.

Hopgood, A. M., Bowes, D. R. and Tonika, J. (1995) Application of structural succession to characterization of the Bohemian Forest tectonic domain and elucidation of geological history in the Central European Hercynides, western Czech Republic. *N. Jb. Miner. Abh.* **169**, 119–56.

Hudson, T. L. (1987) Suspect philosophy? *Geology* **15**, 1177.

Hutton, J. (1795) *Theory of Earth with Proofs and*

Illustrations. I and II. Cadell and Davies, London and Creech, Edinburgh.

Joubert, P. (1971) *The regional tectonism of the gneisses of part of Namaqualand. Bull.* **10,** Precambrian Research Unit, University of Cape Town.

Koistinen, T. J. (1981) Structural evolution of an early Proterozoic strata-bound Cu-Co-Zn deposit, Outokumpu, Finland. *Trans. R. Soc. Edinburgh: Earth Sci.* **72,** 115–58.

Knopf, E. B. (1933) Petrotectonics, *Am. Jour. Sci.* **25,** 460–62.

Lahee, F. H. (1941) *Field Geology.* Fourth Edition. McGraw-Hill, New York.

Laing, W. P., Majoribanks, R. W. and Rutland, R. W. R. (1978) Structure of the Broken Hill Mine area and its significance for the genesis of the orebodies. *Econ. Geol.* **73,** 1112–36.

Lugeon, M. (1949) Unpublished letter to D. B. McIntyre.

MacClay, K. (1987) *The Mapping of Geological Structures.* Open University Press, Milton Keynes.

McIntyre, D. B. (1951) *Alpine Tectonics and the Study of Ancient Mountain Chains.* Doctoral dissertation (Unpublished), University of Edinburgh.

Marques, F. G. and Cobbold, P. R. (1995) Development of highly non-cylindrical folds around rigid ellipsoidal inclusions in bulk simple regimes: natural examples and experimental modelling. *J. Struct. Geol.* **17,** 589–602.

Mehnert, K. R. (1968) *Migmatites and the Origin of Granitic Rocks.* Elsevier, Amsterdam.

Moore, J. C. and Geigle, J. E. (1972) Incipient axial plane cleavage: deep sea occurrence. *Geol. Soc. Am.* (Abstracts) '72. Annual Mtg., p.600.

Naha, K., Srinivasan, R. and Jayaram, S. (1990) Structural evolution of the Peninsular Gneiss – an early Precambrian migmatitic complex from South India. *Geol. Rundsch.* **79,** 99–109.

North American Commission on Stratigraphic Nomenclature (1983) North American Stratigraphic Code. *American Association of Petroleum Geologists Bulletin,* **67,** 841–75, 11 figs., 2 tables.

Odonne, F. and Vialon, P. (1987) Hinge migration as a mechanism of superimposed folding. *J. Struct. Geol.* **9,** 835–44.

Park, R. G. (1969) Structural correlation in metamorphic belts. *Tectonophysics,* **7,** 323–38.

Park, R. G. (1989) *Foundations of Structural Geology.* Blackie, Glasgow.

Passchier, C. W., Myers, J. S. and Kröner, A. (1990) *Field geology of high-grade gneiss terranes.* Springer-Verlag, Berlin.

Phillips, F. C. (1971) *The Use of Stereographic Projection in Structural Geology.* Edward Arnold, London.

Powell, C. McA. (1974) Timing of slaty cleavage during folding of Precambrian rocks, north-west Tasmania. *Bull. Geol. Soc. Am.* **85,** 1043–60.

Ragan, D. M. (1985) *Structural Geology.* J. Wiley & Sons, New York.

Ramsay, J. G. (1967) Folding and fracturing of rocks. McGraw-Hill, New York.

Ramsay, J. G. and Huber, M. I. (1983) *The Techniques of Modern Structural Geology.* **1,** Strain Analysis. Academic Press, London.

Ramsay, J. G. and Huber, M. I. (1987) *The Techniques of Modern Structural Geology.* **2,** Folds and Fractures. Academic Press, London.

Rast, N. (1963) *Structure and Metamorphism of the Dalradian Rocks of Scotland, in The British Caledonides,* Johnson, M. R. W., and Stewart, F. H., (eds). Oliver and Boyd, Edinburgh and London.

Rutland, R. W. R. and Etheridge, M. A. (1975) Two high grade schistosities at Broken Hill and their relation to major and minor structures. *Journ. Geol. Soc. Australia,* **22,** 259–74.

Sander, B. (1911) Über Zusammenhange Zwischen Teilbewegung und Gefüge in Gesteinen. *Tschm. Min. Pet. Mitt.* **30,** 381–84.

Sander, B. (1930) *Gefugerkunde der Gesteine.* Springer, Vienna.

Sander, B. (1970) *An Introduction to the Study of Petrofabrics of Geological Bodies.* Translated by F. C. Phillips and G. Windsor. Pergamon Press, New York.

Schmidt, W. (1932) *Tektonik und Verformungslehre.* Borntraeger, Berlin.

Sederholm J. J. (1967) *Selected Works: Granites and Migmatites.* Oliver and Boyd, Edinburgh.

Sneath, Peter H. A. and Sokal, R. R. (1973) *Numerical taxonomy: the principles of numerical classification.* W. H. Freeman and Co., San Francisco and London.

Sokal, R. R. (1966) Numerical taxonomy. *Scientific American* **215,** 106–16.

Stringer, P. and Treagus, J. E. (1980) Non-axial planar S_1 cleavage in the Hawick rocks of the Galloway area, Southern Uplands, Scotland. *J. Struct. Geol.* **2,** 317–31.

Suo, S., Liu, R. and Ma, X. (1982) Interference patterns of the early Precambrian rock groups in the Songshan area, China. *Geol. Mag.* **119,** 433–61.

Tobisch, O. T., Fleuty, M. J., Merh, S. S. *et al.* (1970) Deformation and metamorphic history of Moinian and Lewisian rocks between Strathconan and Glen Affric. *Scott. Journ. Geol.* **6**, 243–65.

Turner, F. J. and Weiss, L. E. (1963) *Structural Analysis of Metamorphic Tectonites.* McGraw-Hill: New York.

Watkinson, A. J. and Cobbold, P. R. (1978) Localization of minor folds by major folds. *Bull. Geol. Soc. Am.* **81**, 3283–96.

Wegmann, C. E. (1929) Beispiele tektonischer Analysen des Grundgebirges in Finnland: *Bull. Comm. Géol. Finl.* **87**, 98–127.

Whitten, E. H. T. (1966) *Structural Geology of Folded Rocks.* Rand McNally, Chicago.

Whittaker, A., Cope, J. C. W., Cowie, J. W. *et al.* (1991) A guide to stratigraphical procedure. *Journ. Geol. Soc. Lond.* **148**, 813–24.

Williams, P. F. (1967) Structural analysis of the Little Broken Hill area of New South Wales. *Journ. Geol. Soc. Australia,* **14**, 317–32.

Williams, P. F. (1970) A criticism of the use of style in the study of deformed rocks. *Bull. Geol Soc. Am.* **81**, 3283–96.

Williams, P. F. (1985) Multiply-deformed terrains – problems of correlation. *J. Struct. Geol.* **7**, 269–80.

Williams, P. F., Collins, A. R. and Wiltshire, R. G. (1969) Cleavage and penecontemporaneous deformation structures in sedimentary rocks. *J. Geol.* **77**, 415–25.

Wright, J. B. (1981) In *The Cambridge Encyclopedia of Earth Sciences.* University Press, Cambridge.

Wyatt, A., (ed.) (1986) Law of Superposition, in *Challinor's Dictionary of Geology.* University of Wales Press, Cardiff.

AUTHOR INDEX

Bold numbers are figure numbers. Italic numbers are table numbers.

LOCALITY INDEX

Bold numbers are figure numbers. Italic numbers are table numbers.

SUBJECT INDEX

Bold numbers are figure numbers. Italic numbers are table numbers.